CAMBRIDGE LIBRARY COLLECTION

Books of enduring scholarly value

Earth Sciences

In the nineteenth century, geology emerged as a distinct academic discipline. It pointed the way towards the theory of evolution, as scientists including Gideon Mantell, Adam Sedgwick, Charles Lyell and Roderick Murchison began to use the evidence of minerals, rock formations and fossils to demonstrate that the earth was older by millions of years than the conventional, Bible-based wisdom had supposed. They argued convincingly that the climate, flora and fauna of the distant past could be deduced from geological evidence. Volcanic activity, the formation of mountains, and the action of glaciers and rivers, tides and ocean currents also became better understood. This series includes landmark publications by pioneers of the modern earth sciences, who advanced the scientific understanding of our planet and the processes by which it is constantly re-shaped.

A Manual of Scientific Enquiry

Sir John Frederick William Herschel (1792–1871) – astronomer, mathematician, chemist – was one of the most important British scientists of the nineteenth century. Son of the famous astronomer William Herschel, he was persuaded by his father to pursue the astronomical investigations William could no longer undertake; John's subsequent career resulted in a knighthood and a lifetime of accolades. This 1849 publication was commissioned by the Admiralty to encourage and assist naval officers to undertake scientific research while abroad. The work provides instructions in making and recording observations in a wide range of disciplines – astronomy; magnetism; hydrography; tides; geography; geology; earthquakes; mineralogy; meteorology; atmospheric waves; zoology; botany; ethnology; medicine; statistics – written by experts in these fields, including Whewell, Darwin, Hooker and Herschel himself. It was hoped that the instructions could also be used by other travellers to advance scientific knowledge, and the work remained in print for over fifty years.

Cambridge University Press has long been a pioneer in the reissuing of out-of-print titles from its own backlist, producing digital reprints of books that are still sought after by scholars and students but could not be reprinted economically using traditional technology. The Cambridge Library Collection extends this activity to a wider range of books which are still of importance to researchers and professionals, either for the source material they contain, or as landmarks in the history of their academic discipline.

Drawing from the world-renowned collections in the Cambridge University Library, and guided by the advice of experts in each subject area, Cambridge University Press is using state-of-the-art scanning machines in its own Printing House to capture the content of each book selected for inclusion. The files are processed to give a consistently clear, crisp image, and the books finished to the high quality standard for which the Press is recognised around the world. The latest print-on-demand technology ensures that the books will remain available indefinitely, and that orders for single or multiple copies can quickly be supplied.

The Cambridge Library Collection will bring back to life books of enduring scholarly value (including out-of-copyright works originally issued by other publishers) across a wide range of disciplines in the humanities and social sciences and in science and technology.

A Manual of Scientific Enquiry

Prepared for the Use of Her Majesty's Navy and Adapted for Travellers in General

Edited by John Frederick William Herschel

CAMBRIDGE UNIVERSITY PRESS

Cambridge, New York, Melbourne, Madrid, Cape Town,
Singapore, São Paolo, Delhi, Tokyo, Mexico City

Published in the United States of America by Cambridge University Press, New York

www.cambridge.org
Information on this title: www.cambridge.org/9781108029179

This edition first published 1849
This digitally printed version 2011

ISBN 978-1-108-02917-9 Paperback

MANUAL

OF

SCIENTIFIC ENQUIRY;

PREPARED FOR THE

USE OF HER MAJESTY'S NAVY:

AND ADAPTED FOR TRAVELLERS IN GENERAL.

EDITED BY

SIR JOHN F. W. HERSCHEL, BART.

PUBLISHED BY AUTHORITY OF

The Lords Commissioners of the Admiralty.

LONDON:
JOHN MURRAY, ALBEMARLE STREET,
PUBLISHER TO THE ADMIRALTY.
1849.

London: Printed by William Clowes and Sons, Stamford Street.

MEMORANDUM

BY THE LORDS COMMISSIONERS OF THE ADMIRALTY,

Relative to the Compilation of a Manual of Scientific Enquiry, for the use of Her Majesty's Navy.

It is the opinion of the Lords Commissioners of the Admiralty that it would be to the honour and advantage of the Navy, and conduce to the general interests of Science, if new facilities and encouragement were given to the collection of information upon scientific subjects by the officers, and more particularly by the medical officers, of Her Majesty's Navy, when upon foreign service; and their Lordships are desirous that for this purpose a Manual be compiled, giving general instructions for observation and for record in various branches of science. Their Lordships do not consider it necessary that this Manual should be one of very deep and abstruse research. Its directions should not require the use of nice apparatus and instruments: they should be generally plain, so that men merely of good intelligence and fair acquirement may be able to act upon them; yet, in pointing out objects, and methods of observation and record, they might still serve as a guide to officers of high attainment: and it will be for their Lordships to consider whether some pecuniary reward or promotion may not be given to those who succeed in producing eminently useful results.

Their Lordships are aware that in the instructions prepared under the directions of the Royal Society for the Antarctic expedition—in the hints for collecting information given to officers on the expedition to China, in the excellent book by A. Jackson, entitled "What to observe," and in other documents and publications—the fullest directions are to be found; but they are either more voluminous or more closely confined to objects which regard particular localities than is to be desired for a general Manual. Their Lordships are, therefore, desirous that a new compilation should be made, and are satisfied that their wishes would be best met if they could obtain the assistance of some of our most eminent men of Science in the composing, by each, of a plain and concise chapter upon the head of enquiry with which he might be most conversant, and they have been readily and kindly promised the advice and labour of Sir John Herschel in revising the whole and preparing it for publication. The several heads of enquiry are as follows:—

Astronomy.
Botany.
Geography and Hydrography.
Geology.
Mineralogy.
Magnetism.
Meteorology.
Statistics.
Tides.
Zoology.

Independently of matters of exact science, their Lordships would look, in many instances, for Reports upon National Character and Customs, Religious Ceremonies, Agriculture and Mechanical Arts, Language, Navigation, Medicine, Tokens of value, and other subjects; but for these only very general instructions can be given, though valuable Reports may be expected from men of observation and intelligence acting under the encouragement which the notice of whatever is well and usefully done is certain of affording.

It would give additional value to each chapter if the name of him by whom it might be composed should be affixed to it; and their Lordships are anxious that no time be lost in the preparation of this work. They are sending a surveying vessel to New Zealand, and have others in the Torres Straits and in other parts of the world. A new establishment is contemplated at Borneo. Expeditions are proposed in search of Sir John Franklin. They have cruisers in every sea; and where the ships of the navy are not present, it sometimes happens that the vessels of the merchant are conducted with much intelligence and enterprise, and for all of these the work proposed would be valuable.

PREFACE.

THE Memorandum of the Lords Commissioners of the Admiralty, prefixed to the present work, so fully explains its origin and object, that it only remains for the Editor to indicate the reasons which have necessitated what may appear to be in a certain degree departures from the plan as therein sketched out—departures either inevitable in themselves, or sanctioned in their progress by their Lordships' approbation, and which, though involving a material increase in the bulk of the volume beyond what was in the first instance contemplated, can hardly fail to be regarded as adding still more materially to the probability of its proving useful in furthering the collection of scientific information.

On receiving from the eminent authors of the several chapters into which this work is divided, their respective contributions, drawn up by them at the particular and personal request of the late lamented Earl of Auckland (to whose enlightened zeal for the improvement of science the work owes its original conception), the Editor at once perceived that, while, on the one hand, the total amount of matter contributed could by no possibility be compressed into the small compass originally intended (even when reduced by the retrenchment of some portions in his opinion less vitally essential, and which the authors of those portions, on his representation, most readily agreed to abbreviate)—on the other, those contributions bore evidence of having been drawn up with so much care and precision, and elaborated with such consummate ability, that any further attempt to reduce them within those limits would have been at once presumptuous on his part and most injurious to the efficiency of the work itself.

It is hoped and believed that the instructions on each subject treated of are such as completely to fulfil the views of the Memorandum, in so far as it is possible, consistently with the nature of the sciences treated of, that they should do so. But there are some—such as Terrestrial Magnetism—which it is impossible to divest of a certain degree of abstruseness, and in which no observations worth recording can be made without the aid of instruments and methods of observation and reduction decidedly both nice and delicate. The same may be said, so far as the instruments are concerned, in Meteorology. The class of meteorological observations of most importance which require instrumental aid are essentially of some delicacy, and if not executed with a certain precision and nicety, had much better not be recorded at all. But in both these cases particular care has been taken to make the instructions for the use of the necessary instruments so plain and full, that no difficulty can be experienced by any one in acquiring from them, with a little practice, their perfect management.

The time apparently contemplated in the Memorandum for the completion and appearance of the work has been exceeded. The delay, however, has been inevitable. One only of the contributors was aware, until nearly a month subsequent to the date of that document, of what was required of him; and the various professional and scientific engagements of almost all, precluded the possibility of their complying on the instant with the request. The Editor in particular must plead guilty to having been the latest in the field. The subject of Meteorology, originally intended for a far more able hand, devolved upon him, in consequence of the inability from illness, and supervening most pressing engagements, of the party in question to undertake it. His own hands were otherwise full; and he is but too conscious that, although written three or four times over, his contribution falls very far short indeed of what would have occupied its place had the original cast of the subject in this respect been carried out.*

One only of the chapters—that on Ethnology—has not received its final touches and corrections from the hand of its author. The eminent and gifted contributor of that chapter survived only to complete the MS. The correction of the press devolved on his son. Previous to his decease the Editor had received his authority for substituting, as an appendix, a somewhat more extended (though still confessedly incomplete) attempt of his own to facilitate the reduction to writing of foreign native languages, for a system which, on maturely weighing it, appeared liable to serious objections. The subject is confessedly difficult; and it would have been peculiarly gratifying to him to have been enabled to submit this attempt to Dr. Pritchard for revision, and to have received more than the general sanction above alluded to for its substitution.

With the publication of this work the Editor's con-

* In more particular allusion to this chapter the Editor would observe, that though the sympiesometer, and that substitute for the portable barometer to which the name of "aneroid" has been given, are not mentioned among the standard and essential meteorological instruments whose continuous registry is recommended; yet they may be consulted with advantage, and frequent comparisons of the latter with the barometer would be very useful in determining how far its indications at distant times and in different circumstances can be relied on.

nexion with it ceases, beyond a general and earnest interest in its success in forwarding the objects proposed by it. Observations made in any of its departments, in pursuance of the recommendations contained in it, by officers in Her Majesty's service, if officially communicated in the proper quarter, will of course be dealt with officially. But as the work may, and probably will, fall into the hands of voyagers and travellers unconnected with the public service, it may be right to state, for the avoidance of possible misapprehension, that he cannot charge himself with the reception, examination, or discussion of any masses of observations they may accumulate.

CONTENTS.

NAVAL OFFICER'S MANUAL.

SECTION I.

ASTRONOMY.

BY G. B. AIRY, ESQ.,

ASTRONOMER ROYAL.

THE science of Astronomy may occasionally derive benefit from the observations of navigators in the following respects:—

By contributions to Astronomy in general.

By improvement of the methods of Nautical Astronomy.

By accurate attention to Astronomical Geography.

The remarks which follow will be arranged under these heads.

General Astronomy.

1. The first point which calls for attention is the observation of the places of comets or other extraordinary bodies, especially those which can be seen only in low northern or in southern latitudes. In regard to these observations (and indeed to almost all others), one remark cannot be too strongly impressed on the observers: that a bad observation, or an observation which is given

without the means of verification, is worse than no observation at all. In order to make the observations good, the following cautions must be observed :—

The index-error of the sextant must be carefully ascertained. If it has not been found a short time before the observations, it must be found as soon as possible after them.

The distance of the comet from three conspicuous stars in different directions must be measured with the sextant. The point of the comet which is observed with the sextant should be precisely described. It is desirable that the navigator should be possessed of some star-maps or star-charts, by means of which he will be able at once to give the proper names to the stars, and much confusion and loss of time will be avoided.

If the time at the ship and the latitude are very well known, there will be no occasion to make further observations; but if these are not well known, some attempt must be made (by the use of Becher's horizon, or by any equivalent method) to ascertain the altitude of the stars and the comet. The lower these objects are, the greater must be the care in the determination of their altitude.

For affording means of verification, these rules should be followed :—

The observations of distance with the sextant should be entered in the book precisely in the manner in which they are made. The reading of the sextant, *uncorrected*, should be written down: in a column by the side of this should be written the correction for index-error, with a statement whether it is to be added to, or to be subtracted from, the sextant-reading: in the next column should be written a reference to the observations by which the index-error was determined; and in the last column should be written the distance as corrected. For the altitudes, the height of the eye, the depression of the horizon, and the altitude corrected for depression, should also be stated. At some convenient place, either at the beginning or at the end of all, should be written out all the measures by which the index-error was ascertained, exactly in the manner in which they were made, and so that any other person can deduce from them the value of the index-error.

The time of making every observation should be entered exactly as it is read from the chronometer or hack-watch. By the side of this should be placed the error of the chronometer or hack-watch on Greenwich time, or on time at the ship (as may be most convenient); and, after this, the corrected time.

At some convenient place, either at the beginning or the end, must be written out all the observations by which the error of the chronometer is ascertained. If its error on Greenwich time is given, the longitude of the ship must also be given, and the means and observations by which that longitude has been determined must be stated at length.

If a hack-watch is used, the comparison of the hack-watch with the chronometer must be given.

The last observations by which the latitude was determined, and the course and rate of sailing of the ship, must also be given.

All the observations should be sent in this detail to the Admiralty or other body appointed to receive them.

2. Opportunities will sometimes occur, when a ship is

lying in a harbour of which the latitude and the longitude are well known, for observing eclipses of the sun. These observations are almost always valuable. It can seldom be expected that the time of the beginning of an eclipse can be observed accurately, but the time of the end of it can usually be observed with very great accuracy. And if the eclipse is total, the times of beginning and end of the totality can be observed accurately: if it is annular, the times of beginning and end of the annularity can be accurately observed. The observations should be made with the largest telescope which the navigator possesses; and any peculiarity of distortion of the sun s limb or the moon's limb, any light surrounding the moon, &c., should be carefully recorded. [If the eclipse be total, attention should be paid to any coloured or other appendages projecting from the dark edge of the moon, also to the luminous corona surrounding the moon, its apparent breadth, and whether apparently concentric *with the moon* or with the sun near the moments of beginning and end of the total obscuration.] While the eclipse is in progress, but especially near the beginning or the end, measures of the distance between the cusps or sharp points at which the moon's limb crosses the sun's limb may be repeatedly taken. In recording these observations, the observations by which the time is determined, and the observations by which the index-error of the sextant is determined, should be written down in the fullest detail; and the unreduced observations should be given as well as the reduced observations.

3. In similar circumstances occultations of stars by the moon may frequently be observed. Eclipses of Jupiter's

satellites may also be seen; and (if the navigator have a telescope somewhat better than is usually carried in ships, and steadily mounted) the passage of Jupiter's satellites, either behind the planet or in front of the planet, may be seen, and the times at which the centres of the satellites just touch the edge of the planet may be observed. All these observations will be useful: the observations must be recorded with the same fulness which has been mentioned before.

4. It may chance that the navigator is in some climates where the air is much more damp, and in others where it is much more dry, than in Europe. It is possible that in these places he may be able to make observations which will throw some light upon the influence of moisture in atmospheric refraction. It is recommended that repeated observations of the altitude of the sun's upper and lower limb be taken when the sun is very near the horizon. It will be necessary that the time at the ship and the latitude be very well known. The thermometer must be read, as also some hygrometrical instrument, and the barometer, if there is one on board, during the observations. The observations of every kind must be recorded with the utmost fulness.

5. It is certain that some of the stars of the southern hemisphere are variable in magnitude; the most remarkable of these is η Argûs. It is desirable that, on favourable nights, the magnitude of this star should be observed and recorded. The best way of doing it will be, not to state that it looks like a star of the 2nd magnitude, or of the 3rd magnitude, &c., but to compare its brightness with that of some of the stars near it. Thus it will be

easy to say that it appears pretty exactly as bright as one star, certainly brighter than a second, and certainly not so bright as a third.*

6. Much attention has been excited by the appearance, in several years, of meteors in great numbers, on or about the 9th of August and the 12th of November.† It is probable that these appearances may be seen by persons at sea, when, either from the hour at which they occur, or from other causes not yet understood, they cannot be seen in Europe. It is impossible to observe them with accuracy; but very valuable information will be given by counting repeatedly how many can be seen in some fixed interval of time, as five minutes; and by remarking whether they all come from, or go to, one part of the heavens; what is that part of the heavens; whether they usually leave trains behind them; what is their usual brightness (as compared with that of known stars); and by any other remarks which may be suggested by their appearance.

7. Many opportunities will occur of observing the zodiacal light; more especially when the observer is near the equator, where probably it can be seen at all seasons, before sunrise and after sunset; or, if in northern latitudes, after sunset in February and March, and before sunrise in September and October; if in southern latitudes, before

* See a list of variable stars, and some suggestions for observations of brightness of stars, in the Appendix.—(Ed.)

† Humboldt (Kosmos, i. 387) enumerates the following epochs as especially fertile in meteors, viz. April 22—25; July 17—26; Aug. 9—11; Nov. 12—14 and 27—29; Dec. 6—12. Of all these epochs, that of August has hitherto proved to be the most regular. The star B Camelopardali has for several years been their point of divergence about that epoch. Any change in this respect should be recorded.—(Ed.)

sunrise in March and April, and after sunset in August and September. The zodiacal light consists of a pyramid of faint light, whose base is somewhere near the place of the sun, and whose point is at a distance of perhaps 30° from the sun; the axis of the pyramid being usually inclined to the horizon, following nearly the direction of the ecliptic. Although it presents to the eye a considerable body of light, yet the light of any portion of it is so feeble, and the definition of its outline is so imperfect, that it cannot be observed with a telescope. The observer, therefore, should only attempt to observe it with the naked eye, when the sky is very clear, and when the sun is so far below the horizon that no twilight is visible. He should then endeavour, with the assistance of a chart of the stars, to define as accurately as possible its boundary with reference to the stars; remarking especially the place of the point of the pyramid, the width where it rises from the horizon, whether its sides are curved, and in what parts the light is brightest. It will be found that these observations are made most accurately by occasionally turning the eye a little obliquely from the zodiacal light. In registering the observation, in addition to the particulars to be recorded as prescribed above, there should be a statement of the latitude of the ship, the day, the time at the ship (or the Greenwich time and the longitude of the ship), the state of clearness of the sky, and the state of the weather for the day preceding the observation.

Improvement of Nautical Astronomy.

8. So much attention has been given to every detail of Nautical Astronomy, that it is very difficult to fix upon

any part of it to which the attention of navigators should be specially directed with a view to its improvement. Perhaps the principal deficiency at the present time is in the want of well-understood methods of observing (with the sextant) the altitudes of stars at night, and of observing the altitudes of the sun and moon when the horizon is ill-defined. Every endeavour ought to be made to become familiar with the use of Becher's horizon, or some equivalent instrument, and to acquire a correct estimate of the degree of confidence which can be placed in the use of it.

9. It is likewise desirable that efforts should be made to facilitate the observation of occultations of stars by the moon, and the observation of eclipses of Jupiter's satellites at sea. Occultations occur rarely, but the result which they give for longitude is usually so much more accurate than that given by lunar distances, that, in long voyages where little dependence can be placed on the chronometer, the observation of an occultation must be extremely valuable. The eclipses of Jupiter's satellites afford less accurate determinations of longitude, but they occur very much more frequently, and may be very useful where chronometers cannot be trusted.*

* Attempts may laudably be made to devise some available mode of suspending a chair, so as to afford a steady seat to the observer. Hitherto such attempts have failed of practical success, from setting out with the principle of perfectly free suspension, a principle which tends to prolong and perpetuate oscillations once impressed. It remains to be seen what *stiff* suspension, as for example by a rigid rope or cable, or by a hook's joint, purposely made to work stiffly (and that more or less at pleasure), by tightening collars—as also deadening and shortening oscillations, by lateral cords passing through *rings* to create friction—and other similar contrivances may do. In the suspension of a cot, at least, I have found this principle signally available.—(Ed.)

Astronomical Geography.

10. The intelligent navigator, on arriving at any port which has not before been visited, or whose position is not very well settled, ought to consider it his first duty to determine with all the accuracy in his power the latitude and longitude of the port. Supposing him to have determined by the usual nautical methods the approximate latitude, longitude, and error of chronometer, the best method of determining the latitude will be to find the chronometer-time at which the sun or any bright stars of the Nautical Almanack list will pass the meridian, and to observe the double altitude of any such object by reflection in a mercurial horizon, several times, as near as possible to the time of the meridian passage. If the place is in the northern hemisphere, the observation of the double altitude of the pole-star may be made at any time when it is visible: convenient tables for the reduction are given in the Nautical Almanack. For these and other observations the navigator ought to be provided with a proper trough and a store of mercury. For determining the longitude, there is probably no method superior to that of lunar distances (the exactness of which will be increased if the sextant or reflecting circle be mounted on a stand), unless the stay at the port is so long that transits of the moon can be observed. In any case, if there be a transit-instrument in the ship, it ought to be mounted on shore as soon as possible. The instrument ought, on the first evening, to be got very nearly into a meridional position, and then a mark should be set up, and the instrument should always be adjusted to that same mark (even

though it be not exactly in the meridian), and should always be levelled, before commencing a series of observations. One or two stars at least, as near the pole as possible, should be observed every night, in addition to the Nautical Almanack stars necessary for chronometer-error, and the moon-culminating stars which are observed with the moon. The instrument should be reversed on alternate nights; and, if possible, as many transits of the moon should be taken after the full moon as before the full moon.

In the register of all these observations, the same rule should be followed which is laid down under the first suggestion; that every observation should be recorded *unreduced*, exactly in the state in which it is read from the sextant or chronometer; and that the unreduced observations should be accompanied with the elements of reduction of whatever kind; and that (if the navigator has had leisure to reduce them) the reduced results should also be given.

G. B. Airy.

APPENDIX.

BY THE EDITOR.

(A.)

A List of the most conspicuous Variable or Periodic Stars of which observations would be desirable, with their periods of Variation (so far as known) and changes of magnitude.

	Period.			Change of Magnitude.		
	D.	H.	M.			
β Persei . . .	2	20	48	2	to	4
λ Tauri . . .	4	..	..	3·4	to	4
δ Cephei. . .	5	8	37	3·4	to	5
β Lyræ . . .	6	9	..	3	to	4·5
η Aquilæ. . .	7	4	15	3·4	to	4·5
ζ Geminorum .	10	3	35	4·3	to	4·5
α Herculis . .	60	6	..	3	to	4
ε Aurigæ . .	19 months .				..	
ο Ceti . . .	334	..	..	2	to	..
υ Hydræ . .	494	..	..	4	to	10
κ Sagittarii . .	Many years.			3	to	6
η Argus . . .	Irregular .			1	to	4
β Ursæ Minoris .	Unknown .			2	to	2·3
α & η Ursæ Majoris	Do. . . .			1·2	to	2

(B.)

Lists of Fixed Stars in either hemisphere, approximately arranged in order of brightness, down to the fourth magnitude, for the purpose of mutual comparison under favourable circumstances of altitude, and especially in equatorial and tropical voyages, or land stations, with a view to bringing the nomenclature and scale of magnitudes in the two hemispheres to agreement, and to the improvement of this branch of astronomical knowledge. The comparisons to be made by the naked eye among the stars of both lists not differing much (at the time of observation) in altitude, and in the absence of the moon and twilight, and the results arranged in sequences, beginning with the brightest, and ending with the faintest star compared. In each sequence *stars of the two lists should alternate whenever circumstances will allow.*

a. Northern Stars.

Arcturus.
Capella.
Lyra.
Procyon.
α Orionis.
Aldebaran.
α Aquilæ.
Pollux.
Regulus.
α Cygni.
Castor.
ε Ursæ.
α Ursæ.
α Persei.
β Tauri.
γ Orionis.
Polaris.
γ Leonis.
ζ Ursæ.
α Arietis.
β Andromedæ.
β Aurigæ.
γ Andromedæ.
γ Cassiopeiæ.
α Andromedæ.
α Cassiopeiæ.
γ Geminorum.
β Leonis.
γ Draconis.
α Ophiuchi.
β Cassiopeiæ.
γ Cygni.
α Pegasi.
β Pegasi.
ε Pegasi.
α Coronæ.

γ Ursæ.
β Ursæ.
ε Bootis.
ε Cygni.
α Cephei.
α Serpentis.
δ Leonis.
η Bootis.
γ Aquilæ.
δ Cassiopeiæ.
η Draconis.
β Draconis.
β Arietis.
γ Pegasi.
ε Virginis?
θ Aurigæ.
β Herculis.
Cor Caroli.
β Ophiuchi.
δ Cygni.
ε Persei.
η Tauri.
ζ Persei.
ζ Herculis.
z Aurigæ.
γ Ursæ Min.
η Pegasi.
ζ Aquilæ.
β Cygni.
γ Persei.
β Trianguli B.
δ Persei.
ε Aurigæ.
r Lyncis.
ζ Draconis.
π Herculis.

β Canis Min.?
ζ Tauri.
δ Draconis.
μ Geminorum.
γ Bootis.
ε Geminorum.
δ Herculis.
δ Geminorum.
q Orionis.
β Cephei.
θ Ursæ.
Ursæ.
η Aurigæ.
γ Lyræ.
η Geminorum.
γ Cephei.
κ Ursæ.
ε Cassiopeiæ.
θ Aquilæ.
δ Andromedæ.
η Herculis.
ζ Pegasi.
ε Tauri.
ζ Cygni.
α Trianguli B.
ζ Aurigæ.
λ Aquilæ.
μ Herculis.
ι Draconis.
μ Pegasi.
χ Draconis.
η Cassiopeiæ.
θ Pegasi.
ζ Cassiopeiæ.
δ Aquilæ.
μ Herculis.

ξ Geminorum.
κ Geminorum.
ζ Cephei.
η Cephei.
ο Ursæ?
λ Geminorum.
θ Geminorum.
ο Andromedæ.
β Delphini.
ζ Geminorum.
α Delphini.
c Arietis.
ι Geminorum.
λ Tauri.
ο Tauri.
ξ Tauri.
η Piscium.
ι Herculis.
δ Bootis.
γ Trianguli B.
α Draconis.
γ Tauri.
γ Arietis.
τ Cygni.
ι Cephei.
ξ Herculis.
ο Herculis.
ρ Cygni.
ι Pegasi.
ξ Pegasi.
δ Aurigæ.
γ Sagittæ.
γ Ophiuchi.
φ Draconis.
γ Delphini.
α Piscium.

---

b. Southern Stars.

Sirius.
Canopus.
α Centauri.
Rigel.
α Eridani.
β Centauri.
α Crucis.
Antares.
Spica.
Fomalhaut.
β Crucis.
α Gruis.
γ Crucis.
ε Orionis.
ε Canis.
λ Scorpii.
ζ Orionis.
β Argus.
γ Argus.
ε Argus.
α Trianguli A
ε Sagittarii.
θ Scorpii.
α Hydræ.
δ Canis.
α Pavonis.
β Gruis.
σ Sagittarii.
δ Argus.
β Ceti.
λ Argus.
θ Centauri.

β Canis.
κ Orionis.
δ Orionis.
γ Centauri.
ε Scorpii.
ζ Argus.
α Phœnicis.
ι Argus.
α Lupi.
ε Centauri.
η Canis.
β Aquarii.
δ Scorpii.
η Ophiuchi.
γ Corvi.
η Centauri.
κ Argus.
β Corvi.
β Scorpii.
ζ Centauri.
ζ Ophiuchi.
α Aquarii.
π Argus.
δ Centauri.
α Leporis.
δ Ophiuchi.
ζ Sagittarii.
π Ophiuchi.
β Libræ.
γ Virginis.
μ Argus.
δ Sagittarii.

α Libræ.
λ Sagittarii.
β Lupi.
α Columbæ.
ι Centauri.
δ Capricorni.
δ Corvi.
β Eridani.
θ Argus.
β Hydri.
ε Corvi.
β Aræ.
α Toucani.
β Capricorni.
ρ Argus.
π Scorpii.
β Leporis.
γ Lupi.
υ Scorpii.
ι Orionis.
α Aræ.
π Sagittarii.
α Muscæ.
α Hydri?
τ Scorpii.
ζ Hydræ.
γ Hydræ.
β Trianguli A.
σ Scorpii.
τ Argus.
γ Triang. A.
η Serpentis.

ι Scorpii.
υ Argus.
λ Crucis.
γ Sagittarii.
ν Hydræ.
κ Centauri.
N Velorum.
β Columbæ.
ζ Canis.
γ Gruis.
α Indi.
β Muscæ.
λ Centauri.
γ Tubi.
γ Hydri.
ω Argus.
ε Hydræ.
θ Eridani.
ν Argus.
ξ Argus.
o^2 Canis.
π Hydræ.
β Tubi.
α Pictoris.
φ Sagittarii.
α Circini.
σ Argûs.
α Doradûs.
β Phœnicis.
δ Aquarii.
ζ Scorpii.

Section II.

TERRESTRIAL MAGNETISM.

BY LIEUT COLONEL EDWARD SABINE,

OF THE ROYAL ARTILLERY.

1. The magnetic observations which are at present making by naval officers have for their object the determination of the *amount* and *direction* of the *Earth's magnetic force* in different parts of the globe.

2. The amount of the magnetic force at any point of the Earth's surface may either be measured in *absolute* value, or its *ratio* may be ascertained to the value of the force at another station where its absolute measure is already known. No means have yet been devised for measuring absolute values at sea; consequently, all determinations of the magnetic force on board ships are necessarily of the relative class; these give the ratio, or proportion, which the force at the geographical position in which the ship is at the time when an observation is made, bears to its value at some land station which is included in the same series of relative observations, but where an absolute determination has also been made. Ships are therefore supplied with instruments for both absolute and relative determinations; the latter to be used at sea, and on land at times when the ship is in harbour; the former to be used exclusively on land.

Absolute *Measurement of the Magnetic Force.*

3. No satisfactory method has yet been brought into practice for the direct absolute measurement in one operation of the *whole* magnetic force of the Earth (called the "total force") at any particular point of its surface. But that portion of the force which acts in a direction parallel to the surface of the Earth (called the "horizontal component") may be measured with considerable accuracy by a process, of which the following brief description may suffice to give a general idea. If a magnet be suspended horizontally by a few fibres of silk, and made to vibrate in the horizontal plane on either side of its position of rest, the square of the number of vibrations in a given time is a measure of the horizontal component of the magnetic force of the Earth, but is also dependent on the individual properties of the magnet employed. These properties influence the time of vibration in two respects: first, by the greater or less magnetic force which the magnet itself possesses; and, secondly, by the effect which the form and weight of the magnet produce on the time of vibration. The latter effect may be eliminated when the moment of inertia of the magnet is learnt; and this may either be calculated by known rules, or may be ascertained experimentally by vibrating the magnet 1° in its usual state, and 2° with its moment of inertia increased by a known amount. The influence of the magnetic force possessed by the magnet may also be eliminated by ascertaining its magnetic moment. This is accomplished by using it to deflect a second magnet similarly suspended in another apparatus. The deflecting magnet is placed at one or more exactly

measured distances from the centre of the suspended magnet, and perpendicular to it. The deflections thus produced, (*i.e.* the angular differences in the positions of rest of the suspended magnet, 1° when influenced solely by the Earth's magnetism, and 2° when in equilibrium between the Earth's magnetism and that of the deflecting magnet at the distances employed,) furnish the *ratio* of the forces exerted respectively by the Earth s force and that of the magnet; and as the *product* of the same two forces is given by the vibrations of the deflecting magnet when suspended as in the experiments first described, the values of either force may be separately ascertained. The influence of the magnetism of the magnet, and of its form and weight, being thus eliminated, a measure is finally obtained of the force of the Earth's magnetism, independent of the individual properties of the magnet employed in the determination.

4. The numerical expression by which the measure of the Earth's force thus obtained is denoted, depends on the units of time, of space, and of mass employed in the measurements and calculation. In conformity with the Instructions published under the authority of the Royal Society, a second of time, a foot of space, and a grain of mass, are the units so employed; and the horizontal component of the Earth's magnetic force has been found, by the observations hitherto made, to vary at different points of the Earth's surface from 0 to about 8·4 of the scale founded on the units which have been specified.

5. Wherever the horizontal component of the force has been ascertained in absolute measure, there also, if the magnetic direction be known, the "total force" in abso-

lute measure is determined; since it consists of the horizontal component multiplied by the secant of the angle which the magnetic direction makes with the horizon. As ships are supplied with instruments by which this angle, called the dip or inclination of the needle, is measured, the observations on land, when the ship is in harbour, give determinations of the total force, which serve as *base determinations*, to which are referred the *relative* results obtained at sea in the passage from one station of well-assured *absolute* determination to another;—a practice corresponding to that which prevails in determinations of longitude, where stations of well-assured longitude are taken as *base stations*, to which intermediate observations are referred. The total force of the Earth's magnetism, expressed in the scale in which the British units already referred to are employed, has been found to vary at different points of the Earth's surface where observations have hitherto been made, from about 6·4 to 15·8. Before the practice was adopted of determining absolute values, various relative scales were employed, not always commensurable with each other. The one most generally used (and which still continues to be very frequently referred to), was founded on the time of vibration of a magnet observed by M. de Humboldt about the commencement of the present century, at a station in South America where the direction of the dipping-needle was horizontal; a condition which was for some time erroneously supposed to be an indication of the minimum of magnetic force at the Earth's surface. From a comparison of the times of vibration of M. de Humboldt's magnet in South America and in Paris, the ratio of the magnetic force at Paris to what was supposed to be its minimum was inferred; and from the result so obtained, combined with a similar comparison made by myself between Paris and London in 1827 with several magnets, the ratio of the force in London to that of M. de Humboldt's original station in South America has been inferred to be 1·372 to 1·000. This is the origin of the number 1·372, which has been generally employed by British observers, not furnished with the means of making absolute determinations, to express the value of the magnetic force at their base station, viz., London. The essential disadvantage, however, under which any relative scale of the nature referred to labours, is, that the magnetic force of the Earth has been found to be subject to secular variations, so that at no one spot on the surface of the globe can the intensity be assumed to remain constant, and thus to afford a secure unvarying basis for such a scale; whereas by absolute measurements, we are not only enabled to compare numerically with one another the results of experiments made in the most distant parts of the globe, with apparatus not previously compared, but we also furnish the means of comparing hereafter the intensity which exists at the present epoch, with that which may be found at future periods. It is probable from these and other considerations, that the employment of mere relative scales will shortly be entirely superseded by the general adoption of a scale in which the value of the force is expressed in terms of a fixed and unchanging unit.

6. The instrument with which the absolute value of the horizontal component of the force is measured is called the Unifilar Magnetometer; its description, and that of the process by which results are obtained with it, are given in

Appendix No. 1. A tolerably practised observer will complete the process by which a measure of the absolute horizontal force is obtained in about two hours, including the time required for setting up and adjusting the instrument. It is desirable that there should be at least five or six repetitions at places which are to serve as base stations; and also, as a spare magnet is always supplied, that both magnets should be employed. There are certain constants, [such as the moment of inertia of the magnet and stirrup in which it rests, the change which the magnetic moment of the magnet undergoes from an alteration of one degree of temperature, and the coefficient in the correction required for an increase of force which the magnet, in certain positions in which it may be used, may receive by induction from the earth,] which have to be determined for each magnet once for all, and require for their determination apparatus which is not afterwards needed: these constants have hitherto been usually determined at Woolwich before the instrument is put into the hands of the officer who is to use it elsewhere.

Relative *Measurements of the Magnetic Force.*

7. These are the observations which are made at sea, to determine the ratio of the total force in the geographical position of the ship at the time when the observation is made, to its value at some base station where the instrument has been landed and used in precisely similar observations to those made on board ship. The instrument is the well-known apparatus devised by Mr. Fox, which has contributed more to a knowledge of the geographical distribution of terrestrial magnetism than any

other recent invention. The following brief description may serve to give a general idea of the apparatus and of the mode of obtaining results with it, more full directions for its use being given in Appendix No. 2. It consists of a dipping-needle and graduated circle, differing little from the accustomed form of an Inclinometer, except that the needle is supported by the ends of the axle, which terminate in cylinders of small diameter working in jewelled holes. A small grooved wheel is carried on the axle, and receives a thread of unspun silk, furnished at each extremity with hooks to which small weights may be attached, for the purpose of deflecting the needle from its position of rest in the magnetic direction, and causing it to take up a new position, in which it is in equilibrium between the opposing forces of the Earth's magnetism and of the deflecting weight. The weight being constant, and the magnetism of the needle assumed to be so, the intensity of the Earth's magnetic force in different localities is inversely as the sines of the angles of deflection. For greater accuracy, several constant weights are employed on each occasion; and each weight is successively attached to each of the two hooks, a mean being taken of the deflections on either side of the position of rest. The apparatus when used at sea is placed on a gimball table, by which the motion of the vessel is greatly counteracted; and when the weather does not permit the manipulation of the weights, deflecting magnets are substituted, the operation of which may be understood from the detailed instructions in Appendix No. 2. With the gimball table as recently constructed, it is found that but very few days occur in which the angles of deflection, either with weights or deflectors, cannot be satisfactorily ascertained by a careful observer. It is necessary that a spot should be selected for the observations to be made on board ship, which should have as little iron as possible within 5 or 6 feet of it; and that the instrument should always be used in the spot so selected. The mode of investigating and of eliminating (when these precautions are taken) the influence on the results of the iron contained in the ship is explained in Appendix No. 4. It must be carefully borne in mind, that the inverse proportionality of the sines of the angles of deflection to the variations of the earth's magnetism, is only true when the magnetism of the needle has not varied; and although the needles made by Falmouth artists, under Mr. Fox's own superintendence, have generally proved most remarkable in preserving their magnetism unchanged for years and in all climates, it is desirable that reference to a base station should be made as often and with as short intervals as may be convenient; and evidence must always be furnished that the magnetism of the needle has not changed in a certain interval, before the relative determinations made during that interval can have weight. The more frequently references are made to base stations at which the value of the magnetic force is known, the less danger exists that the labour bestowed on observations at sea will prove unproductive; and the more stations are multiplied which afford opportunities of such reference, the greater become the facilities for accurate determinations at sea.

Direction of the Earth's Magnetic Force.

8. The direction of the Earth's magnetic force undergoes every possible variation at different parts of the Earth's surface. For the purpose of determining and representing this direction, it has long been customary to refer it to two planes—the horizontal and the vertical planes—and to take the geographical north as the zero of the horizontal plane, and the horizontal line as the zero of the vertical plane. The *declination* (or *variation*, as it is more usually called by naval men) is the angular difference, measured on the horizontal plane, between the direction of the north end of a magnet or needle and the geographical north point; and the *inclination* (or *dip*, as it is frequently called) is the angular difference, measured on the vertical plane, between the direction of the same north end of a magnet or needle and the horizontal zero point. (The north end of a magnet here spoken of is that end which in Europe points towards the north, and dips below the horizon.) The declination is called West, if the direction of the north end of the magnet or needle is to the west of the geographical north, and is reckoned from 0° to 180°, passing from North through West to South. In like manner, the declination is called East, if the direction of the north end of the needle is to the east of the geographical north, and is reckoned from 0° to 180°, passing from North through East to South. The positive and negative signs are also sometimes applied instead of the terms West and East, in which case + signifies West, and − East Declination.

The Inclination is counted positive, or has the sign

plus prefixed, when the north end of the needle inclines below the horizon; and is counted negative, or has the *minus* sign prefixed, when the north end of the needle inclines above the horizon. Sometimes, instead of the signs + and −, the terms North and South are used, in which case North Inclination or Dip is when the north end of the needle dips below the horizon, and South Inclination or Dip is when the south end of the needle dips below the horizon. Thus an Inclination of − 30° is equivalent to 30° South Dip.

9. The Declination is measured by the azimuth compass, an instrument too well known to naval officers to require any description here, or any directions for the method of observing with it either on land or at sea. As now made, under the superintendence of Captain Johnson, R.N., and on the plan recommended by the Committee for the Improvement of Ships' Compasses, the azimuth compass in the hands of a careful observer, attentive to the practical rules published by the Admiralty for ascertaining the deviations of the compass caused by the iron of a ship, will give results, both at sea and on land, which leave little to be desired. The use of the dipping needle, which measures the inclination, not being so generally familiar to naval officers, full directions for its employment are given in Appendix No. 3.

Local Attraction.

10. It has been found that the results of magnetic observations, whether of the declination, inclination, or of the intensity of the magnetic force, are liable to be influenced by local attraction proceeding from the rocks or soil in the

vicinity of the instrument, and particularly so at stations where the rocks are of igneous character, such as traps, basalts, granites, &c. As a precautionary measure, therefore, magnetical instruments should always be used on stands which raise them 3 or 4 feet above the ground; and those stations are to be preferred of which the geological character is sedimentary or alluvial. Stations of igneous character, though less eligible for obtaining results which show the correct magnetical elements corresponding to the geographical position of the station, may nevertheless be serviceable as stations of comparison between the land and sea instruments; but for this purpose it is essential that the different instruments to be compared should be used precisely on one and the same spot, in which case the local attraction may be supposed to be a constant quantity. And if the station be one frequently resorted to by vessels from which magnetic observations are made, it is desirable that the spot should be susceptible of a definite and well-recognisable description.

At sea, from the quantity of iron which a ship contains, it is scarcely possible that its influence on the instruments should be altogether avoided; but from the circumstance that the greater part of the most influential iron is in fixed positions in the ship, it has been proved by sufficient experience, that by a proper selection of the place in which a magnetic instrument is used on board ship, and by a certain process of observation (repeated whenever the ship has undergone any considerable changes of geographical position), the influence of her iron is susceptible of a sufficiently approximate calculation, and of being eliminated accordingly.

The importance, and in some degree the novelty, of this part of the subject has made it appear desirable that it should be treated somewhat fully in Appendix No. 4.

Summary of the Observations to be made.

11. An officer, therefore, who purposes to make magnetic observations, or to cause them to be made on board his ship, has to attend to the following points:—He must take care that he obtains the instruments some days before the ship is ready for sea, in order that he may assure himself that they are all complete, and that, if inexperienced in their use, he, or the observer whom he selects, may have some preliminary practice with them. He will then have to determine the constants, index corrections, &c. (unless these shall have been furnished with the instruments), and to make the observations required for a base station, with the needles which are to be employed in the relative determinations of the magnetic force and dip at sea. Positions will then have to be selected on board for the standard compass, and for Fox's apparatus, and the pillar for the one, and the gimball stand for the other, fitted accordingly. When the ship is ready for sea, the observations which are directed in the Admiralty Instructions for ascertaining the deviations of the compass caused by the iron of the ship are to be made on board; and when the ship is swung round for this purpose, the deviations of the dip and of the force must be also observed, on the sixteen, or at least on the eight, principal points of the compass, with Fox's apparatus used at the spot selected for it.

This completes the preparations to be made before the ship's departure. Whilst at sea, the observations of dip and intensity described in Appendix No. 2, as well as those of the variation by the standard compass, should be made daily, whenever the weather and other circumstances permit. Whenever the ship is in harbour, and time and opportunities are suitable, it is desirable that the instruments should be taken on shore, to a spot selected as least likely to be influenced by any local attraction; and that the declination, inclination, and absolute horizontal force should there be determined, and the comparative observations made with Fox's apparatus. If the ship has materially changed her geographical position since the last occasion when the deviations were ascertained, or if changes have been made in her equipment by which the deviations may have been affected, it is desirable that the process for their examination should be repeated; and lastly, the harbour observations here described should not fail to be repeated whenever the ship finally returns to England.

Record and Transmission of the Observations.

12. Blank forms are supplied for the entry of observations of all classes, and for the first or uncorrected calculation of those which require that process to be gone through at the time. It is desirable that the forms should be filled up in duplicate, and that one copy should be retained, and the other sent to England from time to time, as soon as circumstances make it convenient. On their arrival they should be immediately examined, and any suggestion to which they may give rise communicated at once to the observer.*

* This has hitherto been done on all occasions when practicable, and it is very desirable that it should always continue to be done.

Application of the Results.

13. The observations when thus received require that the several corrections arising from the influence of the iron, the variations of temperature, the changes in the magnetic force of the magnets, and from various other sources, should be sought out, computed and applied, and the true or corrected results finally derived. These form the materials from which it is intended to construct maps, showing the variations of the magnetic force, and of the magnetic direction in its two co-ordinates of inclination and declination, corresponding to the present epoch, over the whole surface of the globe. The variations of the three elements are shown on these maps, by lines connecting, for example, in the maps of the magnetic force, those points, where the intensity is observed to be the same;—in the maps of the inclination, those points where the inclination is observed to be the same;—and in the maps of the declination, those points where the declination is observed to be the same. These lines are known by the names of Isodynamic, Isoclinal, and Isogonic lines. The Isogonic lines, which form the maps of the declination (or variation charts), have a direct practical importance and value in navigation, which in a notice addressed to naval officers needs not to be dwelt on. In theoretical respects, the Isodynamic and Isoclinal lines are not less essential; the three form the basis of a systematic view of terrestrial magnetism, as it manifests itself to us on the surface of the globe.

The mode in which the results are made to contribute to the formation of these maps is the following:—The results of the three elements finally corrected are entered, each in its proper geographical position, on maps on a large scale, severally appropriated to the force, the inclination, and the declination. Each result has a small characteristic mark denoting the observer. When any portion of the globe is sufficiently covered by the results of observations in proper distribution, the isophænomenal lines are drawn for that portion of the globe in correspondence with the observations, with a free hand, but with a careful judgment, aided occasionally by a process of calculation which it is not necessary here to describe. From these maps tables of double entry are formed, having the latitude at the side, and the longitude along the top of the page, and the values of the magnetic elements corresponding to the several latitudes and longitudes are placed at the points of intersection. By proper care in the process, the step of forming the tables from the maps need involve no additional uncertainty whatsoever. Maps and tables thus prepared will, when completed, form an experimental theory of terrestrial magnetism, in which the facts of nature will be shown with greater or less exactness, in proportion as the observations are numerous, correct, and suitably distributed, and as they are more or less correctly represented in the maps. Mathematical formulæ, based on general mathematical views, having numerical coefficients of which the values are derived from these maps, may also serve for the computation of the magnetic elements at any geographical position on the surface of the globe; and if the points taken from the maps to serve as the basis of the numerical values of the coefficients are sufficiently numerous, and have a proper distribution over the surface of the globe, and if the formulæ are carried to a sufficient number of terms, it may be expected that the elements computed from them will have the same degree of exactness as the maps from which their coefficients are taken.

It may be natural at this stage to inquire what prospect exists of being able to complete a work of this magnitude within a reasonable time; and to this question a more satisfactory reply can be returned than may perhaps be generally anticipated. Nearly three-fourths of the surface of the globe being covered by the ocean, it is chiefly by naval surveys that the materials for such a work can be collected. By the zealous and unwearying assiduity of British officers, acting under the sanction and with the approval of the Lords Commissioners of the Admiralty, and in some instances in expeditions specially appointed for the purpose, magnetic observations designed expressly for the object above-mentioned, and conducted upon a uniform system, have been extended, since their commencement in 1839, over nearly all the accessible parts of the ocean.

Of these surveys, the results of some have been already deduced and published in the Philosophical Transactions, the expense of publication having been borne conjointly by the Government and by the Royal Society; the results of others are undergoing the process of calculation and arrangement for publication; and in a third class, more numerous than either of the other two, the observations themselves are still in progress. In the class first referred to, viz surveys in which the observations have been completed and the results deduced and published, may be named a portion (about the half) of the observations of the Antarctic expedition under Sir James Clark Ross: those of the expedition under Lieutenant Moore, R.N., and Lieutenant Clerk, Royal Artillery, also to the Antarctic Ocean; of Captain Sulivan, from England to the Falkland Islands and back; of Sir Edward Belcher, (in H.M.S. Sulphur,) on the north-west coast of America, and in the Pacific and Indian Oceans; of Lieutenant Alexander Smith, R.N., and Lieutenant Dayman, R.N., in voyages between England and Van Diemen Island. In the second class, viz. surveys completed and in progress of reduction but not yet published, may be enumerated the remaining portions of the observations made in Sir James Clark Ross's Antarctic expedition; a small but valuable collection by the Niger expedition; another by Sir Robert Schomburgk in Guyana; two series on the coast of New Holland, one by Captain Wickham, and the other by Lieutenant (since Captain) Stokes; the observations made in a special expedition to Hudson's Bay and back by Lieutenant Moore; and two extensive series, one by Captain Shadwell, in the hydrographical expedition of Captain Blackwood to Torres Strait and round New Holland; and one by Sir Edward Belcher in H.M.S. Samarang, in the Indian and Chinese Seas.

Amongst the observations now in progress may be named those of the expeditions to the Arctic Polar Sea under Sir John Franklin and Sir James Clark Ross; of the hydrographical surveying expeditions of Captain Barnett in Bermuda and the West Indies; of Captain Graves in the Mediterranean; of Captain Kellett in the Pacific; of Captain Bayfield in the Gulf and River St. Lawrence;

of Captain Owen Stanley in Torres Strait and the coasts of New Guinea; of Lieutenant (now Commander) Moore in the Pacific and Behring's Strait; and of Captain Stokes in New Zealand. To these must be added a very valuable magnetic survey still in progress amongst the islands of the Indian Archipelago by Captain Elliot of the Madras Engineers, at the expense of the East India Company.

When it is borne in mind that several of the above-named surveys include periods of three or four years, and in some instances not only determinations at the several ports and harbours which may have been visited, but also daily observations (weather permitting) of the three magnetic elements at sea in passages from port to port, the accumulation of materials, and their already extensive distribution over the surface of the globe, may in some degree be judged of. No part of the globe has yet however been considered so far completed, that its magnetic curves have been finally drawn, or the tables of the magnetic elements corresponding to geographical positions finally derived from them. The results of each individual series are computed and published as soon after they have been received in a complete state, as can be accomplished by the establishment at Woolwich, which is very limited in comparison with the duties it endeavours to perform. The magnetic lines indicated by each series are also provisionally drawn; but the final co-ordination of the different series, and their combination with the results of the magnetic surveys carrying on at the same time by governments and individuals in various parts of the remaining fourth of the globe which is occupied by land, is deferred until the accumulation of materials in the more frequented portions shall appear sufficient for the purpose, and until in the less frequented portions there shall be no immediate prospect of any further considerable accession. The final period at which the work may be completed cannot but be accelerated by the step which the Lords Commissioners of the Admiralty have taken, in desiring the present notice to be drawn up, and in holding out inducements to naval officers of professional and other advantages, as a recompense for what may in some respects be regarded as extra-professional services. With due hesitation in expressing an opinion on what is future, and contingent on many circumstances, it may be stated that hopes are entertained that the year 1851 may see the work completed.

In prosecuting a work of this general and purely experimental character, unconnected with hypothesis of any sort, the phenomena of all parts of the globe must be viewed in the abstract as possessing an equal importance; and it does not appear desirable, therefore, to name any one of the lines, whether isogonic, isoclinal, or isodynamic, as deserving of special attention in preference to others. There is one direction, however, which may be safely given, and which it may be well to remember at all times, viz., that "the value of each new station is directly proportional to its distance from those where observations have already been made;" and in this point of view it may be useful to notice, that up to the present time fewer observations have been made in the Pacific than either in the Atlantic or the Indian Oceans.

APPENDIX No. 1.

Description and Use of the Unifilar Magnetometer.

The Unifilar Magnetometer consists of two parts—an apparatus for deflection, and an apparatus for vibration: these correspond with the two parts of the process by which the absolute horizontal force is determined; the experiments of deflection consist in observing the angular deflection of a suspended magnet produced by the influence of a second magnet, which is placed on a support at one or more known distances from the suspended magnet, and in a line drawn from its centre perpendicular to its direction: the experiments of vibration consist in suspending the magnet which was used as the deflecting magnet in the experiments of deflection, and observing its time of vibration. By the first part of the process (or the experiments of deflection) we obtain the *ratio* of the magnetic moment of the deflecting magnet to the Earth's horizontal magnetic force at the place of observation; the latter being to the former as 1 to the sine of the angle of deflection multiplied by half the cube of the distance employed;—or if m denote the magnetic moment of the needle, X the Earth s horizontal force, r the distance apart of the centres of the magnets, and u the angle of deflection, the expression is—

$$\frac{m}{X} = \frac{1}{2} r^3 \sin u.$$

By the second part of the process (or the experiments of vibration) the *product* of the same two qualities is obtained: being the quotient of a constant, which we may call π^2 K (see p. 38), divided by the square of the time of vibration: or if T be the time of vibration,

$$m\,X = \frac{\pi^2 K}{T^2}$$

The values of m X and $\frac{m}{X}$ being known, those of m and X may be obtained separately: for if we call $m\,X = \alpha$ and $\frac{m}{X} = \beta$, m (the magnetic moment) $= \sqrt{\alpha\,\beta}$, and X (the horizontal component of the Earth's magnetic force) $= \sqrt{\frac{\alpha}{\beta}}$.

Deflection Experiments.

1. Place the tripod base and copper box upon the stand, and screw on the torsion tube; form a thread capable of carrying the smaller

magnet, *i. e.*, the magnet of 3·00 inches in length (a single fibre of silk is generally strong enough to support it); affix the deflecting rod, and the telescope and scale; level the base circle very correctly, suspend the 3-inch magnet, and raise it until it is in the same horizontal plane with the deflecting magnet when resting on the supports.

2. Place the 3·67-inch (or deflecting) magnet on the graduated support to the west of the suspended magnet, with its centre at the distance determined upon, and with the north end towards the East. Turn the telescope in azimuth until the central division of the scale attached to the telescope is on the vertical wire; write down the readings of the verniers, and the height of the thermometer placed near the deflecting magnet.

3. Reverse the deflecting magnet, placing the south end where the north was, and bringing its centre precisely to the same distance from the suspended magnet as before. Turn the telescope until the central division of the scale is again on the vertical wire, and write down the readings, together with the height of the thermometer.

4. Repeat these observations by reversing the deflecting magnet again and again, until, judging by the agreement of the several results, sufficient accuracy appears to be obtained. Two or three repetitions will generally be found to suffice.

5. Place the deflecting magnet at the same distance on the opposite support, and repeat the same series of observations, commencing as before with the north end of the magnet towards the East. Take a mean of the readings of the circle in the 1st and 3rd positions of the deflecting magnet, and another mean of the readings in the 2nd and 4th positions; half the difference of these means will be the angle u corresponding to the distance r. The distance r should be such as to produce a sufficiently large deflection in the suspended magnet. At Woolwich, where the value of the horizontal component of the terrestrial force is 3·7, the deflecting magnets of 3·67 inches in length, when recently magnetized, will usually produce a deflection of about 10°, when the distance from centre to centre of the magnets is 1·3 foot. But with a constant distance the deflection would increase in localities where the horizontal component is less, and diminish where it is greater: the deflection would also be diminished if the deflecting magnet lost any portion of its original charge. It is not desirable that the deflection should be less than 6°.

When the distance required to produce a sufficient deflection is less than five times the length of the deflecting magnet, the amount of the deflection is liable to be influenced by the distribution of the magnetism in the magnets employed, which does not follow the same law in all magnets. In such cases it is necessary to employ two or more distances for the purpose of eliminating this effect. If two distances are employed, they should be to each other in the proportion of 1 to 1·3 nearly; or for magnets of the specified length and strength, 1 foot and 1·3 foot may generally be found convenient and suitable distances, in which case the lesser distance should be denoted by r, and the greater by $r_{\prime}$, the corresponding deflections being u and $u_{\prime}$.

Where two distances are employed, no alteration is required in the experiments of vibration; but in those of deflection it will be found convenient to adopt the following routine:—

1. Observe the angle of deflection with the north end of the deflecting magnet alternately to the east and to the west (with the proper number of repetitions), the magnet being placed on the eastern arm, and at the least of the two distances.

2. Go through the same process with the magnet still on the eastern arm, but at the greater of the two distances.

3. Repeat No. 2 on the western arm, and

4. Repeat No. 1 on the western arm, the order of the distances in Nos. 1 and 2 being inverted in Nos. 3 and 4.

If the distance required to produce a sufficient deflection should be less than four times the length of the deflecting magnet, it may be desirable to obtain three or more values of u (u, $u_{\prime}$, $u_{\prime\prime}$, &c.) at three or more distances (r, $r_{\prime}$, $r_{\prime\prime}$, &c.); in such case it is convenient for the calculation that the distances should have a common difference; as, for example, (in decimals of a foot) ·9, 1·1, and 1·3; 1·1 being supposed to be the distance ordinarily employed when time and circumstances will not permit more than a single distance; and the other two distances to be employed in addition when there may be full leisure for the purpose.

Vibration Experiments.

1. For these experiments the tripod stand carries a table top, which is furnished with an azimuthal motion, and can be clamped to the tripod. A telescope with a scale attached to it for measuring the arc of vibration is fixed to the table, as is also a rectangular box of wood, with a glazed side, furnished with a tube for the suspension of the 3·67 inch (or deflecting) magnet. The table must be levelled by means of the foot-screws, and turned in azimuth until the mirror attached to the stirrup in which the magnet is supported reflects in the telescope the centre division of the scale attached to the telescope. The magnet rests during the experiments of vibration in a stirrup carrying a mirror, and is suspended by a silk thread of which the ratio of the torsion force to the magnetic directive force must be ascertained, by turning the index of the torsion circle 90°, first in one direction and then in the other, and taking the mean of the corresponding angles through which the magnet

is deflected. Then if H denotes the co-efficient of the torsion force, and F the Earth's directive force,

$$\frac{H}{F} = \frac{\text{the mean angle (in minutes of arc)}}{90^\circ - \text{the mean angle (in minutes of arc)}}$$

2. The magnet, being at rest, must be made to vibrate in small arcs by means of another magnet, and the time of 300 vibrations observed in the following manner:—The arc of vibration being about 60 divisions (or minutes) on either side of the central division, note the time when the central division crosses the vertical wire of the telescope, passing from left to right; do the same when the central division passes from left to right for the third, fifth, seventh, ninth, and eleventh times, corresponding to the completion of vibrations 4, 8, 12, 16, 20, and 24. This will give the approximate interval of time in which 20 vibrations are performed; and will indicate to the observer, without requiring him to count the intermediate vibrations, when he should be prepared to observe the completion of the 60th, 120th, 180th, 240th, and 300th vibrations. At the 300th vibration he should recommence taking the time of the 300th, 304th, 308th, 312th, 316th, 320th, and 324th vibrations. He will then have seven partial results for the time in which 300 vibrations are performed, viz.—

0 to 300	12 to 312
4 to 304	16 to 316
8 to 308	20 to 320
	24 to 324

of which seven partial results he should take the mean.

The temperature of the magnet should be observed by a thermometer placed in the box with the magnet, before and after the experiments by which its time of vibration is determined.

The data required from the observer on each occasion are the following:—The mean time of 300 vibrations, and the particulars from which it is derived (or the partial results); the commencing and concluding arcs of vibration; the temperature of the magnet; the rate of the chronometer by which the times are observed; and the ratio of the torsion force to the Earth's directive force, or $\frac{H}{F}$.

If, then, T′ is the observed time of a single vibration, and T the corrected time, the corrections may be computed by the following formula:—

$$T = T'\left(1 + \frac{H}{2F} - \frac{a\,a'}{16} - \frac{x}{86400}\right)$$

in which a and a' are the commencing and concluding arcs in parts of radius, and x the rate of the chronometer. When the commencing arc of vibration is as small as above directed, the correction for it may be disregarded, except when great accuracy is sought.

It is desirable to commence the observations for the determination of the absolute horizontal force with the experiments of vibration; then to make the series of deflection experiments; and then to repeat the experiments of vibration. The experiments of deflection and vibration thus described will complete a single determination. There should be three such determinations with each of the 3·67-inch magnets at a base station; and all should not be made on the same day, lest it should happen to be a day on which the magnetism of the Earth is disturbed.

The constant π^2 K (page 34) is obtained from π, denoting the ratio of the circumference of a circle to its diameter, and K the moment of inertia of the magnet, including the stirrup (and its appendages) in which the magnet rests during the experiment of vibration. To determine the moment of inertia, the magnet, stirrup, &c. is vibrated a second time with the addition of a metal ring, of which the exterior and interior diameters, as well as the weight, are accurately known. If r and r denote the exterior and interior radii, (in decimals of a foot,) and w the weight, (in grains,) the moment of the inertia of the ring is

$$K' = \frac{1}{2}(r^2 + r'^2)\,W;$$

and if T denote the time of vibration without the ring, and T′ the time of vibration with the addition of the ring, the moment of the inertia of the magnet and stirrup in the vibrations without the ring is given by the formula,

$$K = K'\left(\frac{T^2}{T'^2 - T^2}\right).$$

As this determination requires several repetitions, it is usually made before the instrument is put into the hands of the observer; but as its value must be found afresh if any alteration be made at any time in the stirrup or its appendages, a ring is always supplied for the use of the observer.

If the temperature of the magnet be not the same in the experiments of deflection and vibration, the value of m X as observed at the time of the experiments of vibration, must be reduced to that which it had at the time of the experiments of deflection. For this purpose it is necessary to know the amount of change in the magnetic moment of the magnet occasioned by one degree of temperature. This is called the temperature coefficient, and is usually denoted by q. As its determination requires apparatus with which ships are not supplied, it has been customary to furnish it with the instrument. Then if t and t_0 be the temperature of the magnet at the times of deflection and vibration respectively, m X derived from the experiments of vibration, becomes when corrected to the temperature of the experiments of vibration;

$$m\,\mathrm{X}\left[1 + q\,(t - t_0)\right]$$

The absolute horizontal force determined by the foregoing process will finally require a very small correction, on account of the circumstance, that the magnetic moment of the deflecting magnet is not strictly identical in the different positions in which it is placed in the experiments of deflection and vibration, as it is perpendicular or nearly so to the magnetic meridian in the first case, and coincides with it in the other. In the first case, the magnetic moment is that proper to the bar itself; in the second case it is augmented by the inducing action of the earth. The coefficient for this correction requires a separate apparatus for its determination, and the correction may be applied to the results when finally re-computed at Woolwich, as it requires no special data to be furnished by the observer.

APPENDIX No. 2.

Directions for using Mr. Fox's Apparatus for observing the Magnetic Inclination and Force.

I.—General Remarks.

In fixing the gimball table, it is convenient that it should be so arranged that when the apparatus is placed on it, the zero divisions of the horizontal circle should coincide with the fore and aft midship-line of the ship.

In preparing for an observation at sea, the circle should be turned in azimuth until the vernier of the horizontal circle shows an angle with its zero, corresponding with the difference between the magnetic meridian and the course which the ship is steering. The plane of the circle will then coincide with the magnetic meridian, when the ship is steadily steered. When from circumstances of weather, &c., the steerage is difficult, an assistant is required to indicate to the observer the times when the ship is steady on her course.

The apparatus is usually furnished with three or four needles, one of which is intended to be used on shore for the determination of the true inclination (when no special instrument is provided for the purpose), by the process subsequently described, Appendix 3, in which the poles are reversed The other needles, which are intended for the intensity, are never to have their poles reversed, and care is to be taken not to place them inadvertently near other magnets or iron. Besides the needles, two other magnets are supplied to be used as deflectors. In replacing the needles and deflectors in the travelling box, care should always be taken that the poles of each occupy the places marked for them in the box.

It is desirable to use always the same needle at sea, and to keep it always mounted, clamping it before it is put away for the day; but in case of its undergoing any considerable deterioration from use or accident, one of the other intensity needles may be substituted for it.

When changing the needles at a land station, be very careful not to injure the jewels, or the terminations of the axles of the needles; when a needle is changed it is desirable to hold it chiefly by the grooved wheel; the pivot should first be put into the outer jewelled hole, and the opposite pivot should be carefully guided into the hole at the back whilst the bracket is screwed up.

With respect to the constant weights, it is desirable that the smallest angle of deflection produced by any of the weights employed should not be less than 30°. On account of possible instrumental irregularities it is usual to employ more constant weights than one, with differences between each of half a grain, (as for example, 2 grains, 2½ grains, 3 grains, &c.). Great care is taken that all the weights which have the same nominal value should be equiponderant, but it is desirable if possible to preserve the same identical weights throughout the whole observations of the same relative series.

II.—Observations at Sea.

A.—*Inclination.*

1. *Direct Observation.*—The instrument having been placed on the gimball stand and levelled, and the plane of the circle made parallel to the magnetic meridian, with the face of the circle towards the East, release the needle, which will immediately take approximately the direction of the inclination; rub gently the centre pin at the back with the ivory disk, and read off successively the divisions of the limb indicated by the two ends of the needle; note the readings, which will be +, or positive, when the North pole of the needle dips, and —, or negative, when the South pole of the needle dips: repeat the observation three times, turning the bracket which supports the needle a small quantity between each observation, and being careful to rub the centre pin at the back with the ivory disk whilst reading off. The bracket is turned by means of the screw heads at the back of the circle, and the object of turning it is to cause the ends of the axle of the needle to have different points of bearing on the jewels in each observation.

In reading the divisions on the limb, be careful always to bring the division nearest to the needle to coincide with the corresponding division of the second graduated circle immediately behind it, by which means parallax is avoided.

The mean of the three observations or six readings above described is the apparent inclination by direct observation with the face East.

2. *Observation with Deflectors.*—Having made the preceding observation, screw in the deflector N (or the north pole of a second needle used as a deflector), and adjust the circle at the back by means of its verniers, so that the deflector may be 40° on one side of the division which in the preceding process (§ 1.) was read off as the direct observation with the face East. The needle will then be repelled, and will settle on the opposite side of the dip; read off (always whilst rubbing with the ivory disk) the divisions indicated by the two ends of the needle. Repeat the observation three times, altering the bearings of the ends of the axle between each observation as before directed. Turn the back circle through 80°, so that the deflector may be 40° on the other side of the apparent dip. Move the needle by the bracket, so that it may be deflected on the opposite side of the apparent dip to what it was before, and make three observations. The mean of the six observations or twelve readings is the apparent inclination with a deflector, face East.

Instead of placing the deflector at 40°, another angle, as 45° or 50°, may be taken; or a second angle may be used for the purpose of varying the observations when it may be desired to repeat them; the only essential point being, that the angle at which the deflector is placed should be the same on each side the apparent dip.

Instead of deflector N (or the North pole of a second needle used as a deflector), deflector S (or the South pole of the second needle) may be screwed into the opposite point of the back circle, and six observations taken with it will give as before the apparent inclination with a deflector, face East.

In writing down the observations the following directions must be attended to; if the needle be deflected past the *vertical*, the division of the limb should be read off according to the graduation and noted accordingly, but the mean of the readings must be taken from 180°, in order to give the true arc corresponding to the position of the needle: if it be deflected past the *horizontal*, the readings must be entered as marked on the limb, but with the negative sign prefixed, in which case the mean result will be half the *difference* of the means of the negative and positive readings.

The apparent inclination obtained as above directed, whether by the direct method or with deflectors, requires two corrections to give the true inclination, viz.—1° the index-correction of the particular needle employed, and 2° a correction for the influence of the ship's iron dependent on the direction of her head at the time of observation. The mode of obtaining these corrections will be subsequently explained.

B.—*Intensity.*

3. *With Weights.*—The instrument being on the gimball table and levelled, the plane of the circle parallel to the magnetic meridian, with its face to the East, and the needle shewing the magnetic dip, place the silk carrying the hooks on the grooved wheel; attach one of the constant weights to one of the hooks, and take three readings of the division of the limb at which the needle is in equilibrium, using the precautions already directed of altering the points of support of the axle between each observation, and rubbing with the ivory disk whilst reading off.

If the needle is deflected past the vertical or horizontal, read and enter the angles as already directed under the head of Inclination.

Change the weight to the other hook, when the needle will be deflected to the opposite side of the apparent dip to what it was before, and take three more observations. Half the difference of the mean of the arcs with the weight on either hook is the angle of deflection due to the constant weight employed; or half their *sum*, if one of the arcs was past the horizontal and has therefore the negative sign prefixed.

4. *With Deflectors.*—The instrument being adjusted as already described (and without using the hooks, which are only designed for the observations in which the weights are used), adjust the circle at the back by means of its verniers to the apparent dip, so that the deflectors, when screwed in, may coincide with the line of the dip; the needle will then be repelled to one side; make three observations of the division to which the needle is thus deflected, observing the usual precautions of moving the bracket at the back, reading both ends of the needle, and rubbing with the ivory disk.

Move the needle past the deflector to the other side of the dip by means of the bracket, and take three more observations: if the needle is deflected past the vertical or horizontal, read and enter the angles in the manner already described; half the difference of the arcs on either side of the apparent dip, or half their sum if one be past the horizontal and have the negative sign, will be the angle of deflection produced by the deflector. Instead of the deflectors a second needle may be used as a deflector, either with the end of the needle-case marked N (containing the north pole of the needle) screwed into the arm marked N, or the end marked S screwed into the arm marked S.

The thermometer attached to the circle must be observed at the commencement and close of the observations of intensity, whether with deflectors or weights.

A convenient routine of the observations at sea may be stated as follows:—

1. Take three observations of the apparent dip by the direct observation.

2. Screw in the deflectors N and S, and adjust the back circle to the dip. Make three observations of the angle of deflection produced on either side of the apparent dip; this furnishes one result for the intensity of the magnetic force.

3. Repeat No. 2 with a second needle used as deflector N, which will give a second result for the intensity of the force.

4. Repeat No. 2 with the second needle used as deflector S, which will give a third result for the intensity of the force.

5. Remove the deflector and repeat No. 1, which will give a second result for the apparent inclination.

On days when the weather permits, observe the intensity also by the constant weights.

III.—Observations on Shore.

1. The instrument being adjusted with the plane of the circle coinciding with the magnetic meridian, and the face East, make a complete series of observations of the Inclination with and without deflectors, and of the Intensity with the deflectors and weights, similar in all respects to the observations which have been or which are intended to be made at sea; the needle, deflectors, and weights to be those employed, or to be employed, in the sea-observations.

2. If unfurnished with a separate apparatus for determining the true inclination, substitute in Mr. Fox's apparatus the needle which admits of its poles being reversed (viz. that needle which is not intended to be used in observations of intensity), and obtain the true inclination from the mean of the angles read in four different positions of the instrument, viz. with the face of the circle East and West, and repeating the process with the poles reversed, following the directions for reversing the poles given in Appendix 3. The difference between the true inclination thus obtained, and the apparent inclination with the face East observed with the needle used at sea, ascertained at the several shore stations, furnish one of the data from which the index correction to be applied to the observations made at sea is to be computed.

3. When Mr. Fox's apparatus is furnished with more than one needle for the observations of intensity, each needle must be successively substituted in the shore observations for the needle used at sea, and the inclination as well as the angles of deflection with constant weights observed with it.

APPENDIX No. 3.

Observations of the Inclination, with Barrow's Circle furnished with Microscopes and Verniers.

1. Place the instrument on a tripod stand, and level it by means of the foot-screws; then bring the vertical circle into the magnetic meridian by the following process:—Place the needle designed for the observation of the dip on the agate supports, with the side of the needle on which the letters are inscribed facing the microscopes. Turn the vernier plate so that the microscopes may be nearly in a vertical line; clamp the plate, and set the lower vernier to 90° by the tangent screw. Turn the vertical circle in azimuth, so that its face may be towards the South, and until the North pole of the needle is bisected by the wire of the microscope; raise the Ys and lower gently; if the bisection of the needle has been altered, correct by turning the circle in azimuth. Clamp the horizontal circle, and read off its vernier, calling the reading A. Now set the upper vernier to 90°, unclamp the horizontal circle, and move in azimuth (if required) until the South pole of the needle is bisected by the wire of the upper microscope. Raise the Ys and lower gently; correct the bisection (if necessary) by moving the circle in azimuth; clamp the horizontal circle and read its vernier, calling the reading B. Now unclamp the horizontal circle, and turn the vertical circle 180° in azimuth, so that its face (by which is meant the side on which the microscopes are) which was before to the South may now be to the North. Repeat the process described above, which will give two other readings of the vernier of the horizontal circle, which call C and D. Then

$$\frac{A + B + C + D}{4} = E,$$

the division of the horizontal circle to which the vernier should be set, in order that the plane of the vertical circle may be at *right angles* to the magnetic meridian; therefore, when the vernier is set to $90 \pm E$, the plane of the vertical circle will coincide with the magnetic meridian.

2. The vertical circle being now placed in the magnetic meridian, with its face to the *East*, the needle will direct itself approximately to the inclination; raise it by the Ys and lower it gently on its supports; bring the lower microscope to bisect the North end of the needle, clamp and adjust exactly by the tangent screw, read off the vernier, which will be *Face East, North End.* By means of the tangent screw of the vernier-plate bring the upper microscope to bisect the South end of the needle,

and read its vernier, which will be *face East, South end;* raise the Ys and lower gently; repeat the readings, commencing now with the South end. The mean of the four readings is the inclination with the face *East, poles direct*, or θ_1.

3. Turn the vertical circle 180° in azimuth, and repeat the process in No. 2, taking again the mean of the four readings, which will be $180° - \theta_2$; θ_2 being the inclination with the face *West, poles direct.*

4. The poles of the needle must now be reversed by means of the bar magnets, by the following process:—Take the needle off the agates, holding it by the end which in the preceding observations was a South pole, and which is now to be converted into a North pole; place it with the flat side (which is lettered) uppermost in the wooden frame designed to prevent any injury occurring to the axle, being careful that the end to be made a North pole is placed towards that part of the wooden frame which is marked accordingly; secure the needle by the brass centre-piece, and place the frame with one end towards the right hand and the other towards the left, and let an assistant keep it in that position. Now take the bar-magnets, one in each hand, and let the *North* pole of the bar-magnet be lowermost in the hand which is towards the end of the frame in which that end of the needle is placed which is to be made a *South* pole; and let the *South* pole of the bar-magnet in the other hand be lowermost. Draw the magnet about ten times along the flat side of the needle: the North pole of one bar-magnet being drawn along the end of the needle which is to be made a South pole; and the South pole of the other bar-magnet being drawn along the end of the needle which is to be made a North pole. The needle must then be turned over in the wooden frame, so that its other flat side may become uppermost, which must also be rubbed by the magnets ten times in the manner already described.

The bar-magnets should be held one in each hand, nearly in a vertical position, the lower ends resting on the needle; and must be drawn along the grooves in the wooden frame from near the centre to beyond the ends of the needle. When the process thus described has been gone through, it will be found, on replacing the needle on the agates, that the end which previously dipped below the horizontal line is now inclined above it.

5. The observations in Nos. 2 and 3 must now be repeated, which will give two other mean readings with the face of the circle first East and then West. These readings will be θ_3 (the inclination with the face *East, poles reversed*); and $180° - \theta_4$; θ_4 being the inclination with the face *West, poles reversed.* Then

$$\frac{\theta_1 + \theta_2 + \theta_3 + \theta_4}{4} = \theta,$$

the true magnetic inclination at the place of observation.

6. Two such determinations will generally be found sufficient, but if the results differ from each other more than 3′ or 4′ it is desirable to repeat the observations.

7. On arriving at a new station it is always desirable to magnetise the needle afresh before the observations are commenced. It is indifferent whether an observation is commenced with the end marked A as a North or as a South pole; but it is convenient to call that state of the needle in which the end A is a South pole, and the end B a North pole, "poles direct," and *vice versâ.*

Dr. Lloyd has recently suggested a mode of employing this instrument for measuring the absolute total force in localities where the inclination approaches 90°, and where the usual method of ascertaining the total force from its horizontal component ceases to be satisfactory. It consists in deflecting the dipping-needle from its natural direction in the line of the dip by a second magnet whose magnetic moment is correctly known, placed at one or more accurately measured distances from the centre of the dipping-needle, in a line at right angles to the magnetic axis of the needle, and in observing carefully the amount of the deflections thus produced. For this purpose the vernier-plate of the inclinometer is furnished with two additional arms perpendicular to those which carry the microscopes and verniers; one of these arms is prolonged by an additional piece (removable when not in use) carrying a graduated tube in which the magnet is placed, and in which it can be fixed by binding screws, with its centre at different distances from the centre of the dipping-needle and in the same vertical plane with it. The deflecting magnet being placed in the tube, and fixed at a suitable distance from the dipping-needle, with its North pole towards the needle, and the face of the circle towards the East, the microscopes are brought to coincide with the ends of the needle in the new or deflected direction. The deflecting magnet is then also in its required position, namely, at right angles to the line joining the two extremities of the needle. The arcs are read with the precautions already described in the directions for observing the inclination. The deflecting magnet is then taken out of the tube and replaced at the same distance as before, but with its South pole towards the dipping-needle, and the arcs are again read. This double process is repeated with the vernier-plate turned nearly through a semicircle, so that the deflecting magnet is brought on the other side of the dipping-needle to that which it occupied before; the face of the circle is now turned towards the West, and the observations already described are repeated. The arcs thus obtained give four values for the angle of deflection, the arithmetical mean of which is taken as the deflection (u), corresponding to the distance (r) in decimals of a foot.

If then m be the magnetic moment of the deflecting magnet, and ϕ the total force in absolute measure:—

$$\frac{m}{\phi} = \tfrac{1}{2} r^3 \times \sin u.$$

The Inclinometers supplied to the expeditions under Sir James Ross and Sir John Richardson are furnished with the additional apparatus for this purpose. The dipping-needles are 3·5 inches in length, and the deflecting magnets are those of 3·67 inches, belonging to the unifilar magnetometers, which also make a part of the equipment of these expeditions.

The chief practical difficulty which appears to present itself in the employment of this method consists in the small distance apart at which it is necessary to place the two magnets, viz., the deflecting magnet and the dipping-needle, in order that the angle of deflection may be of a sufficiently large amount. The angle consequently is very considerably influenced by the law according to which the free magnetism is distributed in the particular magnets, and a long process of observation and of calculation is required to eliminate this effect. The formula by which the total force is deduced becomes in effect—

$$\frac{m}{\phi} = \tfrac{1}{2} r^3 \sin u \left(\frac{1}{1 + \frac{P}{r^2} + \frac{Q}{r^4} + \frac{R}{r^6}}, \text{\&c.} \right);$$

P, Q, R, &c., being coefficients depending upon the distribution of the free magnetism in the deflecting magnet and dipping-needle, and of which the values must be determined by experiments at different distances. This process may, however, be gone through in England before the instruments are supplied to the officers, or after they are returned; it has been partially accomplished at Woolwich for the magnets supplied to the Arctic expeditions above-named. The remaining observations required for a determination by this method of the total force in localities where the dip approaches 90° present no particular difficulties, and may be made on shore or on the ice whenever the expedition is stationary for a few hours. It is necessary, however, to be very particular in observing the temperature of the deflecting magnet whenever it is employed in producing deflections, and to know very accurately its temperature coefficient at different parts of the thermometric scale; and it is also necessary to combine in some satisfactory manner the observations (described in App. 1) by which the magnetic moment of the deflecting magnet is examined, with the observations in which it is made to deflect the dipping-needle.

APPENDIX No. 4.

Observations required to determine the Effect of the Ship's Iron on the Magnetic Instruments used on board.

1. When the ship has got her guns, shot, and iron stores on board, and in their places, and has her boats, stanchions, and all other ironwork in the positions in which they are to remain at sea, her head must be successively placed upon each of the sixteen principal points of the compass, as indicated by the standard compass used in its proper position in the ship. When the ship is quite steady, and her head is exactly on the point on which it is first to be placed, the bearing of some distant and well-defined object must be noted by the standard compass, the distance of the object being such that the space through which the ship revolves on being swung round shall make no sensible difference in its bearing; the magnetic inclination by the direct observation with Fox's apparatus must also be noted, and the angle of deflection produced by a second needle used as deflector N, in the manner described in the directions for the observations of the magnetic force with that apparatus (§ 4, page 42).

2. Repeat these three observations (viz., the bearing by the standard compass of the distant object, the inclination, and the angle of deflection for the ratio of the force), when the ship is steady with her head placed successively upon each of the sixteen principal points.

3. Determine the *real* or *true* magnetic bearing of the distant object from the ship, by taking the standard compass to some place on shore where no iron may be near, and from whence the position of the standard compass on board and the distant object shall either be in one, or in directions exactly opposite to each other. The bearing of the distant object observed from this spot will be the true magnetic bearing, or that which the compass should have shown in each of the sixteen observations on board, had it not been for the iron in the ship. The differences between the true magnetic bearing and the successive bearings on board will show the amount of the error occasioned by the ship's iron, when the ship's head was placed on each of the sixteen points. Call the error East, or —, in the cases in which the North end of the needle is drawn to the eastward by the ship's attraction, and West, or +, when drawn to the westward. Observe also and record the magnetic inclination on shore with Fox's apparatus by the direct observation and with the face of the circle to the East, precisely as when on board, as well as the angle of deflection produced by the second needle used as deflector N.

4. The observations which have thus been detailed should be repeated

whenever the ship has so materially changed her geographical position as to have altered the magnetic inclination 30° or 40°, or whenever she has been refitted, or has undergone any other change which may have made a considerable alteration either in the amount or in the distribution of her iron. It is also particularly desirable to repeat them if the ship should be in a harbour where the inclination is very small, or where it is very large; and finally, they should be repeated without fail whenever the magnetic observations made on board a ship are brought to a termination, and before any change has been made in the iron of the ship.

5. If the positions in which the standard compass and Fox's apparatus are used on board are not very far from each other, and if there is no iron within a few feet of either of them, it will usually be found, in sailing vessels built of wood at least, that the effect of the ship's iron is the same, or very nearly so, in the two positions. To prove this, place a second compass (which, like the standard compass, has had its index error, if any, determined) in the gimball-stand of the Fox's apparatus, and observe, generally, whether the two compasses agree when the ship's head is on the different points, and especially on the points of greatest and least error.

6. When the observations thus described have been carefully made and recorded, they furnish the means of calculating approximately all the corrections required to clear the magnetic observations made on board the ship, in her successive passages from port to port, from the effect of the iron upon the needles of the standard compass and of Fox's apparatus.

The calculations for this purpose have hitherto been made at Woolwich, by formulæ derived by Mr. Archibald Smith from the fundamental equations of M. Poisson's theory, in his 'Mémoire sur les Déviations de la Boussole produites par le Fer des Vaisseaux.'

For the full understanding of these formulæ, it is necessary to read Mr. Smith's notices printed in the accounts of the magnetic observations made in the Antarctic expeditions of Sir J. C. Ross, and of Lieutenants Moore and Clerk, published in Nos. V., VI., and VIII. of the Memoirs entitled 'Contributions to Terrestrial Magnetism,' in the Philosophical Transactions; but it may be convenient to reprint here the formulæ which are of most general application. In these formulæ, symbols are used which have the following significations: ϕ is the total magnetic force of the earth observed on shore; θ the inclination observed on shore by Fox's apparatus; ζ the true magnetic azimuth of the ship's head counted from the magnetic North, positive when her head is West of North and negative when East; ϕ', θ', and ζ', are the same elements observed by the needles of the standard compass and of Fox's apparatus on board;

$\zeta' - \zeta = \delta$ is the deviation or error of the standard compass caused by the ship's iron when her head is on the point ζ'; and finally A, B, C, D, E, c, d, and A', are coefficients to be employed in calculating the corrections applicable to the magnetic observations of the declination, inclination, and force made at sea in any part of the globe.

7. For the corrections of the declination A, B, C, D, and E are required; they may be computed by the following rules: let the 32 points of the compass be numbered from 1 to 32, beginning with N. by W. = 1; N. N. W.=2, &c., and going round by S.=16, to N.=32; and let the deviations observed on the different points, when the ship is swung, be designated by δ with the number attached which shows the point to which it refers: as for example δ_1 will be the deviation at N. by W.; δ_2 at N. N. W.; δ_{16} at South; and δ_{32} at North. Then, if the deviation has been observed on the *sixteen* principal points, we shall have the values of δ at all the even-numbered points, as δ_2, δ_4, δ_6, &c. to δ_{32}; and of these values all those in which the North end of the needle is drawn to the westward by the ship's attraction will have the sign + prefixed: and all those in which it is drawn to the eastward the sign —; the signs must be attended to in making the additions which follow:—

$$A = \tfrac{1}{16}(\delta_2 + \delta_4 + \delta_6 \; . \; . \; . \; . \; . \; . \; . \; . \; + \delta_{32})$$

$$\begin{aligned} B = & \cdot 0478 \text{ (its log} = \bar{2}\cdot 6800) \{\delta_2 - \delta_{30} + \delta_{14} - \delta_{18}\}, \\ & + \cdot 0884 \text{ (its log} = \bar{2}\cdot 9464) \{\delta_4 - \delta_{28} + \delta_{12} - \delta_{20}\}, \\ & + \cdot 1155 \text{ (its log} = \bar{1}\cdot 0625) \{\delta_6 - \delta_{26} + \delta_{10} - \delta_{22}\}, \\ & + \tfrac{1}{8}(\delta_8 - \delta_{24}). \end{aligned}$$

$$\begin{aligned} C = & \cdot 1155 \text{ (its log} = \bar{1}\cdot 0625) \{\delta_2 + \delta_{30} - \delta_{14} - \delta_{18}\}, \\ & + \cdot 0884 \text{ (its log} = \bar{2}\cdot 9464) \{\delta_4 + \delta_{28} - \delta_{12} - \delta_{20}\}, \\ & + \cdot 0478 \text{ (its log} = \bar{2}\cdot 6800) \{\delta_6 + \delta_{26} - \delta_{10} - \delta_{22}\}, \\ & + \tfrac{1}{8}(\delta_{32} - \delta_{16}). \end{aligned}$$

$$\begin{aligned} D = & \cdot 0884 \text{ (its log} = \bar{2}\cdot 9464) \{\delta_2 - \delta_{30} - \delta_{14} + \delta_{18} + \\ & \delta_6 - \delta_{26} - \delta_{10} + \delta_{22}\}, \\ & + \tfrac{1}{8}\{\delta_4 - \delta_{28} - \delta_{12} + \delta_{20}\}. \end{aligned}$$

$$\begin{aligned} E = & \cdot 0884 \text{ (its log} = \bar{2}\cdot 9464) \{\delta_2 + \delta_{30} + \delta_{14} + \delta_{18} - \\ & \delta_6 - \delta_{10} - \delta_{22} - \delta_{26}\}, \\ & + \tfrac{1}{8}\{\delta_{32} + \delta_{16} - \delta_8 - \delta_{24}\}. \end{aligned}$$

If the deviations are greater than 7° or 8°, the sines of the angles of deviation should be used in the formulæ instead of the angles themselves. Where time will not allow of observations being made on the *sixteen* principal points, they may be confined to the *eight* principal points. If those observations are carefully made, the results derived from them will have nearly equal value with the results derived from observations made on sixteen points, with the important advantage that the formulæ for calculating A, B, C, D and E become much simplified. Using the same notation as before, and δ_4 δ_8 δ_{12} δ_{16} δ_{20} δ_{24} δ_{28} and δ_{32} being the eight deviations observed, we have

$$A = \tfrac{1}{8}\{\delta_4 + \delta_8 + \delta_{12} + \delta_{16} + \delta_{20} + \delta_{24} + \delta_{28} + \delta_{32}\}$$

$$B = \cdot 1768 \text{ (its log} = \bar{1}\cdot 2474) \{\delta_4 - \delta_{28} + \delta_{12} - \delta_{20}\} + \tfrac{1}{4}\{\delta_8 - \delta_{24}\}$$

$$C = \cdot 1768 \text{ (its log} = \bar{1}\cdot 2474) \{\delta_4 + \delta_{28} - \delta_{12} - \delta_{20}\} + \tfrac{1}{4}(\delta_{32} - \delta_{16})$$

$$D = \tfrac{1}{4}\{\delta_4 - \delta_{28} - \delta_{12} + \delta_{20}\}$$

$$E = \tfrac{1}{4}\{\delta_{32} + \delta_{16} - \delta_8 - \delta_{24}\}$$

The coefficients being known, the deviation on any point of the compass may be computed by the following equation :—

$$\sin \delta = A + B \sin \zeta' + C \cos \zeta' + D \sin 2\zeta' + E \cos 2\zeta'.$$

A, D, and E may be expected to remain constant, or to have the same values in whatever part of the globe the ship may be, whilst no material alteration is made in the distribution of her iron. They may, therefore, be regarded as determined, once for all, by the deviations observed when the ship is first swung, though they may possibly be obtained more exactly by taking the mean of the values obtained on all occasions when that process is repeated. B and C are variable, and depend on the dip, and also on the proportion of the iron which changes its magnetic state cotemporaneously with changes in the geographical position of the ship, to the permanently magnetic iron, or to iron of an intermediate quality to the two which have been named, and of which the magnetism is neither permanent on the one hand nor, on the other, are its changes cotemporaneous with changes of the dip, but are consequent on such changes, and require a greater or less interval to conform to them. It is on account of the uncertainty of the law according to which these two coefficients B and C vary in different ships, that when a ship has changed considerably her geographical position, it is desirable to repeat the process by which the values of the coefficients may be re-determined; and that it is still more desirable that a full and sufficient trial should be made of a very simple method suggested by Mr. Archibald Smith, in No. VIII. of the 'Contributions to Terrestrial Magnetism,' for determining the variable coefficients at any time that may be wished, either at sea or in harbour, by deflections of the compass-needle with the ship's head successively on any two opposite points of the compass. By the addition of a brass bar attached at right-angles to the prism and sight-vane of the azimuth ring of the standard compass, a deflecting magnet or magnets may be temporarily fixed at a convenient distance from the compass-needle, and the deflection observed with the ship's head on opposite points by a process requiring only a very few minutes, and independent of the visibility of the sun or stars, or of any distant object. If the points taken be those which the observations in harbour have shown to be points of no disturbance, and if v_1 v_2 be the angles of deflection on the respective points—

$$\sqrt{B^2 + C^2} = \frac{\sin v_1 - \sin v_2}{\sin v_1 + \sin v_2};$$

$$= \frac{\tan \frac{v_1 - v_2}{2}}{\tan \frac{v_1 + v_2}{2}},$$

and the deviations on the several points may be computed by

$$\sin \delta = A + \sqrt{B^2 + C^2} \sin (\zeta' + \alpha) + D \sin 2 \zeta' + E \cos 2 \zeta',$$

in which α is the easterly azimuth of the line of no deviation.

Should this method of determining the variable part of the correction formula be found to succeed on trial, the correction of the disturbances by the officers of a ship might be still further simplified, by the formation of tables of each term for every probable value of the coefficients, requiring merely the addition of the quantities to be taken out from the tables. In the meantime the calculations may be facilitated by the following table, extracted from Mr. Smith's Memorandum in No. VIII. of the 'Contributions to Terrestrial Magnetism.'

Let B_1, B_2, ... B_7, C_1, C_2, ... C_7 represent the values of B and C multiplied by sin 11° 15′, sin 22° 30′, &c.; and let D_2, D, D_6, E_2, E, E_6 represent the values of D and E multiplied by sin 22° 30′, sin 45°, and sin 67° 30′; we have then—

$$\delta_{32} = A + C + E$$
$$\delta_{16} = A - C + E$$

$$\delta_1 = A + B_1 + C_7 + D_2 + E_6$$
$$\delta_{31} = A - B_1 + C_7 - D_2 + E_6$$
$$\delta_{15} = A + B_1 - C_7 - D_2 + E_6$$
$$\delta_{17} = A - B_1 - C_7 + D_2 + E_6$$

$$\delta_2 = A + B_2 + C_6 + D_4 + E_4$$
$$\delta_{30} = A - B_2 + C_6 - D_4 + E_4$$
$$\delta_{14} = A + B_2 - C_6 - D_4 + E_4$$
$$\delta_{18} = A - B_2 - C_6 + D_4 + E_4$$

$$\begin{aligned}
\delta_3 &= A + B_3 + C_5 + D_6 + E_2 \\
\delta_{29} &= A - B_3 + C_5 - D_6 + E_2 \\
\delta_{13} &= A + B_3 - C_5 - D_6 + E_2 \\
\delta_{19} &= A - B_3 - C_5 + D_6 + E_2
\end{aligned}$$

$$\begin{aligned}
\delta_4 &= A + B_4 + C_4 + D \\
\delta_{28} &= A - B_4 + C_4 - D \\
\delta_{12} &= A + B_4 - C_4 - D \\
\delta_{20} &= A - B_4 - C_4 + D
\end{aligned}$$

$$\begin{aligned}
\delta_5 &= A + B_5 + C_3 + D_6 - E_2 \\
\delta_{27} &= A - B_5 + C_3 - D_6 - E_2 \\
\delta_{11} &= A + B_5 - C_3 - D_6 - E_2 \\
\delta_{21} &= A - B_5 - C_3 + D_6 - E_2
\end{aligned}$$

$$\begin{aligned}
\delta_6 &= A + B_6 + C_2 + D_4 - E_4 \\
\delta_{26} &= A - B_6 + C_2 - D_4 - E_4 \\
\delta_{10} &= A + B_6 - C_2 - D_4 - E_4 \\
\delta_{22} &= A - B_6 - C_2 + D_4 - E_4
\end{aligned}$$

$$\begin{aligned}
\delta_7 &= A + B_7 + C_1 + D_2 - E_6 \\
\delta_{25} &= A - B_7 + C_1 - D_2 - E_6 \\
\delta_9 &= A + B_7 - C_1 - D_2 - E_6 \\
\delta_{23} &= A - B_7 - C_1 + D_2 - E_6
\end{aligned}$$

$$\begin{aligned}
\delta_8 &= A + B - E \\
\delta_{24} &= A - B - E
\end{aligned}$$

If the angles are greater than 7° or 8°, these formulæ give the sines of the angles of deviation instead of the angles.

8. For the correction of the inclinations observed at sea the coefficients c and d may be computed from the disturbances on the several points, shown by the observations of the inclination with Fox's apparatus when the ship was swung in harbour, by the formula:—

$$c \cos \zeta + d \tan \theta = (1 - D) \sin \zeta \operatorname{cosec} \zeta' \tan \theta'$$

for all other points than North and South, and with the ship's head North or South, by—

$$c \cos \zeta + d \tan \theta = (\cos \zeta + B) \sec \zeta' \tan \theta';$$

observing that the values of ζ' employed should be calculated by the declination-coefficients. With the values of c and d thus obtained, tables for the correction of the inclinations observed on different courses at sea in all values of θ may be computed by the formula:—

$$\tan \theta' = \frac{c}{(1 - D)} \left(\cos \zeta + \frac{d}{c} \tan \theta\right) \sin \zeta' \operatorname{cosec} \zeta,$$

$$\text{or } \tan\theta' = c\,\frac{\cos\zeta + \dfrac{d}{c}\tan\theta}{\cos\zeta + \mathrm{B}} \,.\, \cos\zeta';$$

observing that the first must be used when the ship's course is between N. E. and S. E., or N. W. and S. W.: and the second when her course is nearer the N. and S. points; and that the values of ζ' should be calculated by means of the declination-coefficients.

9. The coefficient A$'$, for the correction of the intensities observed at sea, may be computed by the subjoined formulæ, from the observations of the intensity with Fox's apparatus on board with the ship's head on the different points, compared with those of the same instrument on shore, and with the absolute value of the total force on shore (ϕ) obtained in the manner described in App. 3. Tables for the correction of the intensities observed at sea may also be constructed by the same formulæ:—

$$\frac{\phi'}{\mathrm{A}'\phi}\sin\theta' = c\cos\theta\cos\zeta + d\sin\theta,$$

$$\text{or } \frac{\phi'}{\mathrm{A}'\phi}\cos\theta'\sin\zeta' = (1-\mathrm{D})\cos\theta\sin\zeta;$$

the first to be used when the inclination is large, and the second when the inclination is small.

10. If the disturbances of the compass-needle, at the spot where the gimball-table for Fox's apparatus is fixed, differ materially from those at the spot where the standard compass is fixed, the standard compass may be removed temporarily to the gimball-table, and the deviations observed on the eight principal points, which will give the values of B and D for that spot, as well as determine the points on which there is no disturbance: if these are opposite points, the value of the variable coefficient B can always be ascertained experimentally for the correction of the observations of inclination and intensity, by the angles of deflection produced in a compass-needle placed on the gimball-table by the method described in 7, p. 19.

Section III.

HYDROGRAPHY.

BY CAPTAIN F. W. BEECHEY, R.N.

Making a Passage.

The observer's attention is directed first to those objects which affect the passage of a vessel from one part of the globe to another; such as the movement, the duration, the limits, and the periodic occurrences of those great currents of the atmosphere and of the ocean, upon which the speedy and successful issue of a passage mainly depends.

Well recorded and established facts bearing upon the several points connected with these inquiries are highly important to navigation, and may be collected by every assiduous seaman in the ordinary course of his duties.

1. It is well known that in various parts of the globe there exist monsoons, and zones of trade and variable winds; and that these and other disturbances of the atmosphere which influence the surface of the ocean are the principal causes of the many currents which sweep over the face of the earth. The effect of these upon a vessel passing to and fro is one of the most useful inquiries a seaman can make; and as both (wind and current) perform an important part in the economy of nature,

an additional interest attaches to a correct knowledge of them. The seaman should therefore not only carefully note the direction and force of the winds, but should connect with such entries notices as to when and where any continued or periodic wind commenced and terminated; what was its strength and effect upon the passage; whether it came on suddenly, and was furious while it lasted, or otherwise; whether it was preceded by any particular symptoms, and whether it was such as usually occurs at that season; and lastly, whether it be advisable to cross this wind in any particular direction, such as close hauled or large, &c.

2. To detect the current, a more than ordinary attention must be paid to the reckoning of the ship: the compass by which the course is steered should occasionally be compared with that by which the variation is determined, in every position of the ship's head;* and the ship's place should be determined by observation at least once a day. Sights for chronometer morning and evening should both be referred to noon, at which time the latitude will of course be observed; and all observations for latitude at night, or for fixing the ship's place at any time, should be referred to one period of the day, in order that the position of the ship *by observation*, as compared with her place by the *Dead Reckoning*, may give the direction and force of the current, if any, for the twenty-four hours. These observations should all be entered in a table, and at the close of certain obvious and natural periods of a passage, such as that of entering or emerging from the trade-wind, the calm lati-

* See Section 'Terrestrial Magnetism.'

tudes, the commencement or termination of the monsoon, of any positive change of current, or from any continued state of things to another, the whole effect of the current for the period should be deduced, and an average of its daily rate and set be given, together with any remarks which may be considered useful.

3. With the direction of the current thus determined, it is very desirable to connect the temperature of the surface of the sea, for it has been by such observations that we have been able to trace, with a certainty amounting almost to proof, the continuous course of the same body of water for thousands of miles over the troubled surface of the ocean, and that other curious and important facts in physical hydrography have been ascertained. We would therefore urge attention to the subject as one of considerable importance to navigation. As a proof of its influence upon a passage, we need only instance the remarkable phenomenon of the Equatorial and Guinea currents: two streams in contact, but flowing in opposite directions, and having a temperature differing 10 or 12 degrees from each other, and yet pursuing their opposite courses for upwards of a thousand miles; and according as a vessel is placed in one or the other of these currents, will her progress be aided or retarded from 40 to 50 miles a day.*

Could we but obtain a register of the temperature of the surface of the sea from every ship in active service, we should be able in a short time to construct tables showing the normal temperature of the surface of the ocean for every 5° of latitude for every month in the

* Sabine's 'Hydrographical Notices.'

year, and a comparison of these with the actual temperature of the surface at any particular spot, and in any particular month, would at once manifest an abnormal difference, if any existed, and lead to a knowledge of its cause, which might prove of considerable use to the mariner by acquainting him with the movement of the great body of water in which he was sailing; either retarding or accelerating his progress as the case might be, and at all events affecting his reckoning. Or it might lead to a closer determination of the limits and periodical changes of currents which, as before observed, are everywhere running over the surface of the sea as rivers run over dry land.

It is therefore recommended to add to the table of currents a column for the temperature of the open air, and another for that of the surface of the sea, which should be registered frequently during the twenty four hours; but as such observations form an essential feature in the meteorological register of a voyage, they should be made at the times and in the manner indicated under the head of Meteorology.*

4. There should also be noted in the Remark column the occurrence of masses of sea-weed, or of any continued appearance even of small patches of this or of any other floating substances which may be seen; and if opportunity offers, deep-sea soundings should be tried at the spot. "It were much to be wished," says Humboldt, 'Person.

* If passing Cape Horn, or through seas where icebergs may be moving about, these observations cannot be made too frequently in thick weather, especially as a precaution, for the water appears to be influenced to a considerable distance around these masses, particularly in their wake.

Nar.' vol. ii. p. 11, "that navigators heaved the lead more frequently in these latitudes covered with weeds, for it is asserted that Dutch pilots have found a series of shoals extending from the banks of Newfoundland to the coast of Scotland by using lines composed of silk thread." Flocks of birds should also be noted. In many places, the Pacific especially, the tern are useful monitors of an approach to those low specks of coral which endanger the path of the navigator through the labyrinth of the great South Sea. In short everything that may seem to the voyager to be interesting or new, or likely to be useful, should find a place in the Remark column.*

At the end of the passage a summary of these remarks should be given, the whole effect of the current for each particular portion of the passage recapitulated, such as that which was due to the N.E. or S.E. trade-wind, or to the monsoon, as the case might be, and distinguishing each; that which occurred in the calm latitudes or during a period of variable winds, or otherwise, averaging the daily rate; and then might follow any remarks you may wish to make either upon them or upon any other feature of the passage; together with any directions or hints which might be considered useful to those who should follow over the same ground; such as whether any advantage would have been gained by steering more to the east or west, or in any other direction; whether any time would have been saved by making the land on any other bearing than that in which you hit upon it; and in short any remarks which would be instrumental in conveying to

* For the form in which these observations may be tabulated, see Appendix, Table I.

others information which you would have wished to possess yourself at the outset of the passage.

Currents.

5. It is very desirable that observations upon the course of the waters of the ocean should be made without intermission; and that *a continued register* of the temperature of the surface, and occasionally of its submerged strata,* should be kept, as it is only by numerous well-recorded observations of this nature that we shall ever be able satisfactorily to define the limits of the various zones of moving water which sweep over the face of the globe, mingling the waters of the Polar Seas with those of the equatorial regions, and even affecting the climate of extensive districts.† But if from various causes a connected series cannot be continued throughout these great currents, at least an endeavour should be made to commence a register on approaching the limits of such as are now approximately defined, and to continue it while any interest appears to attach to the subject: such as that of the Gulf-stream; the Trade-wind drift; the Guinea and Equatorial current; the Cape of Good Hope current, blending with the south-east trade drift; and the Brazil current—in the Atlantic; the Mozambique and Agulhas current; the Trade drift; and monsoon current of the Arabian and Bengal gulfs—in the Indian Ocean. The remarkable Peruvian current sweeping along the western coast of South America; the

* By means of self-registering thermometers, properly set and carefully lowered and as carefully hauled in (without jerks).

† See Humboldt on the Climate of Peru; Sabine on the Climate of St. Thomas Island, &c.

Trade drift, and Equatorial current; the Mexican current, passing along from Panamá to the Gulf of California, according to the monsoon. The counter-currents north and south of these, and the moving belt along the coast of Japan and Corea to Kamtchatka—in the great Pacific Ocean; particularly noting, as of great importance to navigation, the limits of the outer currents around the Cape of Good Hope and Cape Horn, all of which will be found on a small scale delineated in a general chart at the end of this paper.

Some of these currents maintain a constant difference of several degrees between their own temperature and that of the mean state of the water about them, and all observations which can throw light upon this subject, and upon the limits, course, and velocity of the stream, will be most acceptable.

6. In passing through any of these great currents, the observer should carefully define the extent of the belt of moving water at the parallel in which he crossed it; the limit of the eddy on either side of it; determine the rate and set of both; carefully note every barometrical or thermometrical change of the air, or alteration in the temperature or specific gravity of the sea, and if possible the depths to which these temperatures extend; and record all appearances and changes which may appear of interest or seem to be useful to those who may follow over the same ground.

To detect the motion of the stream the remarks in Art. 2 should be attended to, with the exception that here the position of the ship should be *frequently* ascertained during the day by astronomical observation, and the course and rate of the current deduced for *short intervals* of time instead of for the twenty-four hours. The observations should commence previous to entering the body of moving water, and be continued until after the vessel has quitted it, when it will be advisable to occupy a page of the journal with a graphic delineation of the several courses of the stream, indicated by arrows, and of the several stages of the vessel's progress by the *various* temperatures which have been observed, noting the places where ripples were seen, or where drift-wood, sea-weed, or other floating substances occurred.

The Stream or Surface Drift.

7. Currents have been spoken of under the head of "making a passage," as they affect a ship's route across the ocean, and may have been determined by the position of the ship by DR differing from that by observation. But it will be proper further to *try the set of the surface* of the water on all favourable occasions, by the ordinary method of anchoring, or of sinking a weight, endeavouring if possible to get observations on the same day at about six hours apart, in order that it may be seen whether the stream be due to a tide or not. If the ship be in soundings, and the day be calm, a very simple way of effecting this without the trouble of either anchoring or lowering a boat,* is to drop a heavy lead from the quarter, and after it has reached the bottom, to run out a small quantity of stray line, and then make fast the "nipper," or a billet of wood, to the line; and at the same time to fasten the end

* An objection to trying the current in a boat is the uncertainty of the compass.

of the log line to it, and veer away both together.* Then mark by a watch the time *each knot* is in running out, buoying up the line by a chip of wood; when all the line has run out, take the bearing of the nipper by a compass, and haul all in together. If currents be tried when there are no soundings, the result is merely the relative motions of the upper and lower strata of the water, and it would be difficult to say which way either were going; but if we can possibly determine by astronomical observations the course of the *upper* surface, we shall thence be able to deduce the set of the *lower;* and if there be found any difference of moment, it will be very desirable to ascertain the temperature of both upper and lower strata of the water, and to record them with the other observations. These observations ought always to be made on calm days, and the greater the depth to which the weight be sunk, the better. Bottles thrown overboard with a label inside, containing the date and lat. and long. of the spot where cast into the sea, afford ready means of detecting the current if picked up afterwards, and ships would do well frequently to expend a few empty bottles in this way.† In the event of meeting any such drifting at sea, they should be picked up, their contents copied, and the date and position of the spot added to the label and carefully resealed, they should then be returned to the ocean, and a copy of the label forwarded to the hydrographer.

* If the lead-line be not hitched to the nipper, the tide may drag the line through it, and there will be no result.

† The bottles, before sealed, should be ballasted with a little dry sand, consolidated at the bottom with bees'-wax or pitch run in, that the bottle may be kept upright and not swim too light.

8. If near to any shore, a few points of which are well fixed, and the water be found too deep for anchorage, the course of the stream may still be ascertained by noting the drift of a float—a plank, for instance, weighted at one end, so that the other just floats above the surface; or a weighted *bareca,*—fixing its position from time to time by angles taken in a boat at the several places, and noting the intervals by a watch.

Such methods may of course be resorted to when circumstances do not admit of greater accuracy, but whenever it can be done, the course and rate of the stream should be observed *every hour* during *both* tides, and the times of slack water carefully noted, by anchoring a boat or vessel. Upon an open coast one set of such observations made here and there, well clear of the headlands, will be sufficient; but in channels and straits in which the tide enters at both extremities the tidal phenomena are so varied and full of interest, that it becomes highly important to spread the observations over as large an extent of the channel as possible, and to pursue a regular system of hourly observation throughout both the *ingoing* and *outgoing* streams.

It is desirable to know at each place the time of slack water, the direction in which the stream turns, and the rate and course at which it runs during its several stages. The stations should be numbered, and the times all referred to one meridian. In such channels there will probably be one or more places where the streams meet, and there of course observations will be made; and as one of these places will probably be the *virtual head of the tide wave,* it may so happen that the time of the high and low water there *by the shore* will govern the turn of the stream either along the whole channel or until it reaches a spot where another meeting of the streams occurs. In such a channel also it will probably be found (as in the Irish Channel) that the same stream makes high water at one end and low water at the other at *the same time;* so that the observer must entirely divest his mind of the too often mistaken notion of the turn of the stream being governed by the rise and fall of the water in its immediate locality. As our space does not admit of further detail, I shall leave the subject in the hands of the observer with a remark which, whilst it will put him in possession of what kind of observations are required, will at the same time I think insure his interest in the subject and his hearty desire to co-operate in the matter.

In the 'Philosophical Transactions, 1848,' Part I., it has been shown that in such a channel as that above-mentioned there have been discovered two remarkable spots, in one of which the stream runs with considerable velocity without there being any material rise or fall of the water by the shore, and in the other that the water rises and falls considerably without there being any apparent motion of the stream. Such phenomena are highly curious, and worthy of all the attention that can be bestowed upon the observations. In tracing them it is manifest that they are intimately connected with the height and progress of the tide wave along the shores of the channel, but this properly belongs to another section (see *Tides*).

9. Passing the mouths of great rivers—such as the Amazon, the river Plata, Orinoco, Mississippi, Zaire,

Senegal, Indus, Ganges, Yangtsee or Irawady, &c. &c.—observations on the stream should be more closely made, and discolorations and specific gravity of the water noted.

These and such like stupendous rivers extend their influence to a considerable distance from the coast,* and occasionally perplex and delay the navigator, who finds himself struggling against a difficulty, wholly unconscious of the cause and ignorant of the facility with which he might escape it by changing his route.† River currents of this description vary their direction according to the courses of the stream along the coast, by blending with it, and forming a curve, which vanishes only with their influence upon the ocean current; so that we are not always to look for the outset from the river at a right angle to the coast, nor always in the same locality, but according to the prevailing offing stream.

The limits of the principal currents of the globe have been given (see plate B) in order to apprize the navigator of the places in which he should more closely attend to his observations. If, however, from any cause he may have been prevented continuing the series throughout any of the great currents, and should desire to define their limits, he should begin at least a day's run from the places, and continue his register until he is certain of having passed the boundaries, attending closely to the

* The River Plata, at a distance of 600 miles from the mouth of the river, was found to maintain a rate of a mile an hour; and the Amazon, at 300 miles from the entrance, was found running nearly three miles per hour, its original direction being but little altered, and its water nearly fresh.—*Rennell, Sabine.*

† See the effect of the Equatorial and Guinea current before-mentioned, at p. 56.

temperatures; for although limits have been assigned to these belts of moving water, yet they vary so much according to season, and the data for defining them have hitherto been so insufficient, that it cannot be said they are known with any tolerable degree of precision.

In the China Sea and among the islands of the great Indian Archipelago the tides run strong and are very indifferently known, and observations are especially desired at those places.

In the southern passages it would be well to try during the westerly monsoon, whether the equatorial current may not be found pursuing a subaqueous course to the westward, notwithstanding the surface current be found running in the opposite direction.

Upon the east coast of North America, between the Gulf-stream and the coast, observations upon the set of the stream are also much wanted.

Approaching a Coast.

10. When approaching a coast or any extensive banks in the ocean, the temperature of the surface of the sea should be more closely attended to, for it has been found in many instances that after a certain shoaling of the water, the surface partakes of the temperature of the lower strata of the sea, which are in general colder than the upper. If such should be found to be the case always, and if from well-authenticated facts it should become possible to fix zones of certain temperatures about particular localities, the result would be highly useful to the navigator when out in his reckoning and perplexed with thick and hazy weather.

11. Hydrography requires that the general feature and aspect of every country should be noted from the moment the hills rise above the horizon; that all remarkable objects by which it may be recognised, and by which the position of any port or other locality may be known, either at a distance when the weather is clear, or close in when haze or mist prevails, should be described as graphically as possible; that the extent, direction, and outline of the coast; its capabilities of affording shelter to shipping; its dangers, or freedom from them; its navigable rivers, harbours, and inlets; and the objects adverted to under *Sailing Directions*,—should be fully and carefully recorded; and here it is difficult to avoid infringing upon what properly belongs to *geography*. The two sciences are indeed here so nearly allied, that it is scarcely possible to avoid encroaching upon the province of the sister branch. The observer will, however, do well to describe or delineate the character of the country as far as he can become acquainted with it; the form and elevation of such hills as are visible from the coast; the direction of the valleys and ravines; and to mark the places where they pour their mountain-torrents into the sea; to portray the bold topping cliffs, or low rocky promontories and their reefs; the jutting headlands or deep sinuosities; or the low undulating country with its lagging streams and muddy or sandy fringe of coast; its shallows, bars, and deltas, each as the case may be; with its lighthouses, beacons, buoys, and landmarks, stating the distance which they may severally be seen; with even the forts, towers, churches, and silvery little clusters of cottages upon the inland elevations; with such other varied

features as the coast may present, and as may serve to convey a just idea of what may be expected to meet the eye of the navigator, or be required to keep him clear of danger, and to guide him in safety to his place of destination.

At a distance there is generally some object more remarkable than another which may be singled out as a useful landmark. Note what it is, describe its appearance, and state in what direction the port, or any danger that may lie off the coast, bears from it.

Should the coast be low, buildings will possibly be seen first: large square houses or towers, church-spires, &c.; any of these afford useful guides. Some localities may be distinguished in hazy weather by *patches of white* near the coast, such as masses of sand, chalk cliffs, &c., or one or more large white houses; and these, when viewed against the land or other dark objects, will occasionally afford excellent guides when all other objects are obscured, and at such times are doubly useful. But avoid as marks all *white* objects which have only the sky for a back-ground; such objects are seen only when the sun shines upon the surface presented to the observer, but utterly fail in *hazy* weather, when they are wanted.

Always bear in mind that no description can equal a tolerably faithful sketch, accompanied by bearings. In all your sketches take angles roughly, with a sextant between objects at the extremities of your drawing, and two or more intermediate ones, and affix them to the objects at the moment, and have at least one angular height in the picture; let that be of the highest and most conspicuous or best defined object; thus—

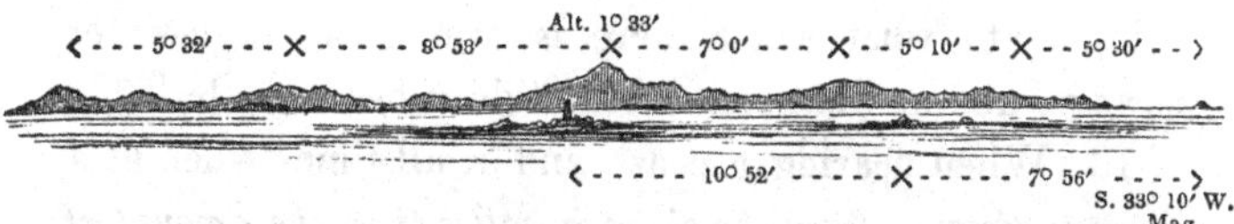

and let your bearing refer to one of the objects between which you have measured angles. Always write under the sketch at the time the name of the place, and especially the native name if you can possibly learn it, and the date; and if you intend any of the objects for leading marks place an arrow at the head of a perpendicular line above and below the objects; thus—

Lighthouse in one with East Peak of Mount Auckland, clears reefs in 4 fathoms, and kept open (S. by E.), leads through the passage, mid channel.

12. Besides marks which are apparent to the eye, the depth of the water and the nature of the bottom are all important, and in all descriptions of a coast as well as in directions for approaching it, these are to be carefully attended to. State as nearly as possible the distances at which certain zones of soundings extend from the shore, and from what part; whether the bottom shelves gradually or abruptly, whether the coast may be boldly approached, or more than ordinary caution be necessary; and whether any peculiarity of the bottom may assist in determining a ship's position or distance from the coast at night, or in thick weather. Always give your depths reduced to low-water spring-tides if possible,* and always give *the least water* upon a reef or shoal; and if it

* See page 81, Art. 24.

dries, state what water there is over it at high-water springs, and at what time of the tide it becomes dry.

13. When nearing a coast, and at all times when at a *greater distance from the shore in miles than the amount of dip in minutes due to the height of your eye*, the height of mountains, or of other objects, may be determined with considerable accuracy if the weather be clear, and proper precautions be taken. To do this, if the distance of the object be not known, it must be found by measuring a base with the patent log. There are various methods given in navigation-books for determining this problem; I shall therefore here merely describe the observations required to be made. At each end of the base measure carefully with a sextant the altitude of the object *off and on;* and if one of Cary's double sextants be on board, measure the terrestrial refraction by bringing the opposite horizons in contact with the arc both above and below the index, and then reading off each time, divide the difference by 4: this will give the dip and terrestrial refraction combined, which is the proper quantity to be allowed in correcting the observed angle.*

In Raper's 'Navigation,' p. 90, 2nd edition, the method of determining a ship's distance from an object by two bearings is briefly explained; and in Belcher's 'Surveying' it is set forth in a manner so clear and ample as to leave nothing to be desired: I shall therefore merely observe here, that according to the accuracy of the observations and the value of the means adopted, will

* If the terrestrial refraction alone be required, take from this quantity the true dip due to the height of the eye, and the remainder is the terrestrial refraction required.

be the correctness of the result. It is clear that the *true* bearing of the object at *each* station should be observed (see Astronomical Bearing); that the course steered should, if possible, be equally well known; (this is effected in the best manner by observing the magnetic bearing of the object with the compass which directs the base, or that which is to be steered by in-running the base; *at the same time that its true bearing is observed;*) that the distance run should be determined by patent log; that the ship *should be on her course at starting when the bearings are observed;* and that the log should be put over and hauled in at the *instant of making* the observations. If the ship should of necessity alter her course during the operation, it should be carefully noted, the log looked at, and fresh bearings of the objects taken.

Two observers are necessary to accomplish these observations nicely, and without hurry.

With these data the height of the object may be found with considerable accuracy,* especially if the dip-sector be used. Having determined the height of a mountain, you may often find it useful to know your distance from it when cruising off the coast; and it will also afford amusement and practice to see how near you can fix the ship by it, as compared with cross-bearings or other observations. For this purpose it will be convenient to make a constant for the height.†

Lighthouses.

14. If lighthouses are erected upon the coast, describe exactly their locality, geographical position, appearance,

* See Appendix No. 2.

† See Appendix No. 3.

height of the lantern above mean water-level, height of the tower, whether the light be fixed or revolving, intermittent, coloured, or otherwise; the distance which it may be seen, and the bearings on which the light is visible. If the light be made use of for the purpose of avoiding any danger, state what the danger is, give its bearing from the lighthouse, and if the light be blinked or changed for this or any other purpose, state what it is, and give the exact bearing on which the change takes place.

If there be a *lower light* in the tower for this, or any purpose of tide work, state as before the bearing on which it opens and obscures, or the times of tide when it is exhibited and extinguished, and how many feet it is below the upper light. State whether pilots are required for the port, and where they are likely to be met with, and the rate of pilotage.

In Port.

15. When at anchor, give the depth of the water, nature of the ground, and whether any precautions are necessary, with respect to protecting the cables if hempen, and whether it be proper to moor in consequence of the difficulty of keeping a clear anchor, or from the treacherous holding or sloping of the ground. Fix the spot by cross-bearings of conspicuous and well-known objects, and note the direction and rate of the tide, and the duration of both ebb and flood stream.

16. The geographical position of a port will necessarily occupy the attention of the persons in whose hands these remarks may be placed, and by assiduity much may be done in a short time with a sextant and artificial

horizon only. But if to these be added a transit and a good achromatic telescope, the longitude by occultations, moon-culminating stars, and eclipses of Jupiter's satellites, will form a valuable addition to that by observations of lunar distances with the sextant.

The earliest opportunity should be taken of determining the error of the chronometers upon mean time at the place, by morning and evening sights, or by *equal altitudes*, which is better. Chronometers will sometimes change their rates on the transition from a passage in which they have been constantly in motion to a state of rest. Besides which, early sights afford a longer interval for rating the watches again.

Survey of a Port.

17. A survey of the port and description of the anchorage will always be desirable if carefully made. If former surveys have been executed, it will afford a useful comparison, and detect alterations of the banks and channels, and the silting up of the port if any. If they have not, such a survey will be doubly useful, and the industrious observer will find very few plans of ports to which he may not usefully add a few soundings or explanatory remarks.

It is not intended in this manual to enter much into the manner of executing a survey, as there are several treatises on the subject, which contain the necessary information; but these works may possibly not be on board, and as "golden opportunities" of acquiring a knowledge of distant ports may thus be lost, from the want of knowing how to construct a rough survey of a place, by persons who probably never contemplated the performance of such an undertaking, it may be useful to describe as briefly as possible the process. Make choice of two stations at as great a distance apart as the survey will admit of, and from which the eye can see over a considerable portion of the ground to be mapped, as A, B, plate A. Put up marks or select objects at all convenient places around the survey, so as to be able to form a network of triangles over the whole space, and include every conspicuous feature around, such as hills, cliffs, rocks, and especially objects at, or near, high-water mark.

18. Having decided upon these marks, as at A, B, C, D, E, F, G, &c., plate A, a base may be measured on shore if there be a convenient spot at hand; but it would be useless to devote much time to this purpose, for the survey of a port in general does not so much require that the absolute distances between places should be *accurately* known, as that the *angles should be carefully observed*, and therefore the relative distances preserved. If it be necessary to measure a short base of a quarter or half a mile, and the ground be uneven, plant staves (boarding-pikes) in the line to be measured, and stretch the lead line along from pike to pike in the direction of the wire of the theodolite when levelled; and measure along the line with a tape, or with rods; then shift the pikes and line on, until the required distance has been measured, the length of which should not be less than 1-7th of the distance between the objects, the distance of which is required.

If a micrometer be on board, a very fair base may be

obtained with it,* or even with a sextant, by measuring the angle subtended by a staff placed at right angles to the observer, and the distance carefully measured between two well-defined marks, one at either end of the staff (such as the clean edge of a sheet of white paper wrapped round each end): the best way of ensuring the staff being at right angles nearly, is to place it upright by a plumb-line. Then treating the figure as a right-angled triangle, the staff will be the perpendicular, and the base the distance between your eye and the station, which may be readily computed, as all the angles and a side are known; but if a micrometer be used, the distance is that between the staff and the *object-glass* of the telescope. If neither of these methods be adopted, or if the field of operations be very extensive, a base by sound, though much less accurate, may be found convenient.

19. The measurement of a base by sound, if several trials are made, and the distance be more than a mile, will, in most cases, be sufficiently exact for the above purposes. If possible land a swivel or small gun upon one of your stations, and go yourself to the other, the more distant the better. Appoint a signal to be shown half or a quarter of a minute before each explosion, in order that the eye may rest between. When the signal is made begin to note the beats,† but not to count until you see the explosion, and then let the next beat be one, and so count up until you hear the report. Let this be done several times, and at the end mean the beats, and turn

* See the Book of Tables and Directions supplied with an instrumen of this kind, by Rochon.

† The stop-watch by Mr. Dent is very convenient for this purpose.

them into seconds of time by the number of strokes your watch makes in a minute. Then, by multiplying this number of seconds by 1090, and adding *one foot* for *every two degrees* of the thermometer *above freezing-point*, you will obtain the length of the base in feet.

To give *a direction* to the base, the readiest way is to observe the passage of the sun's limbs over the wires of the theodolite nicely levelled, and to note the time by a watch, or to take corresponding altitudes, in order to compute the azimuth. If you have only a sextant, astronomical bearing will be found convenient and sufficiently correct, provided the horizon can be seen at a sufficient distance. By either of these methods the angle between the base and the limb of the sun will be known, and hence the true bearing of the object obtained; or if the magnetic bearing of the object be observed by an azimuth compass, and the variation be determined at the same time, it will still be known near enough for the common purposes of navigation. Having arranged the direction and length of the base, at A and B alternately, measure angles between the base and all the stations, taking care that the angles (if measured with a sextant) are as nearly as possible parallel with the horizon, and at all other convenient stations do the same. By this means the relative position of all the stations will be obtained, from any *three* of which the position of another, or of a boat for instance, may be determined by measuring two contiguous angles between them with a sextant. But whenever you have occasion to do this (and in sounding there is no more convenient method of fixing the place of the boat), be careful not to select stations which lie in a curve *concave towards you*, since cases will often arise when stations so situated will give very inaccurate results, as with the objects A H I at P in plate A.

20. While operations on shore are going forward, boats can be sounding out the harbour, and fixing the points of reefs, rocks, &c., bearing in mind that it will always be found more satisfactory to land upon every rock or point, &c., than to lie off in the boat, and fix them by estimated distances or by intersections, either from these or from other stations.

In sounding, fix the boat at starting by two sextant angles; note the direction in which it is intended to run out the line of soundings, and note any two objects distant from each other, that are *in a line* upon that bearing, or if the port be not too extensive make use of staves with flags—shifting them along the coast at the end of each line of soundings the exact distance it is intended to run them apart—the boat showing a signal when the flags are to move; then keep the marks on, and sound at regular intervals 6, 5, 4, 3, 2, or fewer casts in a minute according to the depth; and at given short intervals note the time and fix the position of the boat by two angles as before-mentioned,* as also whenever there is any material alteration in the depth, or whenever the number of casts alters in a given time. When arrived at the end of the line fix the boat's position, and alter the course, sounding all the time until far enough for running back the second line of soundings parallel with the first. Fix the boat's

* See the form in Appendix for entering these angles and soundings, No. 4.

station here again, and take a new leading mark. If the eye cannot catch a leading object at the moment, drop the grapnel to maintain the spot, for much more time is lost by over-running the lines than by coolly waiting for a guide to direct the course. Proceed in this manner, running all the soundings in parallel lines or nearly so, until the anchorage is all sounded out.

Having mapped all that is intended to be comprised in the survey, protract the work carefully on board upon a sheet of drawing paper. Draw in the coast line, rocks, shoals, hills, &c., and every other feature from your rough, attending to the hydrographic method of delineation represented in plate A.

21. The soundings follow next, when reduced to the low water standard of the port by the tide gauge.* If there be no station-pointer on board, protract the angles upon a piece of transparent paper, and mark the stations with their proper numbers. If the soundings have been taken equally, divide the spaces between the △'s in as many parts as there are casts, and fill in the corrected soundings in the order in which they occur. All soundings which may have been taken when the tide was up and by reduction to low water are *dry: draw a line under.* In every chart-box in the service will be found the abbreviations adopted in Admiralty Charts; these are to be strictly followed, and in Appendix No. 5 are given a few symbols which will be found useful in taking angles, &c., and in other surveying operations; and in plate A are given the usual hydrographic delineations of banks, cliffs, shoals, &c.

* See Art. 24.

Lastly, put a meridian line and scale to the plan. Insert the variation, geographical position, time of H. W. F. and C., the low-water standard, *to which the soundings are reduced*, and the range at springs and at neaps; note the duration of the ebb and flood, both by the shore and by the stream; draw leading marks and put in views, heights of mountains, &c.

If time does not permit of a regular survey being executed, still a useful record may be made by an itinerant survey, or even an eye sketch, assisted by sextant angles, a few soundings judiciously taken, the true bearing of one object and thé measurement of a base by sound, or with a Rochon micrometer as before-mentioned.

Sailing Directions.

22. Whenever a survey is executed, *sailing directions* should accompany it, and too much care cannot be bestowed upon this important part of a surveyor's duty.

They should contain a description of the coast (see Art. 11); directions for making the land; for approaching, and sailing into or out of the port both by daylight and with the aid of marks, and also by night or in thick weather, when the lead and the lighthouse, if there be one, must be the seaman's principal guide. How a vessel is to proceed with a leading or a beating wind, and with or against the tide—how far she may stand on either tack—what water she may expect to find at low-water springs—and how she may ascertain the depth by calculation on any other day—within what limits a vessel may safely steer in bad weather and when no pilot is on board—where the best anchorage lies, the depth in which a vessel should anchor, and directions for bringing up. With other particulars which have been mentioned under the heads of approaching a port, especially noting all beacons, buoys, lighthouses, and landmarks, &c. (*see Art.* 11).

Affix to these views of the land and sketches of the leading marks. The geographical position, the time of H. W. F. and C., rise at springs and neaps, the low-water standard of the port, &c., and the variation of the compass, point out the best watering-places, and let all bearings given be *magnetic*, and noted as such.

Port regulations and quarantine laws will not be misplaced at the end of these directions.

Tide Pole.

23. When a survey is determined upon, a tide-gauge should be set up, and from *half an hour before to half an hour after every high and low water* the place of the tide should be registered *every ten minutes.** In addition to this, *whilst the sounding of the port is in progress*, the place of the water must be noted every half-hour to facilitate the reduction of the soundings to the low water standard. The tide-gauge should be fixed in a well-sheltered spot, with its zero such a depth as to ensure its being below the low water at springs. When the pole is properly secured and settled down, paint a mark in the rock corresponding with one of the divisions on the gauge, and *note which* in your book, in case the pole should be washed down. If you remain long enough in port, let your observations be continued at least through an entire

* See Forms, Nos. 7 and 8.

lunar month. When you come away, mean the high and low water heights of each day, and take a mean of them again for the *mean place of the water*, and cut *a mark in the rock corresponding with that mean level of the sea before you remove the pole.* As this is the true scientific level of reference in all matters relative to the tides, refer this level again to some mark in a contiguous building, that a reference may at any time be made to it, by persons who might not be able to find the rock.

Let the watch be always at *mean time* at the place. The high and low water observations should be continued night and day with equal carefulness in order to determine the amount of diurnal tide; and every observation should be recorded, although it may not seem to agree with the others.

If tides are taken at coral islands, or at stations within a belt of coral, it should always be noted in the journal whether the sea or land breeze be blowing and with what strength, and also whether the surf be high upon the reefs and sending its water into the lagoon, filling it faster than it can escape.

In the Appendix will be found two forms, one of which (No. 8) is for registering the tides every half hour, the other (No. 7) is for the high and low water only.

For further information upon *the tides* see that section.

Soundings.

24. Before any soundings are inserted in the chart they should be reduced to *a standard* obtained by meaning the three or four successive lowest waters of each spring-tide, and meaning them again for a general mean. This

standard should be noted in a very conspicuous and *unmistakeable* manner as being so many feet below the mean water level, and recorded as the *low water standard of the port.* It is a quantity which would nearly correspond with half the range of an *ordinary spring* tide, a term often written without any direct reference to the low water standard, and so ambiguous that it is to be hoped it will soon disappear from the face of our charts. With this standard, and the known daily height of the tide above mean water level, soundings taken at any hour may be prepared for comparison with the depths upon the chart by the simple formula

$$R + r \cdot \text{cosine}\left(\frac{t}{D} \cdot 180^\circ\right)$$

Where R = the low water standard to which the chart is adapted.
r = the height of tide for the day above mean water level.
D = the duration of the tide.
t = the time from high water previous.

Or, enter the traverse table with the time from the nearest high water as *a course* (allowing 5° of arc to every 10 minutes of time), and with $r =$ (half the range of tide for the day) as a *distance ;* in the *latitude column* will stand a quantity which applied to the *low water standard of the port* + or −, according as the arc is *less* or *greater* than 90°, will give the reduction required. If the arc exceeds 90° take its supplement. But it is to be observed that all these corrections, although preferable to the old method of reducing soundings, are but approximations. In many places, especially in such as have great tides, it is necessary to distinguish between *rising* and *falling*.

If in a country subject to earthquakes, carefully watch

the tide-pole during and after the shock, and if any undulations of the water are observed, note them, and the direction whence they proceed.

Be careful never to place the tide-pole at the mouth of a river, and especially guard against having it within *a bar*, sand-bank, or any such impediment to the free action of the water.

The Bore.

25. If any place should be visited by that peculiar phenomenon, the bore, a wave which in some places comes rolling in with the first of the flood, with a crest foaming and rushing onward, threatening destruction to boats and even to shipping; note the time of the tide at which it begins, whether there be *one wave* only or more, the height to which it rises, and *where it first* appears with respect to any alteration in the feature of the river; and especially note the situation and extent of shoals at or below the spot. The bore is said to occur only at spring tides; note particularly whether this observation be correct.*

Freshes.

26. Connected with the rise and fall of the water is that periodical elevation of the surface of rivers by "freshes," occasioned by heavy and continued rains in the interior of the country. These torrents not only raise the general level of *the river,* properly so called; but where a bar

* The remark made in "How to observe [p. 35] that either rocks or shoals, or great depth of water secure a river from the inconvenience of the bore," is not always correct; for the Severn is encumbered with shoals, and has a bore which has proved destructive to vessels grounded upon the sands.

exists, also raise the level there, so that vessels which cannot enter during the dry season are at such times able to pass over the bar. The time when the water begins to rise, when it attains its maximum, when it begins to subside and regains its mean or ordinary level, should be carefully noted, and with it the elevation of the water in feet, both in its ascent and descent.

Discovery of Land.

27. On the discovery of any unknown lands or dangers, the first endeavour, after the vessel is placed in safety, should be to fix the position of the place as accurately as the means of observation admit, and not to quit the spot until the danger is satisfactorily placed upon the chart.* Describe it as accurately as you can; determine its extent, height, and configuration; the adjacent soundings, and the quality of the ground; and give a sketch of its outline. If it be extensive, a running survey will be desirable.† If it be within sight of other land, its position must be fixed by bearings or angles between known points of the coast, and some conspicuous objects upon the land selected, which being *brought in a line* will lead ships clear of the danger. Do this for *both sides*, and give correct bearings of the transits, and, if possible, sketches of the objects.

* See Raper's 'Navigation,' 855, p. 328; and 856, p. 329. "No commander of a vessel," observes that talented officer, "who might meet unexpectedly any danger (before unknown), could be excused, except by urgent circumstances, from taking the necessary steps both for ascertaining its true position and for giving a description as complete as a prudent regard to his own safety allowed."

† See Art. 29.

As regards coasts, and islands which are but little known, I have given in the Appendix a list of such as are most deserving of attention, extracted from a return made by the able and indefatigable officer at the head of the Hydrographic department to an order of the House of Commons, 1848, and all general directions for acquiring information which may have been already given must be considered to apply with double force to these countries. The limits of this paper do not permit of our entering into particulars as to the probable position of places which may be imperfectly determined, nor of the reported position of islands which are considered doubtful. In the Atlantic alone, for instance, there are islands reported continually where none could possibly exist; and the islands of the Pacific have been multiplied by the errors of the longitudes of persons visiting them; but wherever the charts place any islands as doubtful, which you wish to seek (as it is always more probable that the latitude is correct than the longitude), the *parallel* of the supposed latitude should be gained, at a meridian sufficiently distant from that given to exceed the probable limit of error in longitude, and a due east or west course pursued until a similarly distant meridian is gained on the other side; and if there should be any change in the colour of the water, sounding ought by all means to be tried; and especially we call attention to soundings upon the site near the equator marked as the seat of volcanic action from about 3½ S. and 15° to 24° W., and also to the vicinity of the great bank S. and S.E. from the Falkland Islands, called Burdwood Bank, on which there has been found recently as little as 24 fathoms; the Agulhas Bank, and the sites of any volcanic islands which may have risen and disappeared.*

Sailing along a Coast.

28. When sailing along a coast or islands which may even be known and charted, it is advisable, as a general practice, to *verify* the position of the points and headlands as the ship sails along; and when the coast is new, or but indifferently explored, no opportunity should be omitted of determining as accurately as possible the position of every part within your power.

The position of places is determined from a ship with the least disadvantage, by being brought to bear east or west when the latitude is taken, and north or south when longitude is observed. And as these observations may be made during several hours of the day,† much may be done in a single day's run, especially if patent log bases connect the stations, and astronomical bearings be employed. And upon all occasions the noting of transits, or the coming in a line of remarkable objects and of points of interest, should form a necessary portion of our duty, although we may believe them to be already satisfactorily determined, as they afford the most critical test of the accuracy of former surveys, and are especially useful in cases where longitudes of contiguous places may have been had by different observers.

If time admits of more than this being done, and in some of the countries which are but little explored, it is

* See also Art. 4.

† See Raper's Navigation, 830 et seq., p. 320; also 834, p. 321, second edition.

extremely desirable that no opportunity should be lost of perfecting their outline, the heavy boats may be hoisted out and sent in shore of the ship to run in the coast line and the detail whilst the ship carries on a triangulation and continuation of bases in the distance, making what may be termed a running survey.

Running Survey.

29. Whenever this can be done, send the boats to a distance of 4, 5, or 6 miles at starting, and let them and the ship anchor, if possible, to measure a base by sound (Art. 19), and to get astronomical bearings and angles to the *same points.* Fix the ship's position by repeated observations for the latitude and by chronometer; then weigh and put the patent log over and steer a steady course along the land (sounding, if the depth of water admit of it, without stopping). One boat now runs along the land from point to point, putting in the coast line and its detail, getting astronomical bearings and angles as she proceeds, especially of all transits of points and headlands, and measuring her distance between them by patent log, and sounding, but without stopping. The other boat attends principally to the soundings, fixing herself as she requires, by angles and bearings between the points determined by the other boat and the ship.

At the end of a few miles' run, or at noon, or when necessary to renew the angles and bearings, a signal is to be shown and the logs are then to be hauled in and read off, but not reset, fresh angles and bearings to be taken and a new base commenced, the distance between the ship and boats being again measured by sound. The log is then again put over and the course of the vessel resumed. In this manner the day passes, the bearings and observations all being worked out at the moment—the outline run in, views taken, and every particular mapped and booked at the time so as to leave nothing to memory. At the close of the day's operations anchor in position, measure a base by sound, and repeat operations as at starting, recall the boats, and in the grey of evening get the ship's position by stars and planets, which may at this time be observed with great accuracy before the horizon becomes too obscure. If the ship can remain at anchor, she will observe the set of the stream and the rise and fall of the water, however roughly it may be done.

As early as possible commit the triangulation to paper that the vessel may start in the morning with some points of land well fixed so as to enable the ship to continue her triangulation throughout the day without the aid of the boat—although her co-operation as before should be renewed.

If there be no anchorage, the ship will maintain her position during the night under canvas, and in the grey of the morning picking up the place where she left off on the preceding evening, send the boats away, get altitudes of stars for latitude and longitude, measure a base by sound; get astronomical bearings and angles, &c., and putting over the patent logs continue along the coast as before.*

Thus far we have considered the observations as being wholly confined to the vessels, but it will add considerably

* For further information, and a more extensive application of this method, see Belcher, Mackenzie, and other works on nautical surveying.

to the accuracy of the survey if landings be occasionally made, and the stations be critically determined by astronomical observation, *i. e.*, by latitudes and chronometers, and the positions connected with the rest of the work.

30. It is not necessary to be provided with a regular chart for this purpose; the projection may proceed as you advance. Thus, consider how the coast runs, and draw a line along the paper to represent the meridian at starting; set off on this a degree of latitude according to the scale on which the survey is to proceed, 1 inch or 1½ inches to the mile, or more or less according to circumstances, and begin at once to lay off the bearings and angles. As you take up other stations proceed to throw out meridians and parallels in the manner described in Appendix No. 10. A chart upon this projection will be found easy of construction and more satisfactory than any other; and when the survey does not extend over more than 8 or 10 degrees of latitude is sufficiently correct. In laying off bearings upon it, it must be borne in mind that they are to be projected with reference to the meridian passing through the spot. Mercator's projection, in which the meridians are all parallel, and which is in such general use in the navy, except in very low latitudes, is not adapted to the purposes of a survey, as the bearings and the protraction will never agree together nor with the observed latitude and longitude of the stations.

With reference to the longitude I may remark, that the absolute longitude of the place is not required, but it is necessary to determine the *difference of meridians* as you proceed; and these should afterwards be compared with some well-determined meridian. I may observe here, once for all, that the longitude of a place, by chronometer, from Greenwich, should never be given without the *accompanying longitude from which the deduction of the meridian was made;* in short, that chronometers should be referred to only as a *measure of* DIFFERENCES.

Coral Islands.

31. Should coral islands be fallen in with, determine their position, extent, and map their outline; fix the openings into the lagoons, and describe their general appearance, whether wooded or not, and whether any high clumps of trees (distinguishing the palms) be conspicuous upon them, and at what part; you should then particularly notice the slope of the coral on both the outside of the island and the inside, and run off lines of soundings in various parts from the water's edge to as great a depth as you can reach, and at each cast particularly note the bottom, whether it be living or dead coral; note the greatest depth at which live coral is brought up: the existence of living coral at great depths is a point of interest. A swab fixed to the lead will often bring up specimens of coral which might otherwise be missed.

Point out the place of the anchorage in the lagoon by an anchor, and state whether vessels can sail in with the trade wind or not, and the best time for going in, for in many of these islands there is so strong a current running out through the channel after the trade wind has set in, in the morning, as to render it imprudent to attempt the passage; and in some it is only after the sea wind subsides, and the land breeze has commenced, that the passage can be effected. It is the sea getting up with the breeze and beating over the reefs into the lagoon that occasions such a current through the opening. Inquire into this on the spot, and do not commence any tide observations in the lagoon if the reefs are low and the channel small: if, however, the lagoon be open on one side and sheltered on the side of the prevailing wind, these spots in the ocean afford excellent places for observations upon the tides.

Currents occasioned by the trade-wind prevail about all the islands situated in those latitudes; their direction and force should be ascertained and stated in your remarks

Rivers.

32. All rivers should be traced to the furthest possible point that time will allow, for although it is the usual practice to limit hydrographic inquiry to the vanishing point of tidal influence, yet there are many reasons why we should not here so circumscribe our views. Rivers are the great arterial features of our globe; they define the valleys, give boundaries to the hills and mountain ranges, and if traced to their source enable us, with the aid of a few well-determined culminating points of contiguous ranges, to trace upon our charts the general feature of the country through which they flow. Besides which they are so far connected with the navigation of our ports and harbours that their aid is often indispensable to a free access and egress, by affording a powerful means of scouring channels and removing impediments to shipping, which would otherwise be denied admission. They may therefore be said to be of almost equal importance

to hydrography as to physical geography. In all cases then where rivers approach or flow into any of the ports under examination, you should acquire as extensive a knowledge of them as you possibly can, map as much of the windings and feature as is practicable, and especially of such parts of those that are not navigable as may be made available to the improvement of the navigation of the port, or in any way be converted to hydrographic use, particularly noting the depth, extent, and variations of surface, of all widenings of the stream, or basins affording back water and capable of being retained, or converted to a scouring power, carefully determining the elevation of the surface above the *mean level of the ocean*, and. if the river does not run into the port, whether it could not be conveyed to it, and with what facility. These inland basins are occasionally greatly affected by mountain torrents, melting of snow, and rainy periods, raising their surface to an extraordinary height even in a few days; while, on the other hand, long dry seasons depress them as much below the mean level. Our endeavour should be to ascertain these variations and the mean level of the water of the basin; we may often see, for weeks after the event, the mark of the wash of the water around the lake or basin far above the existing level; this may be measured and compared with the place of the mean level, and be coupled with the place of the water according to the best information to be procured at the place (noting the informant).

Note the depth and capability of transport or of inland navigation, and the power of traversing the stream for military purposes; also the nature and peculiarity of construction of the vessels employed and the means they have of advancing against the stream, &c., and the distance to which navigation is practicable, severally for vessels, boats, or barges.

In large rivers communicating with the sea note the facility of access and egress, the depth of water on the bar, if there be one,* the position and nature of shoals or rocks, and the navigable capabilities of the stream, the rate and duration of flood and ebb, that is, of the *ingoing* and *outgoing* stream. The distance to which the stream runs up, and the extent to which the rise and fall of the water is felt, or what may properly be called the end of the tide; and here always, if possible, determine the elevation of the high-water line above the *mean level* of the ocean.

Lastly, in speaking of rivers, let it be understood that the *right* or *left* bank should have reference to the *downward direction* of its course, so that, when descending the stream, the *right* bank is on the *right hand* and *vice versâ.* It is better to adopt this phrase than to say east or west, which might at the least be ambiguous, for it is clear that if a stream meander much, its course being always of necessity downwards, it might be successively diverted to every point of the compass.

Lakes.

33. Lakes, properly so called, or which have no rivers running through them, can scarcely ever be turned to the uses of hydrography, except when they are upon a level

* What has been already said on leading marks, lighthouses, beacons, buoys, &c., &c., of course applies here also.

with the sea, when a communication has been or may be made and a scouring power obtained by the admission of the tide through the port. However, what has been said of river basins may be applied to these enclosed sheets of water. The principal points are, their distance from the port, height above mean water line of the ocean, depth, dimensions, and fluctuation of surface, the quality, temperature, and sweetness of the water, the nature of the bed and borders, inland navigation, if any, &c.

Artificial Harbours.

34. In all harbours, but especially in the vicinity of those which are formed by piers carried out into deep water, it is proper to notice whether there are shoals formed about the piers, and the pier-heads especially. If there are, obtain information as to the probable cause, when they were first noticed, carefully note their extent and direction, and connect with them the direction of the tide, ebb and flood, and if there be any stream through the piers out, or in, note its rate, direction, and the distance it extends. The form and construction of artificial harbours, piers, and breakwaters, does not properly belong to hydrography; but it may be well to describe and record the form of the breakwater, the pitch or slope of the stonework, *the depth* in which it is erected, the material of which it is composed, the nature of the work, and how it has resisted the sea. Or if there be an opportunity of seeing it in a gale of wind, the power any peculiar form or construction of breakwater may have in repelling a heavy sea, or the effect any peculiar form of pier may have in diverting the sea at the entrance from the

anchorage within. The position of the entrance with regard to the offing stream and prevailing wind, the width of the channel, the protection of the anchorage, the number of square acres enclosed. If there be any backwater, state its extent, how the scouring, if any, is managed, at what time of tide and what is its apparent effect—and at all places wherever backwater is used, it may be as well to sound off the mouth of the port to as great a distance as the effect of the scouring action can possibly extend—for occasionally injurious effects have been produced by this powerful agent at a distance scarcely contemplated. State all deposits, siltings up, and at what rate it proceeds.

Foreign Ports.

35. In visiting foreign ports, a particular account should be given of the resources of the place in the event of vessels requiring either a repair or a refit. Such as whether there are any docks, wet or dry? what sized vessels they are capable of receiving, and how many at a time, is there a patent slip or gridiron, &c. How near vessels, of particular dimensions, can approach the wharfs, or at what time of tide lie alongside of them; whether there are sheers for removing masts, and of what size, or cranes for lifting machinery and boilers; whether there be a dockyard or arsenal, or whether stores can be procured from other sources. Whether there is a steam-yard, and to what extent they cast and manufacture machinery or boilers, or can repair steamers?

Whether there is a coal depôt, and what quantity of

coal can be generally relied upon as at hand; nature and quality of the material? &c.

Are there any piers, jetties, or wharfs for landing passengers, or cranes for carriages, and at what time of tide available? If the country be low, are there any sea walls, and would the country be flooded by their removal?

Waves.

36. Lastly, the attention of the observer should be directed to the measurement of the height, the extent, and the velocity of the waves of the ocean. Not only of those high swelling seas which are common to every gale, but especially of those gigantic ridges which are occasionally met with off Cape Horn, the Cape of Good Hope, and even in the Atlantic; coming in couplets and triplets in the course of a gale, and occasioning fearful lurches which are long remembered. Opinions differ greatly as to the dimensions of these stupendous bodies, and any observations which will assist in determining their limits cannot fail to be acceptable. The inquiry is, *first*, as to the height of the solid wave above the mean water level. *Secondly*, the distance of the ridges apart. *Thirdly*, the rate at which the wave travels, and whether the height and distance of the ridges vary with the velocity. *Fourthly*, what is the greatest estimated extent of any one of those ridges.

The most simple way of measuring the height, is, when the vessel is in the lowest part of the trough between two following seas, to ascend the rigging to such a height as will bring the top of the wave on with the horizon, to put a mark, note the inclination of the vessel, and *at leisure* to measure the perpendicular height of the eye above the water line, which we may presume will be double the height of the wave above the mean water level. It will necessarily require several observations to be made before any satisfactory conclusion can be arrived at. The distance of the waves apart may possibly be tested by actual measurement, by means of the lead-line and a float veered out to such a distance that the float shall be on the crest of one wave when the ship is on the top of the other. And the rate may be determined by the time occupied by the wave in passing from the float to the ship: the rate of the ship through the water and the angle her course makes with the route of the wave being known. There are other methods of determining this interesting problem which will no doubt occur to the intelligent observer, and they are sufficiently numerous to afford ample exercise of his ingenuity, but all are attended with difficulty owing to the circumstances under which the observations are required to be made.

APPENDIX No. 1.

DATE.	LATITUDE.	LONGITUDE.	CURRENT.		VARIATION.	WIND.		TEMPERATURE.		BAROMETER.	REMARKS.
	N. or S.	E. or W.	Direction.	Rate.	E. or W.	Direction.	Force.	Air.	Sea.		

Remarks upon the Currents which prevailed in the passage across the Atlantic.

"From the time we quitted Teneriffe, with the N.E. trade-wind, until we lost the breeze in lat. 7° 40′ N., long. 26° 40′ W., the current set on an average S. 54° W. true, at the rate of 11¼ miles per day. On losing the trade and entering the calm latitudes, the westerly current ceased, and the next 24 hours the ship was set N. 83° E. true 23 miles. The meeting of the opposite currents was marked by a strong ripple, which was traced to a considerable distance. The four succeeding days, in which we changed our position from lat. 7° 20′ N., and long. 26° 58′ W. to lat. 3° 56′ N., long. 26° 44′ W., the current ran S. 70° E. true 13 miles per day. Here we met the S.E. trade, and with it experienced a strong current, which carried us N. 62° W. true 23 miles per day, until we made Fernando Norhona. Hence to a position 100 miles due east of Cape Ledo the current set between S. 78° W. and S. 21° W. (true), on an average daily rate of 27 miles," &c.

"While in Rio Janeiro H.M. ships A. and B., the packet C., and a fast-sailing schooner the D. arrived, and we learnt that the A. had crossed the equator in 18° W., the B. in 25° W., the C. in 29¼° W., the D. in 39° W., whilst we crossed in the E. in 30° W.; and upon inquiry it appeared that the passages from England were as follows, viz.: the A. was 49 days, B. 40 days, C. 38 days, the E. 36 days, and the D. 110 days, having got *so far to the westward that she could not weather Cape St. Roque*, and was obliged to stand back to the variable winds to regain her easting. Thus it appears that, with the exception of the D., *the passages were shortened in proportion as the equator was crossed to the westward*," &c. &c.

Plate A. *To illustrate Hydrographic delineation.*

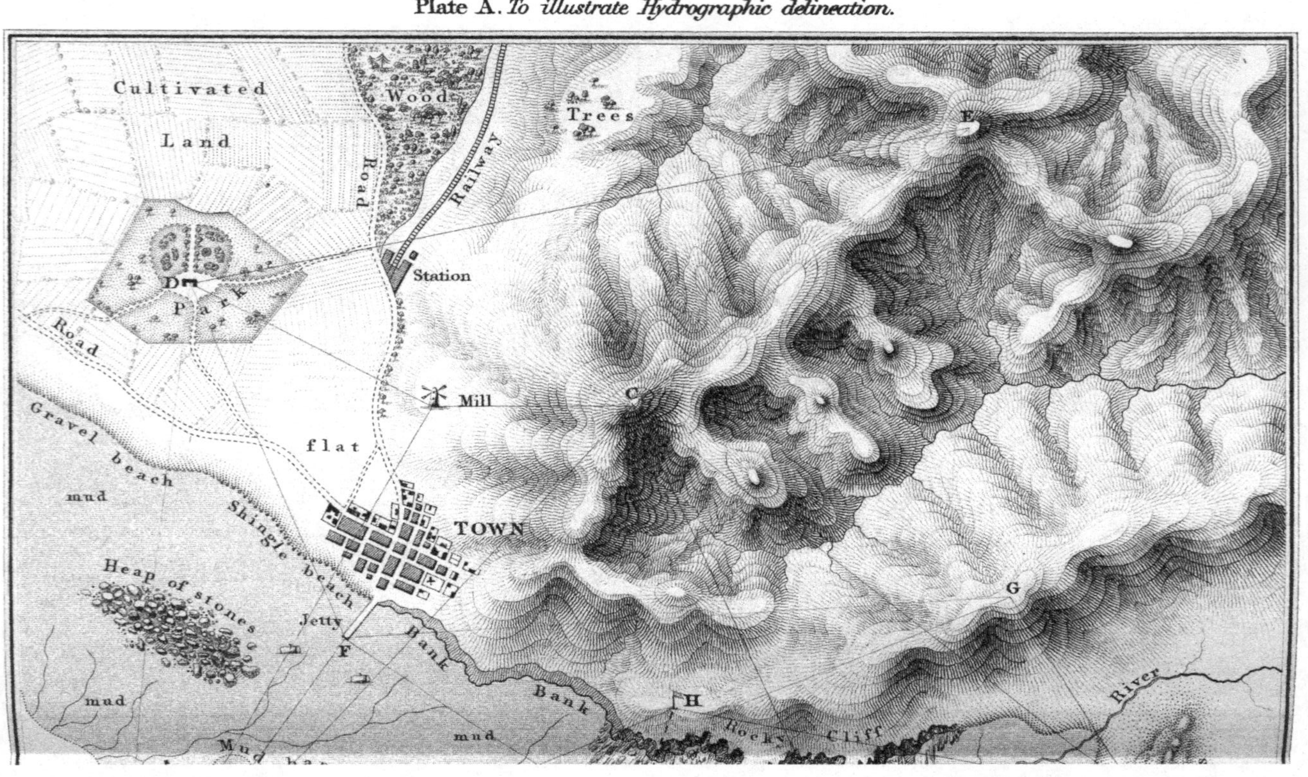

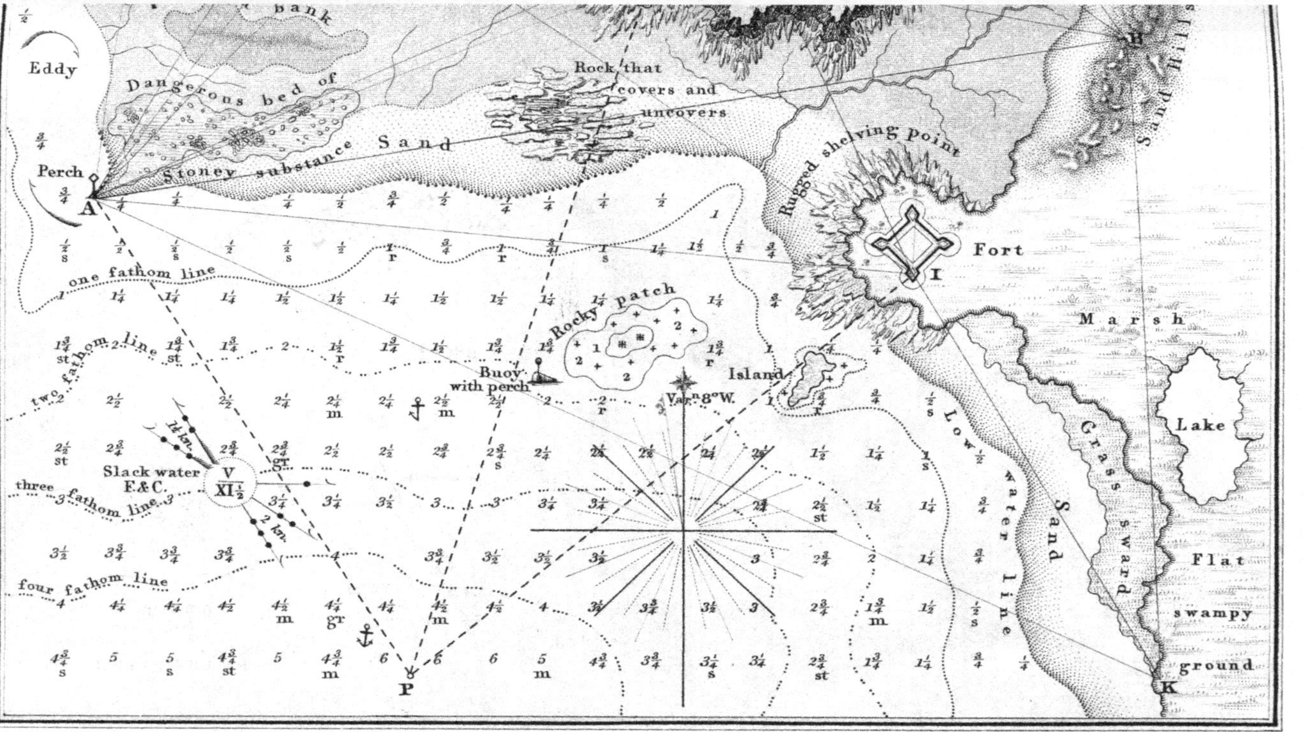

J.&C. Walker Sculp[t].

Published by John Murray, Albemarle Street, Piccadilly 1849.

APPENDIX No. 2.

To find the Height of an Object the Distance of which is known.

Rule.—To the observed altitude apply the true dip, less the terrestrial refraction.* The result call corrected altitude; to the log. of the distance in yards add the constant 8·073007, and find the log. of the sum, which turn into arc and add to the corrected altitude; then to the log. tangent of this sum add the log. of the distance in yards as above-mentioned, the result will be the log. of the height of the object in yards.

Example.—Mount Etna was seen at 57 miles distance, and subtended an angle of 1° 30′ 00″ with the horizon; elevation of the eye 20 feet, required the height of the mountain?

	°	′	″		
Altitude . . .	1	30	00	Distance 57′, in yards 115650 log	5·063157
† Dip . . .		− 4	43	Constant . . .	8·073007
	1	25	17	60)1368″ log =	3·136164
† $\frac{1}{11}$ of Dip . .		+	26		
Corrected Altitude	1	25	43		
		+ 22	48	. . . . . Correction	. 22,48
True Altitude . .	1	48	31	Tangent . . .	8·4993668
				Constant . .	5·0631570
					3·5625238 log 3652 Yards. height required.
					× 3
					10956 feet.

APPENDIX No. 3.

To find the Constant for a Height, in order to compute its Distance readily from its observed Altitude.

Rule.—From the log. of the height in yards subtract the constant log. 6·5424481, halve the sum—find its sine, and take out the corresponding co-sine, which is the constant required. (*a*)

* The terrestrial refraction varies from $\frac{1}{8}$ to $\frac{1}{14}$ part of the arc.

† If the Dip Sector had been used, the observed Dip should be substituted for these two quantities.

To find the Distance.

RULE.—From the observed altitude subtract the dip less the terrestrial refraction,* and call the remainder corrected altitude. To the constant above mentioned 6·5424481 add the cosine of the corrected altitude, and from the cosine of the sum subtract the corrected altitude. The remainder is the log of the approximate distance in arc. Divide the approximate distance so found by the proportion of terrestrial refraction allowed, and subtract the quotient from the before found corrected altitude for the true altitude.

Lastly, add the cosine of the true altitude to the constant due to the height of the object (a); find the cosine of the sum, and subtract from it the true altitude; the remainder is the distance in arc required.

EXAMPLE.—Observed the altitude of Snowdon to be, On 45′ 00″
its height being 3565 feet = 1188 yards, Off 45 10
required its Constant and its Distance, height of eye
being 14 feet. 45 5 mean.

Log of height 3·0722499
Constant . . 6·5424481
2)16·5298018
Sine = 8·2649009
Cosine = 9·9999265 Constant required.

To find the Distance.—

Constant for Snowdon 9 9999265
Cosine corrected Alt. 9·9999686
1° 15′ 33″ = Cosine 9·9998951
− 41 20 Alt.
$\frac{1}{10}$)34 13 Approx. dist.
3 25 Correction.

Observed Alt. . . . 45′ 05″
Dip for 14 feet . . − 3 45
41 20
Terrestrial ref. $\frac{1}{10}$ of Dip + 22
Corrected Alt. . . . 41 42
Correction . . . − 3 25
True Alt. 38 17

Constant . . . 9·9999265
Cosine true Alt. . 9·9999731
1° 13′ 55″ Cosine = 9·9998996
True Alt. − 38 17
35 38 = 35·6 miles. = distance of object.

* The terrestrial refraction varies from $\frac{1}{9}$ to $\frac{1}{14}$ of the arc.

APPENDIX No. 5.

Useful Abbreviations in Surveying.

Two Objects in a line (conjuction) .... ☌

Right tangent .... › .... as ....

Left tangent .... ‹ .... as ....

Leading mark in a view or sketch ....

Angle subtended .... ‹-- 20° 30′ --›

Sun's right limb ....

D° left „ ....

D° centre „ ....

D° upper „ ....

D° lower „ ....

No. of angle .... ∠

Station .... △

Wind Mill ....

Church ....

Direction of stream ....

flood or ingoing — 1st Qr, 2nd „, 3rd „

ebb or outgoing — 1st Qr, 2nd „, 3rd „

Necessary to moor ....

Whirl of tide ....

Races and overfall ....

Low water standard .... L. W. S.

Mean water level .... M.W. L.

Rock which covers and uncovers .... ⁂

Rocks always under water ....

+ less than 1 fm.

1 .... 1 fm.

2 .... 2 „

3 .... 3 „

APPENDIX No. 4.

TUESDAY, 24th. Sounding in the 1st Cutter, WILLIAM REEDER, Leadsman. Sextant used, D 56. Line correct at starting.

AUGUST 24, 1847. (Officer's Name, Mr. D. HALL.)

Red to L.W.	Mean Time.	SOUNDINGS. High Water 5·36 P.M., Low Water . 11·42 A.M. h. m. Range 33·9 ft. in.	Course.	P. Log.	DATE AND REMARKS. Spring Range 44 ft.	Objects.	Angles.	Objects.	Angles.	Objects.
fm.	h. m.	fms.		m.			°		° ′ ″	
$\frac{5}{6}$	11·45	∠1 $2\frac{4}{6}$ $2\frac{4}{6}$ $2\frac{4}{6}$ $2\frac{4}{6}$ $2\frac{3}{6}$ $2\frac{3}{6}$ $2\frac{4}{6}$ $2\frac{4}{6}$						∠1		
		Reduced $1\frac{5}{6}$ $1\frac{5}{6}$ $1\frac{5}{6}$ $1\frac{5}{6}$ $1\frac{4}{6}$ $1\frac{4}{6}$ $1\frac{5}{6}$ $1\frac{5}{6}$	Pulling in the direction of Clevedon church, tree in one with house on the side of the hill.			Denny Island	63·46	Clevedon church	79·11	Worle mill
	11·55	∠2 $1\frac{5}{6}$ $3\frac{4}{6}$ $3\frac{3}{6}$ $3\frac{3}{6}$ $4\frac{5}{6}$ $6\frac{1}{6}$ 7 $5\frac{2}{6}$						∠2		
		Reduced 1 $2\frac{5}{6}$ $2\frac{4}{6}$ $2\frac{4}{6}$ 4 $5\frac{2}{6}$ $6\frac{1}{6}$ $4\frac{3}{6}$				Do. .	61·12	Do. .	84·18	Do.
1	12·0	∠3 $3\frac{3}{6}$ $3\frac{3}{6}$ $2\frac{1}{6}$ $2\frac{2}{6}$ $2\frac{4}{6}$ 3 $3\frac{2}{6}$						∠3		
		Reduced $2\frac{3}{6}$ $2\frac{3}{6}$ $1\frac{1}{6}$ $1\frac{2}{6}$ $1\frac{4}{6}$ 2 $2\frac{2}{6}$	Altered course, and pulled towards Denny Island, in one with church.			Usk light house	89·4	Walton castle	103·22	Worle mill.
$1\frac{1}{6}$	12·5	∠4 $2\frac{1}{6}$						∠4		
		Reduced 1				Do. .	109·21	Do. .	71·32	Do.
		At 6 o'clock left off sounding and measured the line over, found it correct.								

NOTE.—The objects are here placed according to their observed relative positions, and should always be so written down: the *right-hand* angle being invariably read off *first*.

APPENDIX No. 6.

The number of Miles or Minutes of the Equator contained in a Degree of Longitude under each parallel of Latitude for the Spheroid. $\frac{1}{304}$ Compression.

Lat.	Length of Degree.	Lat.	Length of Degree.	Lat.	Length of Degree.
°	′	°	′	°	
0	60·000	31	51·475	61	29·161
1	59·991	32	50·930	62	28·240
2	59·964	33	50·370	63	27·310
3	59·918	34	49·793	64	26·372
4	59·854	35	49·202	65	25·426
5	59·773	36	48·596	66	24·471
6	59·673	37	47·975	67	23·509
7	59·556	38	47·339	68	22·540
8	59·419	39	46·688	69	21·564
9	59·266	40	46·021	70	20·581
10	59·094	41	45·346	71	19·592
11	58·905	42	44·654	72	18·596
12	58·697	43	43·948	73	17·595
13	58·472	44	43·229	74	16·588
14	58·229	45	42·495	75	15·577
15	57·968	46	41·750	76	14·560
16	57 690	47	40·992	77	13·539
17	57·394	48	40·220	78	12·514
18	57·081	49	39·437	79	11·485
19	56·751	50	38·642	80	10·452
20	56·403	51	37·834	81	9·416
21	56·038	52	37·015	82	8·377
22	55·657	53	36·185	83	7·336
23	55·258	54	35·343	84	6·292
24	54·842	55	34·400	85	5·246
25	54·410	56	33·627	86	4·199
26	53·962	57	32·754	87	3·150
27	53·496	58	31·870	88	2·101
28	53·015	59	30·977	89	1·050
29	52·518	60	30·074	90	0·000
30	52·004				

APPENDIX No. 8.—Daily Tide Journal.

☽'s Transit, of 18 h. m. A.M. ☽'s Transit, h. m. P.M. { Place, Lat. Long. Var. of Compass.

Time. Mean.	Height.	Direction of Stream Magnetic.	Velocity of Stream	Winds. Direction Magnetic.	Winds. Force.	Weather	Barom.	Therm.	REMARKS.
UPPER TRANSIT.									
H. M. 0 30 A.M.	Ft. In.		Knots.						
1 0 ,,									
30 ,,									
2 0 ,,									
30 ,,									Additional Observations.
3 0 ,,									H. M. F. I.
30 ,,									. 0 . .
4 0 ,,									. 10 . .
30 ,,									. 20 . . From ½ hour
5 0 ,,									. 30 . . before
30 ,,									. 40 . . to ½ hour
6 0 ,,									. 50 . . after
30 ,,									. 0 . . H. W.
7 0 ,,									
30 ,,									H. M. F. I.
8 0 ,,									. 0 . .
30 ,,									. 10 . . From ½ hour
9 0 ,,									. 20 . . before
30 ,,									. 30 . . to ½ hour
10 0 ,,									. 40 . . after
30 ,,									. 50 . . L. W.
11 0 ,,									. 0 . .
30 ,,									
12 0 ,,									
LOWER TRANSIT.									Moon's Age.
0 30 P.M.									
1 0 ,,									
30 ,,									
2 0 ,,									
30 ,,									Additional Observations.
3 0 ,,									H. M. F. I.
30 ,,									. 0 . .
4 0 ,,									. 10 . .
30 ,,									. 20 . . From ½ hour
5 0 ,,									. 30 . . before
30 ,,									. 40 . . to ½ hour
6 0 ,,									. 50 . . after
30 ,,									. 0 . . H. W.
7 0 ,,									
30 ,,									H. M. F. I.
8 0 ,,									. 0 . .
30 ,,									. 10 . . From ½ hour
9 0 ,,									. 20 . . before
30 ,,									. 30 . . to ½ hour
10 0 ,,									. 40 . . after
30 ,,									. 50 . . L. W.
11 0 ,,									. 0 . .
30 ,,									
12 0 ,,									

RESULTS		H. M.	H. M.		feet.	inches.
	High Water at.............	A.M.	P.M.	Rise, A.M..		
	Low Water at..............	,,	,,	Fall.......	,,	,,
	Flood Stream terminated at	,,	,,	Rise, P.M..	,,	,,
	Ebb ditto ditto	,,	,,	Fall.......	,,	,,

APPENDIX

Register of TIDES observed at Portished in the Month

Date.	High Water.			Barometer at High Water.	Low Water.			Barometer at Low Water.	Range of Tide.	High Water.		
	A.M. or P.M.	Time.	Height.		A.M. or P.M.	Time.	Height.			P.M. or A.M.	Time.	Height.
1	A.M.	h. m. 5 53 6 3 6 13 6 23 6 33	ft. in. 43 2 43 8 43 11 43 8 43 2	29·95 A.M. 29·74 P.M.	A.M.	h. m. 0 20 0 30 0 40 0 50 1 00	ft. in. 2 10 3 2 3 00 3 2 2 10	29·92 A.M. 29·70 P.M.	ft. in. 40 11	P.M.	h. m. 6 20 6 30 6 40 6 50 7 00	ft. in. 44 3 45 2 45 2 44 8 44 0
2												
3												
4												
5												
6												
7												
8												
9												
10												
11												
&c.												

No. 7.

of October, 1847. (Observer Wm. Quin, Qr.-Master.)

Low Water.			Range of Tide.	Interpolated High Water.	☽'s transit.	Diff.	Mean Water Level.	Wind.		Stream.	
P.M. or A.M.	Time.	Height.						Direction.	Force.	Flood turned.	Eb turnl.
	h. m.	ft. in.	ft. in.	h. m.	h. m.	h. m.	ft. in.			h. m.	h.
	0 50	3 0	43 0	6 13 A.M.	11 19 A.M.	7 16	23 5	North.	2	6 0	12
	1 00	2 5									
P.M.	1 10	2 2		6 38 P.M.		7 19	23 8	N.N.W.	1		
	1 20	2 4								6 20	1
	1 30	2 11									

APPENDIX No. 9.

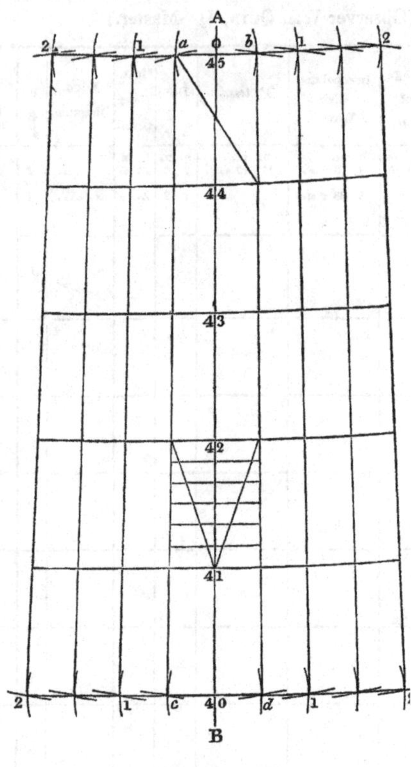

APPENDIX No. 10.

To construct a Chart for a Running Survey of a Coast.

Draw the meridian line A B (Appendix No. 9) through the centre of the chart, and set off the degrees of latitude upon it, of equal lengths, according to the scale which it is intended to construct the chart upon—draw short lines at right angles to each of these.

At the extremities of the degrees of latitude, as at 40° and 45°, set off right and left upon the perpendiculars distances equal to half a degree of longitude in those parallels respectively (taken from Appendix No. 6), as at *a b c d;* then with the diagonal distance *a d* or *c b;* and with the foot of the compass at 40° sweep the small arcs 1 1 at top, and likewise the arcs 1 1 at bottom from 45°. Again, with the length of a degree of longitude in 40° cut the small arcs 1 1 before described: these intersections will be corners of parallelograms, each of a degree of longitude, extended over as many degrees of latitude as your chart contains. Repeat the process for other meridians right and left of A B, and connect the points 1 1, &c. by meridian lines. Set off upon these from either the top or bottom the degrees of latitude before laid off upon A B, and connect them all throughout the chart by straight lines, as at 40, 41, 42, 43, 44, 45, &c. For a scale, divide the degree of latitude into sixty equal parts, or into such equal portions of it as the scale admits of, and it will give a similar proportion of geographic miles of distance; and for longitude, if each degree of latitude be so divided by lines extending from meridian to meridian, and the corners of the parallelogram be connected by a straight line, as is shown in the plan between the 41st and 42nd degree, a scale of miles will be given for that parallel.

When bearings are taken they must be laid off from *the Meridian passing through the station.*

APPENDIX No. 11.

Coasts and Islands of which our Hydrographical Knowledge is imperfect.

Abstract from a Return made to the House of Commons 10*th February,* 1848, *from the Hydrographic Department of the Admiralty.*

There is wanted a critical examination of "the eastern islands of the Mediterranean, along with the coasts of Syria and Egypt, and as much of the northern shore of Africa as would meet the French survey, which, having commenced with Algiers and Morocco, will very probably be continued along Eastern Barbary and Tunis.

"From the Strait of Gibraltar the western coast of Africa has been sufficiently surveyed and published as far as Cape Formosa, in the Bight of Benin; but as there is much legitimate traffic in the eastern part of that great Bight, as well as further to the southward, both it and many of the ports and anchorages on this side of the Cape of Good Hope require a more careful and connected examination.

"The charts of the whole of the Cape colony are exceedingly defective, and from thence to the Portuguese settlements of Delagoa we know scarcely anything.

"From Delagoa to the Red Sea and the whole contour of Madagascar are sufficiently represented on our charts for the general purposes of navigation, though many further researches along the former coast might still be profitably made.

"The Red Sea, part of the coast of Arabia, the Gulf of Persia, and many detached portions of the East Indies, have been already executed by the Company's officers; and no doubt it is intended that the coasts of Malabar and Coromandel shall soon be undertaken by the same hands. The long Malay Peninsula and the Strait of Malacca will require much time and skill to complete, and to combine with each other those parts that have been surveyed.

"With the China Sea we are daily becoming better acquainted, but much is still to be done there; for probably not one of the multitude of rocks and shoals with which it is almost covered is put exactly in its right position; and while some are repeated two or three times, others have been omitted.

"On the coast of China the charts are excellent from Canton round to the mouth of the great river Yang-tse-Kiang; but of the Yellow Sea we know very little, and still less of the Corea, Japan, and the coast of Tartary, and up to the confines of the Russian empire.

"The southern passages into the China seas have never been examined with the care they deserve; and all that is known of what are called the eastern passages through the Great Malay Archipelago are only the results of the casual observations and sketches made years ago by industrious seamen.

"The islands and surrounding shores of the Arafura Sea, if better known, would offer many ports of refuge, and probably an increased opening to commercial enterprise.

"The Strait of Torres has been satisfactorily surveyed; but before it becomes the great highway for steam-vessels to and from Sydney, its approaches, and also its contiguous coasts of New Guinea, should be more intimately known.

"The whole circuit of the great island of Australia has been well explored, and the general characteristics of its several shores are sufficiently known for all general purposes; but far more minute surveys of its immediate waters and maritime resources must precede their being inhabited, beginning with the eastern coast, along which the tide of colonization seems to be already creeping.

"The shores of Tasmania, in like manner, are but very roughly laid down, and even to this day there is no chart of the harbour and entrance to Hobart Town, its capital and principal seat of trade.

"A full survey of New Zealand has just been commenced, and will no doubt answer all the wants of both the settler and navigator.

"In advancing to the eastward across the Pacific Ocean, there are

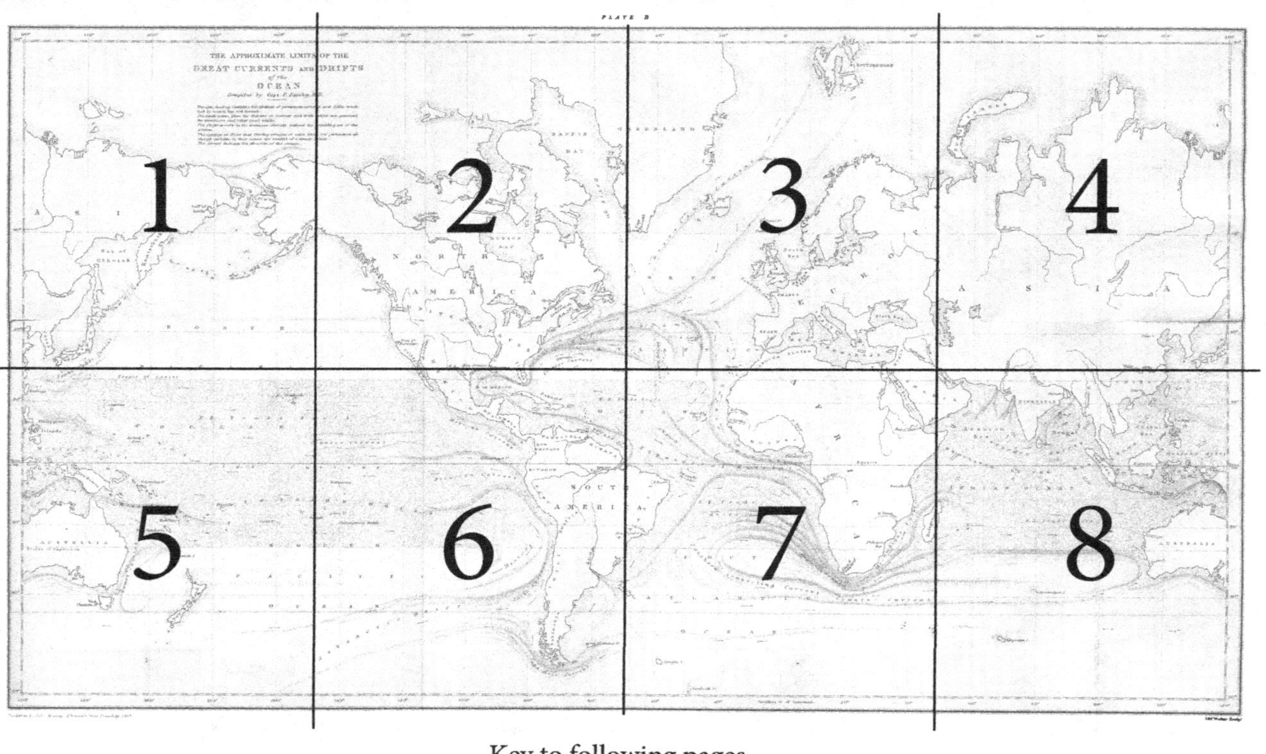

Key to following pages.

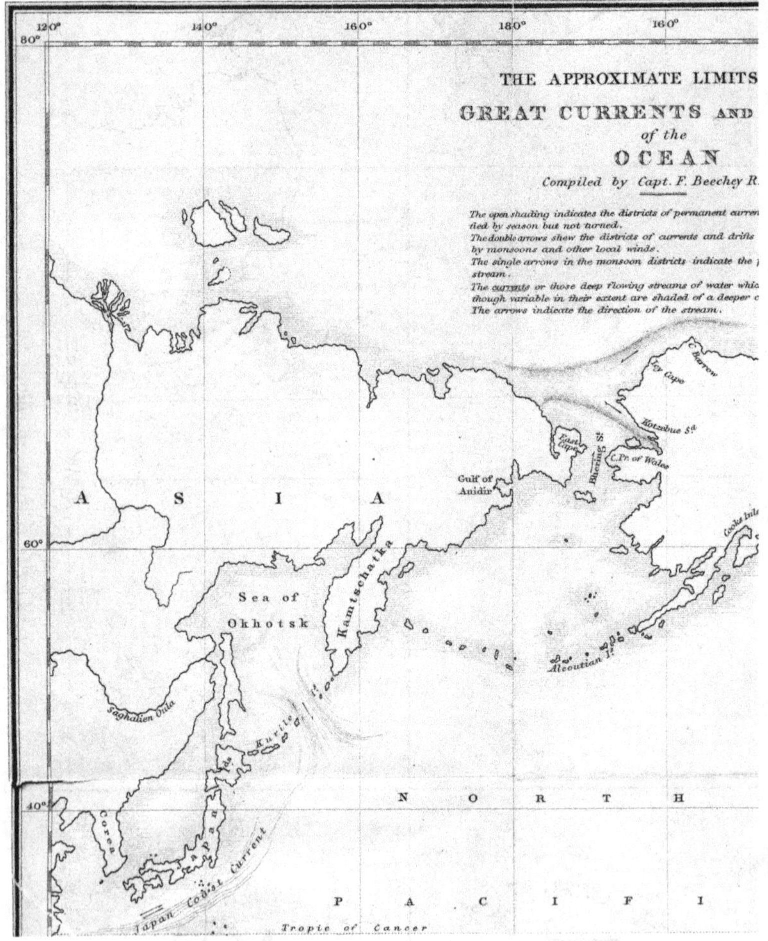
THE APPROXIMATE LIMITS
GREAT CURRENTS AND
of the
OCEAN
Compiled by Capt. F. Beechey R.
The open shading indicates the districts of permanent curren
fied by season but not turned.
The double arrows shew the districts of currents and drifts
by monsoons and other local winds.
The single arrows in the monsoon districts indicate the
stream.
The currents or those deep flowing streams of water whic
though variable in their extent are shaded of a deeper c
The arrows indicate the direction of the stream.
A S I A
Sea of
Okhotsk
Kamtschatka
Gulf of
Anidir
East Cape
Bhering St
C. Pr. of Wales
Kotzebue Sd
Icy Cape
C. Barrow
Cooks Inlet
Aleoutian Is
Saghalien Oula
Kurile Is
Corea
Japan
Japan Coast Current
N O R T H
P A C I F I
Tropic of Cancer
120°
140°
160°
180°
160°
80°
60°
40°

1

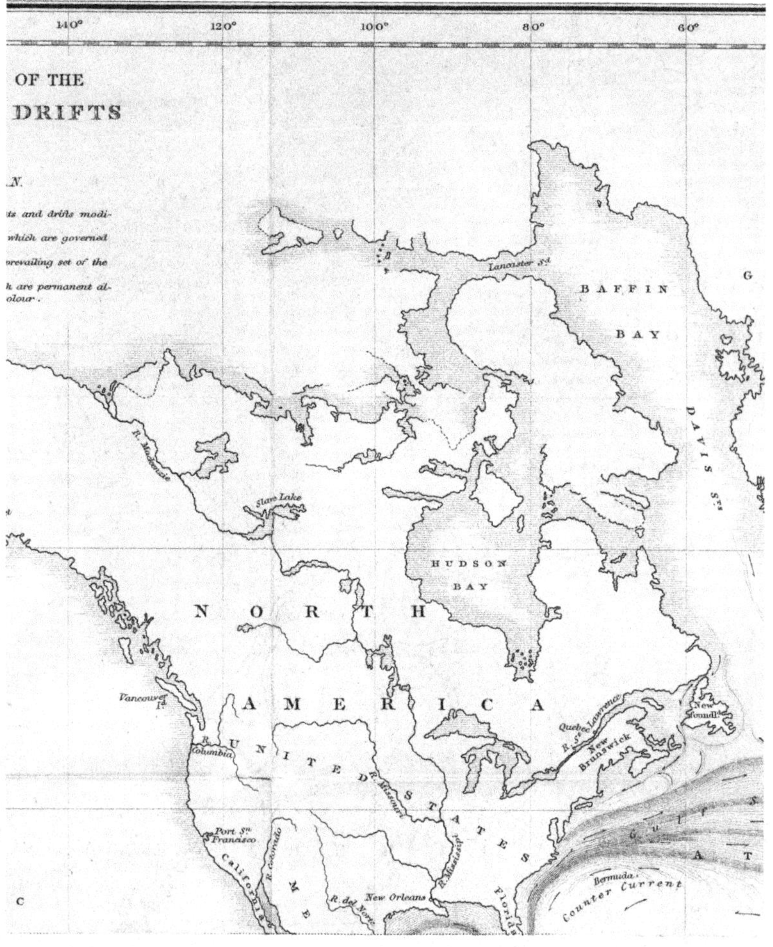
140°
120°
100°
80°
60°
OF THE
DRIFTS
.N.
ts and drifts modi-
which are governed
prevailing set of the
h are permanent al-
olour .
Lancaster S^d
BAFFIN
BAY
DAVIS S^ts
G
R. Mackenzie
Slave Lake
HUDSON
BAY
NORTH
AMERICA
Vancouver I^d.
R. Columbia
UNITED STATES
R. Missouri
R. Missisipi
Quebec
R. St Lawrence
New Brunswick
New Foundl^d
Port S^n Francisco
California
R. Colorado
R. del Norte
New Orleans
Florida
Bermuda
Counter Current
Gulf S
A T
C

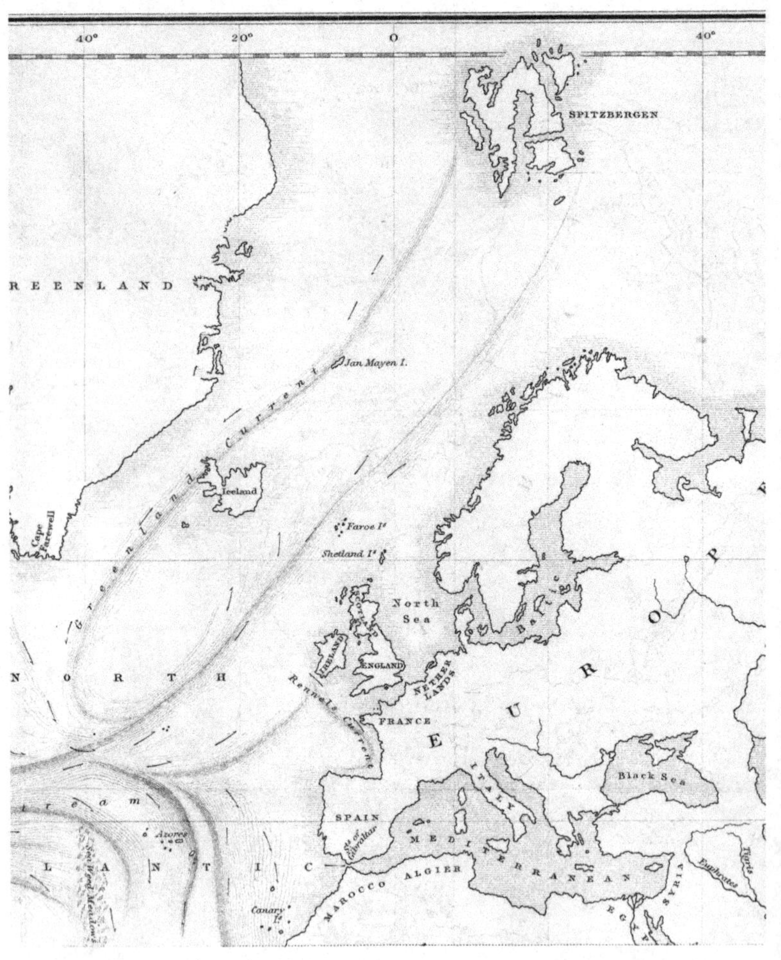
40°
20°
0
40°
SPITZBERGEN
REENLAND
Jan Mayen I.
Greenland Current
Iceland
Cape Farewell
Faroe I^s
Shetland I^s
North Sea
Baltic
IRELAND
SCOTLAND
ENGLAND
NETHERLANDS
N O R T H
Rennels Current
FRANCE
E U R O P
Black Sea
ITALY
SPAIN
Azores
Str. of Gibraltar
MEDITERRANEAN
L A N T I C
ALGIER
MAROCCO
SYRIA
Canary I^s
Euphrates
Tigris

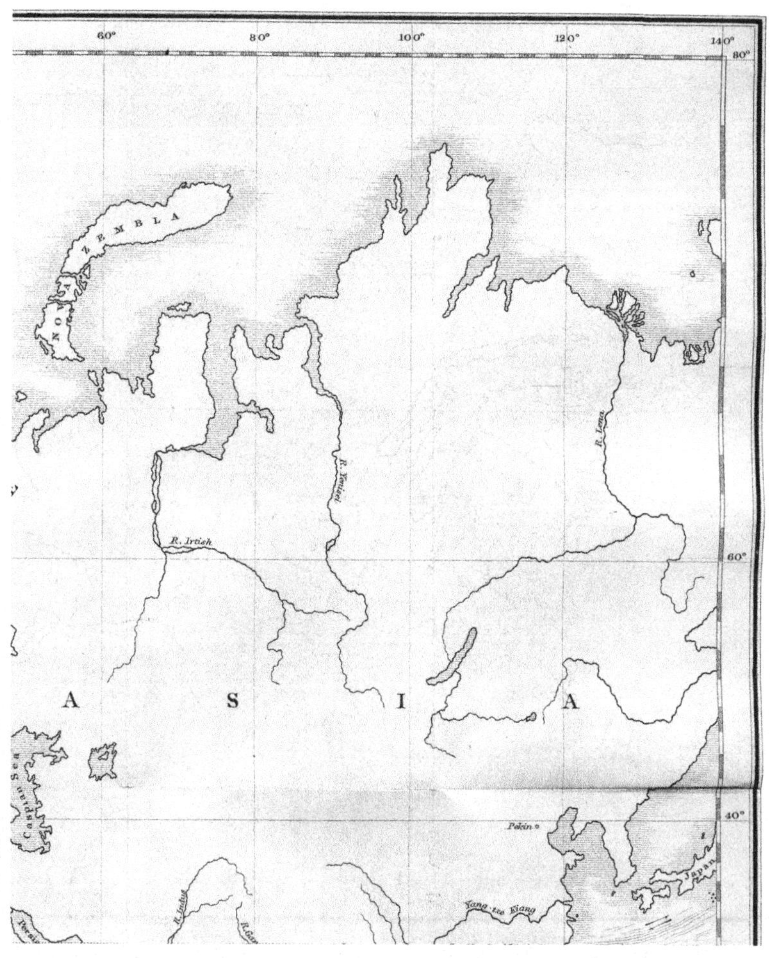
60°
80°
100°
120°
140°
80°
60°
40°
NOVA ZEMBLA
R. Irtish
R. Yenisei
R. Lena
A S I A
Caspian Sea
Pekin
Yang-tze Kiang
Japan
R. Indus

4

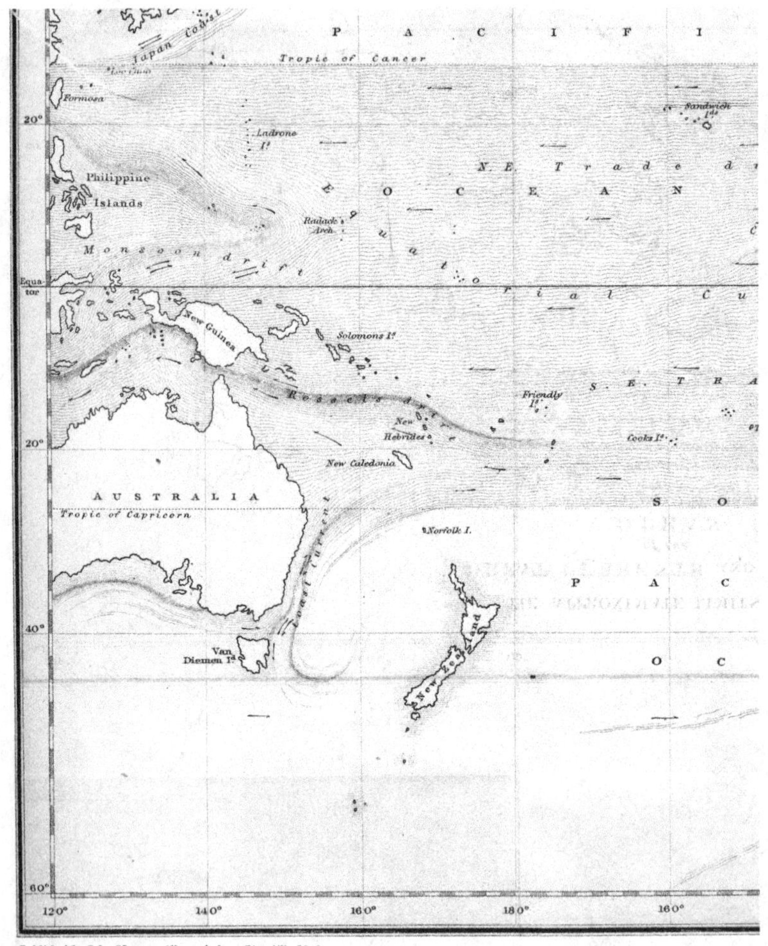

Published by John Murray, Albemarle Street, Piccadilly. 1849.

5

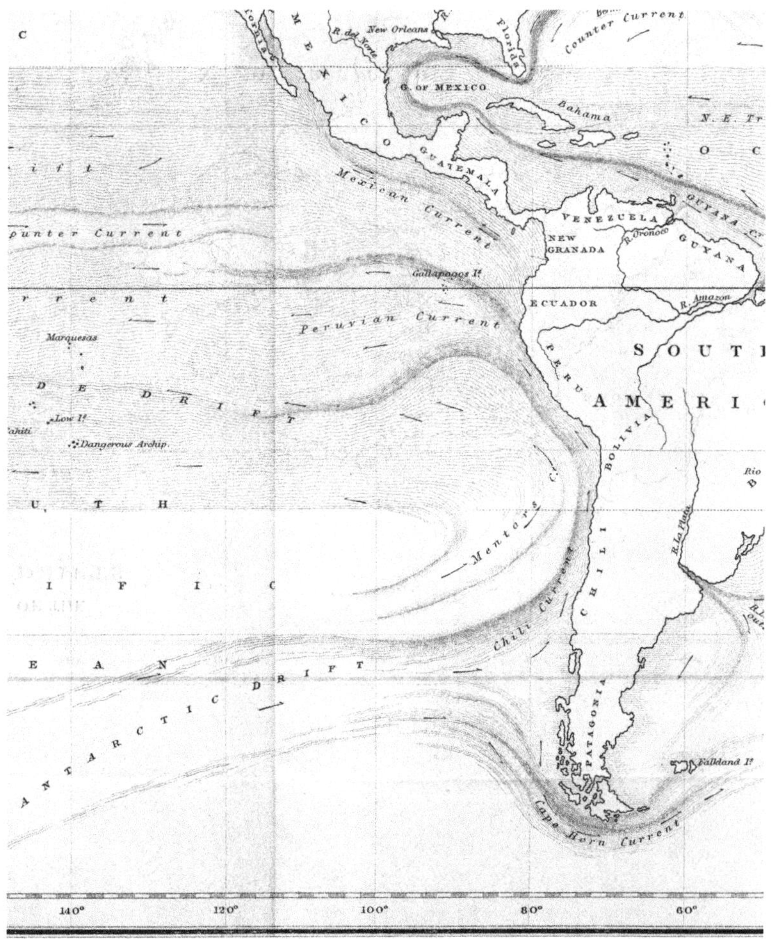
New Orleans
R. del Norte
Florida
Counter Current
G. OF MEXICO
Bahama
N. E. Tr
GUATEMALA
Mexican Current
VENEZUELA
NEW GRANADA
R. Oronoco
GUYANA
Counter Current
Gallapagos I.
ECUADOR
R. Amazon
Peruvian Current
Marquesas
SOUTH
AMERIC
PERU
BOLIVIA
D E D R I F T
Low I.
Dangerous Archip.
Mentors
CHILI
R. La Plata
Chili Current
ANTARCTIC DRIFT
PATAGONIA
Falkland I.
Cape Horn Current
140°
120°
100°
80°
60°

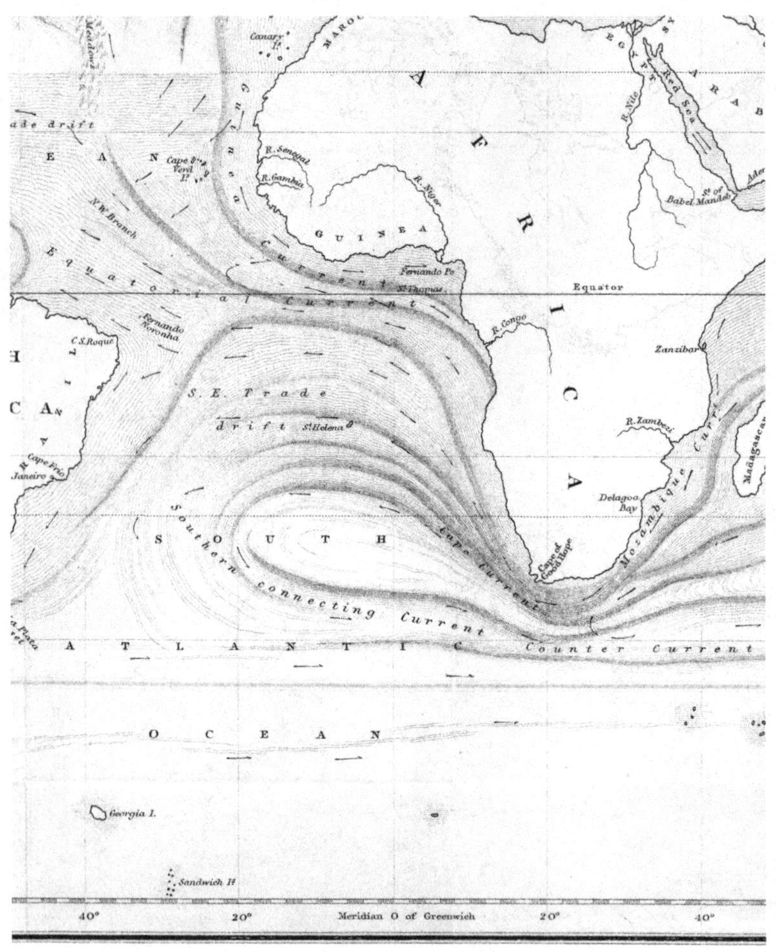

Canary Is
A F R I C A
Red Sea
R. Nile
St of Babel Mandeb
R. Senegal
R. Gambia
R. Niger
G U I N E A
Cape de Verd Is
Guinea Current
N.W. Branch
Equatorial Current
Fernando Po
St Thomas
Equator
R. Congo
Zanzibar
Fernando Noronha
C.S. Roque
S. E. Trade drift
St Helena
R. Zambezi
Mozambique Current
Delagoa Bay
Cape of Good Hope
Cape Current
Cape Frio
Janeiro
S O U T H
Southern connecting Current
Counter Current
A T L A N T I C
O C E A N
Georgia I.
Sandwich Is
40°
20°
Meridian 0 of Greenwich
20°
40°

7

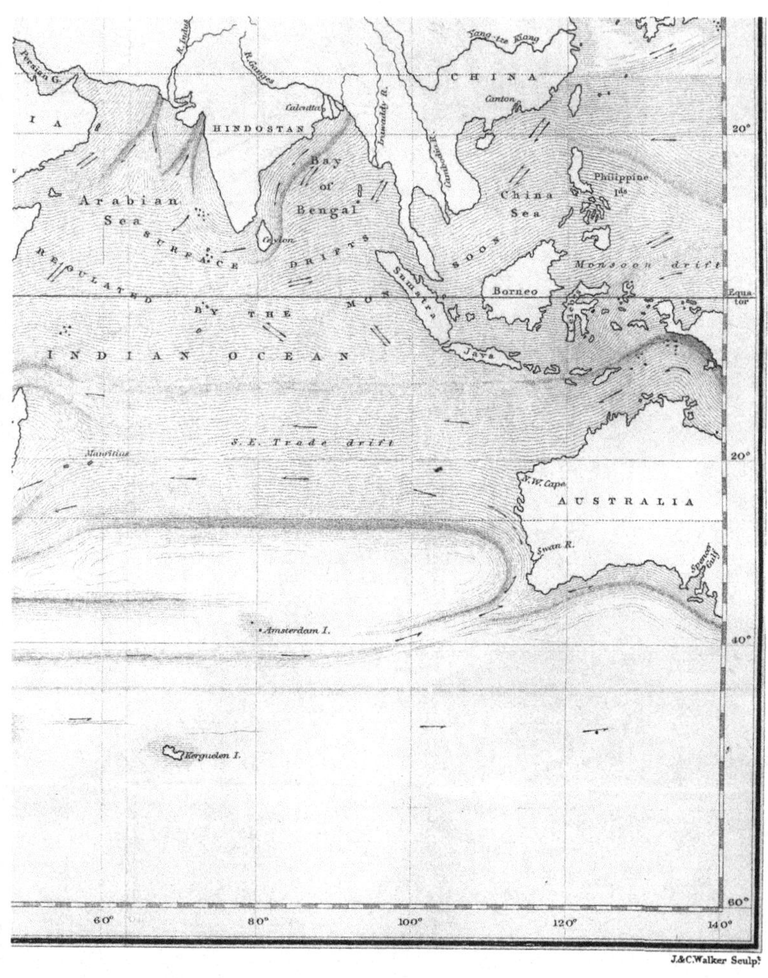

J.&C.Walker Sculpt.

8

many groups of islands with which our merchant-vessels have occasional traffic, or in which the whaling vessels refit, and which ought, therefore, to be more efficiently examined.

"On the opposite side of the Pacific some progress has been made in surveying the coast between the Russian territory and the Strait of Juan del Fuca; but with the long interval between the Oregon district and the entrance of the Gulf of California we are very superficially acquainted, and but little is known of the interior of that extensive Gulf. In the present state of those countries it does not appear necessary to push our survey into their inner waters; but there can be no doubt that the coasts of Mexico, Guatemala, and New Granada, which contain many valuable harbours and innumerable trading ports, ought to be minutely and connectedly surveyed.

"From the Equator to Cape Horn, and from thence round to the river Plata, on the eastern side of America, all that is immediately wanted has been already achieved by the splendid survey of Captain Fitzroy.

"Some parts of the great empire of Brazil we owe to the labours of Baron de Roussin and of other French officers; but there is much yet to be done on that coast between the Plata and the Amazon rivers, and again along Guyana and Venezuela up to the mouth of the Orinoco.

"The shores of the main land between Trinidad island and the Gulf of Mexico have been charted and published by the Admiralty; but many of the West India islands are still wanting to complete a wholesome knowledge of those seas.

"The United States are carrying on an elaborate survey of their own coasts; and to the northward of them a part of the Bay of Fundy has been done by ourselves, as well as all the shores of Nova Scotia, Canada, and Newfoundland; and when these surveys are finished, we shall only want to complete the eastern coast of America, those of Labrador and of Hudson Bay, which, being in our possession, ought to appear in our charts with some degree of truth."

As it is impossible here to open the question of the positions of the multitude of islands, of the Pacific especially, the apparent number of which has been so greatly increased by the errors of observation of navigators who have reported them, we can only recommend to the observer the propriety of fixing astronomically every island which he may fall in with, and to note any peculiarity by which it may be identified hereafter.

Section IV.

TIDES.

BY THE REV. DR. WHEWELL.

Directions for Tide Observations.

1. In making tide observations, the main object is, in the first place, to refer the tides to the motions of the *moon*, by which they are, in most places, mainly governed.

For this purpose, the *time* and *height* of *high water* (and of *low water*) at each place must be obtained; and this *time* will have to be compared with the *time* of the moon's passage across the meridian of the place.

The latter time (the *time of the moon's transit*) may be known by the common table given in the Nautical Almanac, or in other books of the same kind.

2. The time of high water (and low water) may sometimes (when the sea is calm) be observed with sufficient accuracy by observing the surface of the sea, where it washes a vertical scale fixed in the open water, and divided into feet and inches. The moment when the water is highest (and lowest) must be observed by a watch or clock, well regulated, or corrected for its error.

3. In general, the waves will make it difficult to observe the moment of the highest (and lowest) open water with much accuracy. The following methods may be

used to make the observations more accurate:—An upright tube, open below and above, may be placed in the water, reaching above the high water, and below the low water (or two tubes, one for high water and one for low water, if this mode be more convenient). In this tube must be a float (a hollow box or ball, for example), which must carry an upright rod, or else must have attached to it a string which passes upwards over a pulley and is stretched by a weight; and the part of the rod or of the string which is outside the tube must carry an index, which shall mark on a vertical fixed scale the rise and fall of the float.

By making the tube close below, except one or more small openings, the motion of the waves will very little affect the float, and the true rise and fall of the surface may be observed with much accuracy.

4. It may happen that the moment of the highest or lowest water is difficult to determine, either with or without the tube, on account of the water, while near the highest or lowest, stopping or hanging still, without either rising or falling, or else rising and falling irregularly.

If there is a considerable time during which the water neither rises nor falls decidedly, note the moment when it ceases to rise, and the moment when it begins to fall, and take the time half way between these for the time of high water.

5. Another method is the following:—At certain intervals of time near the time of high water, for example, every ten minutes, or every five minutes, let the height of high water be observed, say for half an hour or an hour, and from the height so observed pick out the

highest for the high water, and note the height and the time; and in like manner for low water.

6. But the following is a better mode of dealing with observations thus made every five or ten minutes. Let a number of parallel lines (*ordinates*) be drawn at intervals corresponding to the intervals of observations, and bounded by a line perpendicular to them on one side (the *abscissa*), and on these lines (the ordinates) let the observed heights of the surface be set off (from the abscissa) and let a line be drawn through the extremities (of the ordinates). This line, if it be tolerably regular, will give the time of high water; and if it be somewhat irregular, it can be smoothed into a curve, and then the time and height of high water read off. And in like manner for low water.

Suppose, for example, that we have the following observations of the height of the water made every five minutes for an hour:—

Times of Observation		h. 0 m. 0	m. 5	m. 10	m. 15	m. 20	m. 25	m. 30	m. 35	m. 40	m. 45	m. 50	m. 55	m. 60
Heights observed	ft.	6	6	6	6	6	6	7	6	6	6	6	6	5
	in.	0	6	6	9	10	11	0	11	11	9	5	2	10

The selection of the greatest height (as in Art. 5) would give high water at 0h. 30m.; but the general run of the height (Art. 6) would give the high water two or three minutes later, as appears by drawing the dotted curve in fig. 1.

7. It is easy to draw such curves, if we have, ready prepared, *paper ruled* into small squares, the divisions in the horizontal line representing hours and minutes, and

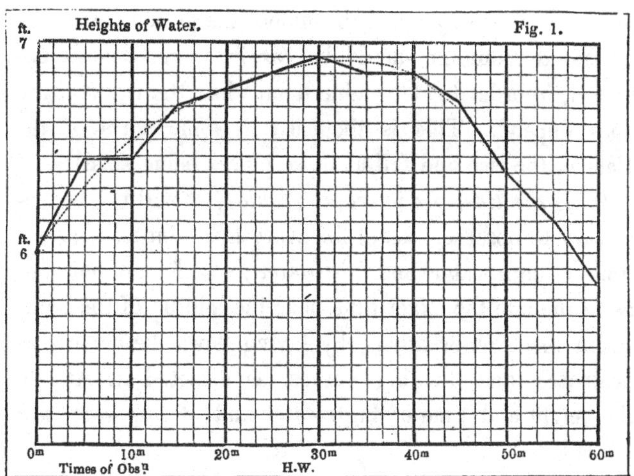

the divisions in the vertical line representing feet and inches.

8. It is well to begin a series of tide observations at any place by observing the height of the water during the *whole of the day and night* every half-hour or every quarter of an hour. For if the rise and fall be very irregular, or have any features which make it differ much from the common rule, it will, by this means, be seen that the case is a peculiar one, and that peculiar methods must be used: but if there is nothing peculiar in the case, the common methods may be used.

For instance, if, instead of there being two tides in every (lunar) day, there be one only, or four (both which cases occur at several places), these peculiarities will be discovered by observations continued during the day and night, in the way just recommended. If there be a periodical rise and fall of the sea's surface not depending in any obvious way upon the moon, the periods of maximum and minimum should be carefully and exactly observed, in order to determine upon what the rise and fall does depend. This is the case in some parts of the Pacific, the rise and fall at those places being small.

9. If *the tides are tolerably regular,* it will not be necessary to observe, except for every five minutes near the time of high water and low water; say for an hour, so as to include the exact time near the middle of the hour. From these observations, by laying down the heights as ordinates, and drawing curves, as directed in Art. 6, the height and time of high water and of low water will be deduced.

10. It is desirable to compare the observations of the time of high water and low water with the time of the moon's transit (see Art. 1) *while the observations are going on:* for if the tide follow this transit at very irregular intervals, the common modes of observation will probably be of no use, and the time and trouble employed in making them will be lost.

11. The time of high water at any place on the day of new or full moon is commonly called the *establishment* of the place; because, this being established, the time of high water on any other day may, in most cases, be known.

12. But if the tides are very irregular, this is not the case; and then the establishment of the place is of no use; or, rather, there is no proper establishment. And if the tides be regular, the establishment may be got from observations made *on other days*, just as well as from those made on the day of new or full moon. See Note A.

13. To compare the times of high water with the times of the moon's transit (see Art. 10), we must take the moon's transit from the tables (see Art. 1), and reckon how much the time of high water is *after* the time of the moon's transit, and put down these intervals, which are called the *lunitidal intervals.**

Suppose, for example, that we have the observations of high water contained in the following table: we add to them the other columns, containing the moon's transit and the lunitidal interval calculated therefrom. The alternate transits are interpolated midway between the others, which are given by the table. The A.M. transit which happens on the 14th is given in the table as 12h. 32m. on the 13th, the hour of the table being reckoned from noon.

1847. Jan.	Times of H.W.	Times of Moon's Transit.	Luni-tidal Interval	1847. Jan.	Times of H.W.	Times of Moon's Transit.	Luni-tidal Interval
	h. m.	h. m.	h. m.		h. m.	h. m.	h. m.
11 A.M.		[10 33]	2 34	16 A.M.	3 54	2 6	1 40
P.M.	1 7	10 57	2 32	P.M.	4 9	[2 29]	1 34
12 A.M.	1 29	[11 21]	2 26	17 A.M.	4 26	2 52	1 28
P.M.	1 53	11 45	2 20	P.M.	4 43	[3 15]	1 24
13 A.M.	2 11			18 A.M.	5 3	3 39	1 20
P.M.	2 29	[0 9]	2 16	P.M.	5 23	[4 3]	1 18
14 A.M.	2 48	0 32	2 8	19 A.M.	5 47	4 27	1 18
P.M.	3 3	[0 55]	2 2	P.M.	6 9	[4 51]	1 18
15 A.M.	3 21	1 19	1 52	20 A.M.	6 34	5 16	1 20
P.M.	3 36	[1 42]	1 48	P.M.			

14. To see whether the lunitidal intervals follow the

* It is not necessary, for the purposes considered in these directions, to calculate the time of the moon's transit *at the place* of observation by differences of days. It is sufficient to take the time of the moon's transit at Greenwich, and to *add* two minutes for every hour of *west* longitude of the place. For the moon (on the average) moves away from the sun so that her distance from the sun is increased 48 minutes in time for every 24 hours, and therefore the transit of the moon is later at every other place by two minutes for every hour.

regular law, the best way is to put them into a curve, setting off the lunitidal interval belonging to each tide as an ordinate, as in fig. 2.* If the curve drawn through

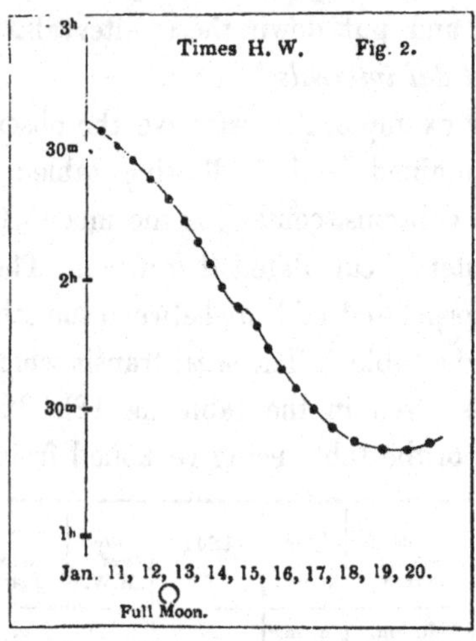

the extremity of the ordinates be tolerably regular, the tides may be presumed to be so.

15. In the observations given in Art. 13 we may see how loose a term the "*establishment*" is. The 13th is the day of full moon, for in the course of that day the moon is 12 hours from the sun. The time of high water on the 13th is—A.M., 2h. 11m.; P.M., 2h. 29m.; and either of these might, in the common use of the term, be called "the establishment."

16. If the lunitidal intervals be set off for a fortnight

* In actual practice it will be better to draw the figures on a larger scale than those here given.

or more, the curve (Art. 14) will descend and ascend alternately every fortnight, as in fig. 3.

This curve is the curve of the *semi-mensual inequality;* and when this curve has been determined by observations at any place, the hour of high water at any time at that place may be predicted.

17. But the curve will be better determined if, instead of taking for the abscissa the day of the month, as in fig. 2, we take for the abscissa the time of the moon's transit as in fig. 3.*

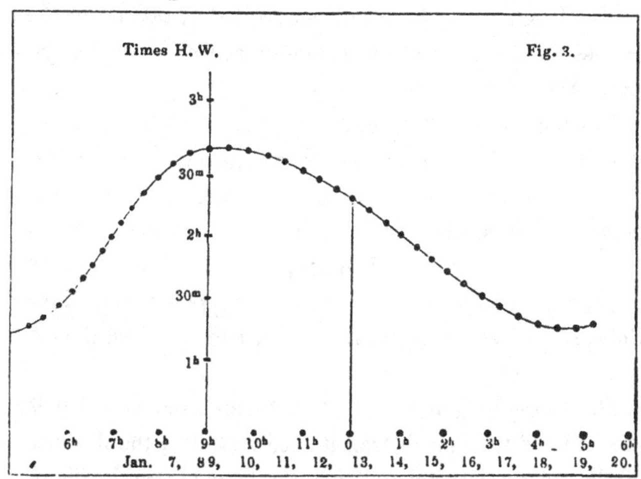

In this case *the establishment* is the ordinate of this curve which corresponds to the time of moon's transit 0h. or 12h. In the figure it is 2h. 16m.

The mode. of calculating the hour of high water on

* Since the moon's transit is about 48 minutes later every day, there will be along the abscissa five days of the month for every four hours of moon's transit.

any day, when the establishment of the place is known, as in Art. 17, is given in Note A.

The establishment of the place may be known by observations made at any age of the moon, as well as at new and full moon, by the same kind of calculation.

18. It is also advisable to set off the *height* of high water as ordinates, and to draw a curve through the extremities. This curve also will ascend and descend every fortnight (ascending at spring tides and descending at neap tides).

The heights may be set off as ordinates, taking for the abscissas equal intervals to represent successive half-days, as in Art. 16.

But the curve will be better determined if we take for the abscissas the hour of the moon s transit, as in Art 17.

19. The *maximum* or greatest ordinate of this curve of heights (that is, the spring-tide height) follows the day of new and the day of full moon, by one, two, or three days; and as the new or full moon is supposed to produce the spring tide, this interval of one, two, or three days is called *the age of the tide.*

20. If the heights be set off from an abscissa which is the hour of the moon's transit (see Art. 18), the distance of the maximum ordinate from the hour of transit, 0h. or 12h. (which are the same thing), will give the *age of the tide* more exactly than Art. 19.

21. The lunitidal intervals and heights of *low water* may be laid down in curves in the same manner as those of high water.

22. The curve of the semi-mensual inequality of times and heights should be determined, when opportunity

allows, for several weeks or months in succession: for from such observations we can obtain other scientific results (the effect of the sun, of the moon's parallax, and the like).

23. Besides the changes which are produced from day to day by the semi-mensual inequality of times and heights, there are at many places other considerable changes produced between the two tides of the same day by the *diurnal inequality*.

For example, there are many cases in which the height of high water is alternately lower and higher in successive tides.

24. In this case, if we set off the successive heights of high water as ordinates at equal intervals, and draw a line through their extremities, this line will have a zigzag form, as in fig. 4.

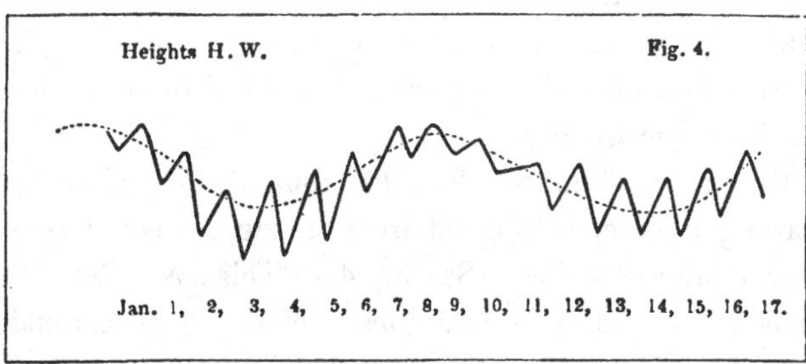

The width of the zigzag increases from nothing to a maximum, and then diminishes to nothing again, generally in the course of a fortnight; and so on perpetually.

25. In consequence of the diurnal inequality, it sometimes happens that the afternoon tides are higher than the forenoon tides, or the reverse, for many weeks together.

And hence, it has sometimes been stated as a rule, at such places, that the afternoon tides are always the highest, or the reverse. But this is not the rule. If the afternoon tides are the highest at one time of the year, they are the lowest at another.

The rule of the diurnal inequality depends on the moon's declination, and will be given in Note B.

26. There is often a diurnal inequality of the height of low water, and at some places it is greater than the diurnal inequality of high water (as at Sincapore, and at Port Essington in Australia).

27. Also there is often a diurnal inequality in the times.

When this is the case, if we set off the lunitidal intervals as ordinates (see Art. 14), the line drawn through their extremities will have a zigzag form, like that of the heights in fig. 4.

28. When this is the case, we cannot determine *the establishment* (see Art. 17) without making allowance for the diurnal inequality.

We make allowance for the diurnal inequality by drawing a curve, cutting off from the zigzags equal portions above and below. (See fig. 4.) This *mean line* will be of a wavy form in consequence of the semi-mensual inequality; and the ordinate corresponding to the new or full moon, or to the hour 0 or 12 of moon's transit, will give the establishment.

But if we apply this establishment to predict the time of tide on any day, we must also apply the diurnal inequality predicted according to its rule. (See Art. 25, and Note B.)

29. The diurnal inequality sometimes becomes so large that there is *only one tide in 24 hours* (and then we have *single-day tides*). But this does not generally happen through a whole lunation; it happens only for a few days in each semi-lunation; and at other times there are two tides as usual. Cases of one tide in 24 hours should be particularly observed, making the observations every half-hour, or, if possible, oftener, say every 5 minutes.

30. In some places the tide rises and falls *four times* in the 24 hours. The cases where this occurs are to be particularly observed.

They may be observed, as in Art. 29, by making observations every half-hour, 10 minutes, or 5 minutes.

These may be called *double half-day tides.*

31. Where double half-day tides exist, they do not commonly extend over any considerable length of coast. If there be time and opportunity, it will be well to examine, by observation, how far they do extend. But if the object be to determine the laws of the tides in a larger area, it is better to make the observations out of the region of these anomalies.

32. It is well to observe the *direction* of the stream of flood and of ebb, and the *time* at which the stream turns.

We must take care not to confound the time of the *turn of the tide-stream* with the time of high water. Mistakes and errors have often been produced in tide observations by supposing that the turn of the tide-stream is the time of high water. But this is not so. The turn of the stream generally takes place at a different time from high water, except at the head of a bay or creek. The stream of flood commonly runs for some time, often for hours, after the time of high water. In the same way, the stream of ebb runs for some time after low water.

33. The time at which the stream turns is often different at different distances from the shore; but the time of high water is not different at these points. In general, what is wanted in tide observations is the time of *high water*, not the time of *slack water*.

With regard to the streams of flow and ebb, they are often not merely two streams in opposite directions at different times of the tide; they generally turn successively into several directions, so as to go quite round the compass in one complete tide; either in the direction N. E. S. W. (with the sun), or N. W. S. E. (against the sun). It is desirable to note which of these ways the tide-stream goes round, as this fact may help to determine which side the tide-wave comes from.

34. One important object to be answered by means of tide observations is to trace the progress of the tide from one place to another.

This may be done in some measure by determining the *establishments* of a series of places in the region which we have to consider. For these establishments, reduced to Greenwich time by allowing for the longitude, give the time at which the tide is at each place, and hence its progress.

35. The progress of the tide may be conceived as the progress of a very wide *wave* which brings the high water to each place in succession.

But the motion of this *tide*-wave is not that motion of the water which makes the stream of flood. Nor does the motion of the wave coincide with any motion of the parts of the water. The tide-wave may be going one way when the water is going another, as happens in some rivers when the tide is travelling upwards in them.

36. The *establishment*, which is wanted in order to determine one progress of the tide-wave (see Art. 34 and 35), may be known from observations made at any age of the moon, as well as at new or full moon. (See Art. 17 and Note A.)

37. In tracing the progress of the tide-wave, instead of using the *vulgar establishment* hitherto spoken of, it is better to use the *mean establishment*, namely, the mean of all the lunitidal intervals.

For the vulgar establishment is affected by the age of the tide (Art. 20), which the mean establishment is not.

The mean establishment is (say) 10m., 20m., 30m., or 40m. less than the vulgar establishment, according to the age of the tide. (See Note A.)

38. When the tides are regular, good observations, made for a few days or a week at each place, may give the establishment (either vulgar or mean) with sufficient exactness to determine the progress of the tide-wave.

39. But the progress of the tide-wave may be much better determined by means of *simultaneous observations*· namely, observations made at different places on the same days for a few days or a week.

For such a purpose persons must be posted at different points of the shore or shores where the motion of the tide-wave is to be traced; say 10, or 20, or 40, or 80 miles from each other, as may be convenient. They must observe the tides at these places *on the same days,* morning and evening, by the methods already described. The times of high water at the different places on each half-day, being compared, will give the progress of the tide-wave.

40. In order to trace the progress of the tide-wave still more widely, the observers described in the last article, after having made the observations there spoken of, may be removed to new positions of the same kind, and thus trace the tide farther.

When this course is adopted, it will be well to have one (or more) fixed or standard station, at which tide observations are constantly made; and the observations made at any time at any other place may be compared with those made at the standard station.

41. The tides which take place far up deep bays, sounds, and rivers, are *later* than the tides at the entrance of such inlets; but they are not more irregular: on the contrary, the tides in such situations are often remarkably *regular*.

42. The progress of the tide-wave up inlets may be determined by the method described in Art. 39.

43. The tide in its progress up inlets and rivers is often much magnified and modified by local circumstances.

Sometimes it is magnified so that the wave which brings the tide at one period of its rise advances with an abrupt front of broken water. This is called a *bore* (as in the Severn, the Garonne, the Amazons River).

Sometimes the tide is divided into two half-day tides in its progress up a river (as in the Forth in Scotland).

In all cases, after a certain point, the tide dies away in ascending a river.

44. The tide observations made at any place, when the times and heights of high water (and of low water) have been deduced in the way directed in Articles 2, 3, 4, 5, 6, may be entered in a table of which the form will be given, and must then be sent to the Hydrographer's Office in the Admiralty.

45. It is to be remarked that, though there is generally an A.M. and a P.M. tide, there is one day in every half-lunation on which there is only one tide.

(Because the interval of the two tides is, on the average, about 12h. 24m.; so that if there be a tide at 11h. 50m. A.M., there will be no other tide till 12h. 14m. P.M., that is 0h. 14m. A.M. of the next day.)

46. *Self-registering tide-machines* are used in several places, and may be constructed at no great expense. (They are made by Mr. Newman, of Regent Street, for about 30*l.*: they are constructed so as to work with a tube and float, as described in Art. 3.)

These machines give the whole course of rise and fall of the tide; and record several successive tides on the same paper.

47. The wind often produces a considerable effect upon the tides, especially upon the height, and should be noted, although it is difficult to give any general rule for the effect.

48. The surface of the sea rises and falls as the barometer falls and rises; namely, about 1 inch for every $\frac{1}{20}$ inch of mercury. This may be applied as a correction when very exact observations are made.

NOTE A.

Note to Articles 17, 19, and 37.

To find the Hour of High Water on any day, at any place, when the Establishment of the place is known.

The rule is different (as to amount) according to the *tidal force* of the sun; for though the tidal force of the sun in theory is the same at all places, it is found by observation to be different at different places.

This difference appears in the different ratio of the rise of spring-tides to the rise of neap-tides: (this difference is the semimenstrual inequality of heights.) In general the rise of spring-tide is about double that of neap-tide, which gives the solar tide *one-third* of the lunar tide. But in some cases the spring-tide exceeds the neap-tide only by one-third, which gives the solar tide only *one-seventh* of the lunar tide.

Also the difference of the greatest and least lunitidal intervals (that is, the semimenstrual inequality of times: see Art. 13 and 16) shows the difference of the solar tidal force at different places. The difference of the greatest and least intervals is 1 h. 28 m. at London and Liverpool, but at Plymouth it is 1 h. 36 m., and at Portsmouth 1 h. 21 m. On the coast of North America it is generally less than 1 h. 20 m., while at some places on the coasts of France and Ireland it is above 2 h.

We may take 1 h. 28 m. as the mean value of this difference, which agrees with the supposition that the solar tide is about one-third the lunar tide.

In finding the hour of high water on any day when the *vulgar* establishment is known, the rule will also be different according to the age of the tide. We shall give the rule when the tide is a day and a quarter old, and also when the tide is two days and a half old. In general, the tides will be between these limits.

(1.) *Tide a day and a quarter old.* Minutes to be added to or subtracted from the establishments, according to the hour of the moon's transit on the half-day in question:—

Hour of the Moon's Transit after Sun	h. 0	h. 1	h. 2	h. 3	h. 4	h. 5	h. 6	h. 7	h. 8	h. 9	h. 10	h. 11
Correction of the vulgar Establishment to find the Lunitidal Interval . . .	m. 0	m. −16	m. −32	m. −47	m. −57	m. −60	m. −47	m. −16	m. +15	m. +28	m. +25	m. +15

For example—if the establishment be 2 h. 27 m., at what hour will the high water come after a moon's transit which takes place at 4 h. A.M.? The minutes to be added to 2 h. 27 m. for 4 h. transit are, by the table, − 57 m.; therefore the high water will be at 1 h. 30 m. after the moon's transit, that is, at 5 h. 30 m.

(2.) *Tide two days and a half old*:—

Hour of Moon's Transit	h. 0	h. 1	h. 2	h. 3	h. 4	h. 5	h. 6	h. 7	h. 8	h. 9	h. 10	h. 11
Correction of the Establishment . . .	m. 0	m. −15	m. −31	m. −47	m. −62	m. −72	m. −75	m. −62	m. −31	m. 0	m. +13	m. +10

This table is to be used in the same way as the other.

Hence we see that the age of the tide most affects the lunitidal interval when the time of moon's transit is between 7 and 8 hours.*

The mean lunitidal interval, or *mean establishment*, is 16 minutes less than the former, and 31 minutes less than the latter establishment supposed in the above tables. (See Art. 37.)

If the tides are observed for a semilunation, or any complete number of semilunations, the mean lunitidal interval, or mean establishment (see Art. 37), will be found by taking the mean of all the lunitidal intervals observed.

The lunitidal interval *corresponding to* any given distance of the moon from the sun may be found by the following table. But the tide corresponding to the given distance may not really *occur* till one, two, or three days later, according to the age of the tide.

(3.) Correction of *mean* establishment.

Hour of Moon's Transit (1, 2, 3 days preceding)	h. 0	h. 1	h. 2	h. 3	h. 4	h. 5	h. 6	h. 7	h. 8	h. 9	h. 10	h. 11
Corresponding Correction of Mean Lunitidal Interval . .	m. 0	m. −16	m. −31	m. −41	m. −44	m. −31	m. 0	m. +31	m. +44	m. +41	m. +31	m. +16

This table may be used when we know the age of the tide. Thus, let the age of the tide be a day and a quarter, and the mean lunitidal interval 2 h. 11 m.; let the moon's transit take place at 4 h.; then at the *birth of the tide*, a day and a quarter earlier, the transit took place at 3 h.; therefore the correction of the lunitidal interval is, by the table, − 41 m., and the interval so corrected is 1 h. 30 m., which, added to 4 h., the time of moon's transit, gives 5 h. 30 m. as the time of high water.

To find the Establishment at any place when the Hour of High Water on a given day is observed.

On the given day, the time of moon's transit is known, and hence the lunitidal interval; and, by the above tables, the correction by which this differs from the establishment is known.

Hence it is desirable to make tide observations in the first and fourth quarters of the moon, rather than in the second and third quarters.

Thus, if high water occur at 5 o'clock when the time of moon's transit is 3 h., the lunitidal interval is 2 h.; and the correction (if the first table be applicable) is − 47 m.; hence the establishment is 2 h. 47 m.

NOTE B.

Note to Article 25.

The Rule of the Diurnal Inequality.

The Diurnal Inequality depends upon the moon's declination, as has been said already. It increases from 0 up to its maximum, and decreases to 0 again, as the declination does so; following these changes at an interval of one, two, or three days, according to the age of the tide. The rule is expressed in this way:—

For *north* declination of moon,

Add to the tide following moon's *south* transit;

Subtract from the tide following moon's *north* transit.

For *south* declination of moon,

Subtract from the tide following moon's *south* transit;

Add to the tide following moon's *north* transit.

The south transit is the superior transit in the northern hemisphere, and the north transit the inferior. The contrary is the case in the southern hemisphere.

Form for Tide Observations.

Tides observed at , *Lat.* , *Long.* . *By*

Mode of observation . . . . . { Fixed scale in open water? Tube with float? Self-registering gauge?

Mode of deducing H. W. and L. W. { Mere looking? Ordinates every 5 m. near max.?

1848.		High Water.		Low Water.		Wind. Barom.	Moon's* Transit.	Lunitidal* Interval. H. W.
Month.	Day.	Height.	Time.	Height.	Time.			
	1 A.M.							
	P.M.							
	2 A.M.							
	P.M.							
	3 A.M.							
	P.M.							

* These columns to be filled at leisure (see Art. 13, 41).

SECTION V.

GEOGRAPHY.

BY W. J. HAMILTON, ESQ., PRES. R.G.S.

BEFORE alluding particularly to the individual objects to which, in reference to Geographical observations, the attention of travellers should be more immediately directed, it may be, perhaps, expedient to mention a few general points which should be constantly borne in mind as the basis of all observations, inasmuch as without them, all individual remarks, however carefully made, must be desultory and unsatisfactory.

Most prominent amongst these general points is the necessity of acquiring a habit of writing down in a note-book, either immediately or at the earliest opportunity, the observations made and information obtained. Where numbers are concerned, the whole value of the information is lost, unless the greatest accuracy is observed; and amidst the hurry of business or professional duties the memory is not always to be trusted. This habit cannot be carried too far. A thousand circumstances occur daily to a traveller in distant regions, which from repeated observation may appear insignificant to himself, but which may be of the greatest importance to others, when brought home in the pages of his note-book, either as affording new information to the scientific inquirer, or as corroborating the observations of others, or as affording the means of judging between the conflicting testimonies of former travellers.

It is also important, in order to secure accuracy, that the observations should be noted down on the spot. It is dangerous to trust much to the memory on such subjects; and if the observation be worth making, it is essential that it be correct. And here it may not be inappropriate to hold out a caution against too hasty generalization. A traveller is not justified in concluding that because the portion of a district, or continent, or island which he has visited is wooded or rocky, or otherwise remarkable, the whole district may be set down as similarly formed. He must carefully confine himself to the description of what he has himself seen, or what he has learned on undoubted authority.

Again, to the geographer, the constant use of the compass is of the greatest value. No one attempting to give geographical information should ever be without an instrument of this kind, as portable as is consistent with correctness. The bearings of distant points, the direction of the course of a river, however they may be guessed at by the eye, can never be accurately laid down without the compass; and these should be immediately transferred to the note-book. This and his compass should on all occasions be his constant and inseparable companions. In using the former, he should not forget that slight sketches of the country, and of the peculiar forms of hills, however hastily and roughly made, will often be of more assistance in recalling to his mind the features of the district he has visited than long and elaborate descriptions. Let him then acquire the habit of never quitting his ship without his note-book and pencil and his pocket-compass, and although at times it may seem irksome to have to remember and to fetch these materials, the traveller, if he acquires the habit of constantly using them with readiness, will never have reason to regret the delay or the inconvenience which may have temporarily arisen in providing himself before starting with such useful companions.

Having made these few introductory remarks, I shall proceed to describe as briefly and succinctly as possible some of the principal features to which the attention and the inquiries of the young geographer should be chiefly directed. For this purpose I propose dividing the subject into two heads, which, without straining the use of words, may be not inappropriately called Physical and Political Geography. By physical geography I mean everything relating to the form and configuration of the earth's surface as it issues from the hand of nature, or as it is modified by the combined effects of time and weather, and atmospheric influences. By political geography I would wish to imply all those facts which are the immediate consequences of the operations of man, exercised either on the raw materials of the earth, or on the means of his intercourse with his fellow creatures.*

* An Italian writer of considerable eminence, Count Annibale Ranuzzi, in a little work published at Bologna, 1840, entitled 'Saggio di Geografia Pura,' divides geography into two branches, which he calls pure and statistical geography: the former professes to describe the results of physical forces, the latter the effects of moral force; the former is expressed by measurement, the latter by numbers.

I. Physical Geography.

The principal heads under which this branch of the subject may be divided, and respecting each of which it will be necessary to say a few words, are—

1. Form of country; whether consisting of hills, valleys, or plains.

2. Mountain ranges; their direction, height, spurs, woods, and forests.

3. Rivers; their sources, obstacles, size, affluents, and confluents.

4. Springs; whether hot or cold or mineral, their localities, temperature, &c.

5. Lakes, marshes, lagoons; how surrounded, &c.

6. Coast line, mouths of rivers, their beds and banks, harbours, nature of shore; sandy, rocky, or muddy.

1. *Form of country; whether consisting of hills, valleys, or plains.*—The general configuration of a country is the first object which engages the attention of a traveller on entering a new locality, and this may be described in general terms as flat, undulating, hilly, or mountainous; or the country may be divided into districts, to each of which one of the above terms of configuration may be applied. Each of these, however, is susceptible of great modification. A *flat* country may be a sandy desert, a rich alluvial plain, or a marshy, boggy tract; it may be well watered by rivers and streams, or arid and parched up; it may contain numerous lakes; it may be barren or wooded, or cultivated as arable or grass land: each of these features may be of importance, or at least of interest; nor must the nature of its soil be omitted, whether sand, or

marl, or clay, as the appearance of the country will often depend greatly on this circumstance. Another important characteristic is its general form and extent, and the natural features by which it is bounded, whether mountains, rivers, or seas; how many miles wide, and how many long; whether extending parallel with the coast, or running up between hills into the interior.

Many of these characteristics, it will be observed, belong equally to the other forms which constitute the character of the district. An *undulating* country may be barren, wooded, or cultivated; it may be arid, or watered by streams, &c. The undulations may be abrupt, or only gently swelling, and this may be in a great measure owing to the nature of the subsoil, whether it consists of gravel, or sand, or rock; but a country of this description is easily described. A *hilly* country, on the other hand, is more complicated. Not only is the term vague and uncertain, but other features have to be considered. Neither hills nor mountains can exist without valleys, and these also deserve to be considered and described. Then, again, the hills themselves may be of various forms and characters; do they extend in long parallel chains or ranges, or are they detached and isolated? Do they radiate or converge? Do they rise abruptly or gradually from the low country? and how are they wooded? What do the rocks which constitute their nucleus consist of? If possible, it is desirable to ascertain their height, which, in the absence of complicated instruments and barometers, may be very fairly obtained approximatively by marking the exact point at which pure fresh water boils. Of course the same accuracy cannot be obtained as with the barometer, but much may be done with the help of well-graduated thermometers.

2. *Mountain ranges.*—The most important features in the configuration of a country are the mountain ranges by which it is traversed. The exact point of distinction between a hill and a mountain is difficult to describe; in some cases it will be purely comparative, in others it will depend on the general character of the country, and in some it will be arbitrary. But in all cases it will be desirable to endeavour to ascertain the height of the principal points, the direction of the main ranges or chains, and whether they are parallel or not. The ridges also may be *serrated* (jagged like a saw), or smooth and even, and the summits themselves will be either pointed, or dome-shaped, or flat. Is the mountain insulated or not? and if so, is it conical and sloping on all sides to the surrounding plains, or does it consist of a detached ridge? Many of these points will be found to depend on the geological formation of the country, and this branch of our subject is very closely connected with that science. It is also desirable to ascertain how far the mountain tops are covered with perpetual snow, and how far down their sides snow lies during the whole year. Is there any marked difference in the slope on the one side or on the other? Does vegetation abound more on one side than on the other? *e. g.* in Asia Minor all the mountain ranges which extend from E. to W., and this is their principal direction, are covered on their northern flanks with luxuriant vegetation and magnificent forests, while the southern flanks, exposed to the rays of an almost tropical sun, are void of vegetation, barren, and generally rocky. Here,

again, we trench on the province of the botanist; and yet the geographer should inquire how far vegetation extends up the mountain side, and what are the changes which it undergoes. How far is it influenced by the change of soil, or the abundance or absence of springs? Nor can we complete our information respecting a mountain chain, unless we know the length to which it extends, and the breadth of country which it covers.

Valleys are a necessary complement to mountain masses, and there are many peculiarities connected with them well deserving observation. Are the sides precipitous or sloping? are they wide or narrow? well watered or arid? wooded or barren? Do the rocky sides correspond with each other in their salient and re-entering angles? How far do they extend into the bosom of the mountains? and how are the subordinate valleys connected with the principal one? But there is another peculiarity of valleys not to be lost sight of. There are some which convey to the traveller the impression that he is passing through a mountainous or hilly country, so steep, rugged, and lofty are the hills by which he is surrounded. It is only on reaching their summit that he becomes aware that the country through which he has been passing is an extensive plain, or table-land, intersected by deep chasms and valleys, cut through the soft soil by the constant efforts of the streams by which it is traversed; such valleys of excavation as these have been sometimes not unaptly called negative valleys.

3. *Rivers.*—Scarcely less important than that of mountains is the effect of rivers in modifying the geographical configuration of a country. From their sources in

the mountain recesses to their final disemboguing in the sea, their course, their currents, and their shores afford an endless variety of remarks and observations. The depth and colour of the water, the rate at which it flows, the eddies and currents by which its course is marked, are all deserving of notice, as are also the rocks and shoals which obstruct its uniform progress, either interfering with its navigation, or, by projecting beyond its ordinary banks, throwing back the rushing torrent on the opposite shores, as has been so eloquently described by the Latin poet:—

> Vidimus flavum Tiberim, retortis
> Littore Etrusco violenter undis,
> Ire dejectum monumenta regis,
> Templaque Vestæ—

thus causing the gradual fall of the cliffs by undermining their precarious foundation. Nor in describing the size or extent of rivers should we neglect to state how far up they are navigable, to what vessels, and by what means, whether the mouth is constantly free, or whether closed by a bar, and how much water there generally is over it. Some rivers, however, are not only closed by a bar, but, as in the case of Western Australia, are, during periods when the water is low, completely masked by the sandhills or dunes which are blown up, forming a continuous bank with the hills which skirt the shores, and only when freshets of more than ordinary force come down are these sandy barriers overthrown, and the rivers enabled to find an uninterrupted outlet. In other cases the effect of beaches thrown up by the constant set of currents in one direction is not so absolutely insurmountable, the streams are only partially deflected from their proper course, and instead of flowing into the sea in a continued line, are compelled to run for some distance parallel to the coast, until the accumulated backwater has acquired sufficient power to overcome the diminished resistance of the sea-beach: this, however, more properly belongs to the consideration of the coast line.

But the description of a river will be imperfect, unless we also state the number and character of the streams which fall into it. And here we have to consider the angle at which the rivers join each other, whether the direction of the main stream is altered or not by the junction, the relative size of two confluent streams, and which of them may be said to preserve its former course with the smallest deviation. On the true description of these details must depend the question as to which of two confluent rivers should be considered as the main or parent stream. Rivers are said to be confluent when both branches are nearly equally deflected from their former direction, and that of the united streams may be said to be the resultant of two contrary forces. An affluent is a stream which falls into another called the recipient without changing the direction of the latter, and entirely losing its own.

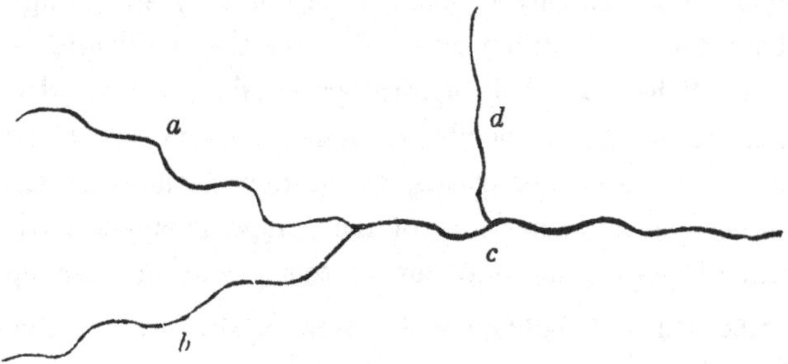

a and *b* are confluent streams, *d* is an affluent falling into *c*, the recipient.

An affluent, too, may generally be said to be smaller than its recipient, and may often be more correctly called a rivulet or a torrent; and here it may be remarked that there is great advantage in attending to the true and proper use of these relative terms, rivers, torrents, rivulets, or brooks, the two latter being more or less synonymous, and a torrent being generally applied to a rapid mountain stream; all these, more or less, bring down detritus from the hills, which is deposited at the mouths of the streams, or wherever other natural causes retard the rapid flow of water. In these cases deltas are formed, which deserve examination, and are either fluviatile, lacustrine, or marine, according as the river empties itself into another river, a lake, or the sea.*

But there are other important characters which deserve attention in the description of a river; and chiefly the *name* is of importance. Does it change during its course, and where and when? How far up from the mouth is the same name preserved? and is it the same on both banks? What is its origin, and by whom was it first given? Then we must inquire what islands are met with in its course? Where are they situated? Are they low? subject to inundation? marshy or rocky? or do they stand high above the level of the stream? Are they cultivated or not? What are their natural productions? By what creatures are they inhabited? Again, is the river at all affected by rapids, or shoals, or cataracts? and what are the peculiar characteristics of these impediments to navigation? Does the tide flow in them, and how far up

* See Col. J. J. Jackson's work, 'What to Observe.' London, 12mo., 1841.

is it felt? Does the river abound with eddies or whirlpools, and how are they occasioned? Do they interfere with navigation or not? Are they accompanied by rocks or shoals? Again, we must ascertain what fords a river offers, and what depth of water is generally found over them: the nature of the bed of the river, particularly in the case of a ford, should also be carefully ascertained.

In addition to these remarks, many other important peculiarities will often be discovered by the careful observer. In some countries, particularly in secondary limestone districts, the rivers are remarkable for their subterranean courses.* Suddenly emerging in large volumes from the base of a lofty mountain, they flow across rich alluvial plains, and are then as suddenly lost in the cavities of another mountain, again to issue forth to the light of day in a distant region, after their subterranean course. Nor should the traveller omit to notice, when crossing a river, the direction in which it flows as regards his own course, whether to the right or to the left. Several distinguished travellers have been unable to connect their observations from not having attended to this point.

4. *Springs.*—The phenomena connected with the outbursts of water from the surface of the earth are not only of the greatest interest, but a correct observation of them is attended with the greatest practical advantage. The traveller should state, approximatively at least, their size or volume, and the nature of the rock or soil out of which they rise; also whether they are pure or mineral, and what deposits are formed about the orifices through

* Styria and the neighbourhood of Trieste.

which they issue; how they are affected by different seasons; whether they are of ordinary temperature or thermal, and if the latter, it is desirable to ascertain the degree of heat by means of a thermometer: the touch alone is a very vague and uncertain guide. It is also desirable, when it can be done conveniently, to procure specimens, in closely sealed bottles, of the water of such springs as appear to possess mineral properties, or to contain salts in solution, for the purpose of analysis at home. Naval officers whose ships are at hand have in this respect great advantages over those whose only means of transport is on horseback or on camels.

5. *Lakes.*—These sheets of water varying greatly in size, form very important features in the geographical description of a country, and the traveller should carefully remark their connexion with the other hydrographical features of the district. Whether they constitute the sources of rivers, or are their ultimate recipients, whether they are or are not connected with the ocean or other great seas, their levels with regard to the ocean, particularly when at a lower level, what rivers flow into or out of them, and whether they are fresh water or salt.

I cannot here do better than quote the following remarks from Colonel Jackson's work, who says, "With regard to lakes in general, the observations to be made upon them may be comprehended under the following heads:—

"Name; geographical and topographical situation; height above the level of the sea, and as compared to other neighbouring lakes; subterranean communication; form, length, breadth, circumference, surface, and depth;

the nature of the bed and of the borders; the transparency, colour, temperature, and quality of the water; the affluent streams and springs; the outlets, the currents; the climate, soil, and vegetation of the basins; the height and nature of the surrounding hills when there are any; the prevailing winds; the mean ratio of evaporation compared with the quantity of water supplied; and any particular phenomena; the navigation and fisheries of the lake; formation and desiccation of lakes." This latter point, depending as it mainly does on the relative elevation or subsidence of the country, may indeed be said almost to belong to the kindred science of geology, and yet it bears so immediately on the physical configuration and geographical features of the country, that it may fairly be mentioned in this place.

Connected with the question of lakes, are the scarcely less important features of lagoons and marshes, and smaller hollows called ponds; the extent of these marshes and lagoons should be ascertained, also whether connected with the sea or not; and what portions of them become dry and passable during the summer or other periods of the year. Peat bogs, in many cases the remains of former lakes, may also be classed amongst these features, and their extent and depth and qualities should be ascertained.

6. *Line of coast, &c.*—This may be indeed said to be the peculiar province of the naval officer; but as forming one of the chief boundaries of those great geographical subdivisions, the details of which we have been here alluding to, we must not omit a brief allusion to some of its most important features. And 1st, with regard

to the actual line of coast itself, the traveller should remark the various headlands jutting out into the sea, as well as the deep bays and recesses running up into the land, and affording refuge from the dangers occasioned by the neighbouring headlands; all gaps and breaks in the continuity of hills or cliffs, or mountain ranges, the occurrence and nature of rivers and streams emptying themselves into the sea, the character and extent of their mouths, the nature of the detritus and alluvial matter brought down by them, and whether or not deltas are formed near their mouths. In another aspect he should inform us whether the coast is bold or flat, whether formed by cliffs or by sloping plains, and whether the rivers enter the sea by one or by numerous channels; whether the coast is clear from danger, or whether sunken rocks and reefs render more than usual precaution necessary in approaching it; whether the sea deepens gradually or suddenly, and whether there are any extensive shoals or sand-banks. Soundings also may be given when practicable, as well as the nature and colour of the sand, clay, or other substances brought up from the bottom by the lead. Do these appear to belong to the same formation as the adjacent mountains, or to have been carried thither by tides or currents, &c.

The nature of the shore, also, should be carefully ascertained, whether it consists generally of sand or mud, or rocks, either in the shape of reefs, or occurring as detached blocks, also whether the landing is easy or not on the beach, and whether this consists of sand or shingle. What bays or coves occur along the line of coast to serve as harbours of refuge? What is the nature of the anchorage? Are there any harbours along the coast? and how far have natural harbours been rendered more available and safe by the erection of breakwaters or piers?

In concluding this portion of the first division of physical geography, I would also mention a few points connected with the physical features of the country which deserve notice, but which, being of an accidental rather than of a normal character, did not easily find a place in any of the natural subdivisions of the subject. The traveller should always pay particular attention to those phenomena in the physical structure of the country which may be called by some persons natural curiosities. Amongst the principal of these are grottoes, caves, and caverns; some of them are not only strikingly beautiful, but of great scientific interest. They are more usually met with in limestone districts than in any other; it is interesting to ascertain their size and extent, and the distance to which they have been traced. Are they traversed by subterranean streams, and if so, do these streams enter or escape by known channels or mouths, as is frequently the case in Istria and Carniola, and in the west of Ireland? Natural bridges present another instance of this kind of phenomena. How have they been formed, and what is the nature of the rock of which they consist? Are they stalactitic, or of a more compact nature? Mines are also to be noticed, although they come more directly under the head of geological observation. All volcanic phenomena and earthquakes are also deserving of notice. Springs of fresh water rising up in the sea are not of unusual occurrence; and any information respecting them is always desirable, such as the depth of water and the effect of the fresh water on the surrounding ocean. Any instances of that remarkable phenomenon observed in Cephalonia, where the sea-water flows inland into a hollow in the rocks, should also be carefully described. In short, it may be safely asserted that there is no single fact connected with the physical structure of the earth, falling under the notice of an intelligent observer, which may not be of value or importance either to himself or others, if he will only give himself the trouble of carefully noting down the main facts on the spot itself, with as much accuracy and detail as circumstances will permit. With this view we must again urge what was stated at the beginning, and would add in the words of Mr. Darwin, "Trust nothing to the memory; for the memory becomes a fickle guardian when one interesting object is succeeded by another still more interesting."

II. Political Geography.

We now proceed to notice some of the principal features to which attention should be directed on the subject of political or statistical geography. In many respects this branch of our subject approaches very closely to that of statistics, to the consideration of which a distinct and separate article will be devoted; we will however endeavour to steer clear of collision, by confining ourselves to the definition already given, and by avoiding those questions of detail which are more peculiarly the province of the statist. Nor can it be expected that the casual visitor should devote to the examination of documents and books the time that is necessary to arrive at any important results in reference to these questions, or to make much progress in the investigation of a subject, however important, the whole value of which depends on the extent and minute accuracy of its detail; but yet there are many matters connected with man's social state which the traveller may easily elucidate by availing himself of the opportunities thrown in his way, and carefully preserving the information he obtains.

This branch of our subject may properly be divided into the following heads:

1. Population; different races of inhabitants.
2. Language; words and vocabularies.
3. Government; ceremonies and forms.
4. Buildings; towns, villages, houses.
5. Agriculture; implements of labour and peculiarities of soil.
6. Trade and Commerce. Roads, and other means of communication.

1. *Population.*—One of the most interesting inquiries on visiting new countries relates to the people by whom they are inhabited. It is not enough to ascertain the mere amount of population although even this is by no means easy; and unless obtained from official documents, it cannot always be relied on. The oral information first obtained by a stranger is almost invariably incorrect, and particularly so in barbarous countries and amongst an ignorant population, where truth and accuracy are equally disregarded. Various sources must be referred to before we can venture, in such cases, to place confidence in our information. Another and more interesting

question, as regards the population of a country, is the nature and character of the races by which it is inhabited; we wish to know whether they all belong to one of the great races of the human family, or to a mixture of several; how far the national character has been affected or modified by such mixture; whether it took place long ago, or is an event of recent occurrence. In many instances, casual intercourse with the natives will lead to information on this subject; local traditions will be found to have been preserved, which, after making due allowance for exaggeration and prejudice, will generally give a clue to the details required. It is also worth noticing, when the population consists of various races, whether one race or nation is more confined to a rural or a town life than the other; whether there exists any feeling of hostility or jealousy between them; whether any particular trades or occupations are more exclusively practised or followed by one race than the other; whether one race is kept down or oppressed by the other, or whether they enjoy a state of comparative equality.

When the population of a country has up to a certain period consisted of one race, and a mixture has subsequently taken place, this change may have been occasioned in three different ways. The new race may have come down with force and violence on the original inhabitants, and having gained possession by right of conquest, may have constituted themselves the masters of the country; or, secondly, they may have been introduced as slaves in the first instance, captured in war or taken by stratagem by their more successful neighbours; or, thirdly, they may have come gradually, few at a time, with the free consent of the inhabitants, seeking to make their fortunes in a new country as settlers or as colonists. Any information on these points, where a mixture of races does exist, will be interesting. Not only will the moral character of the united people be differently influenced, but even their political rights, their institutions, and form of government, will have been greatly modified according to the different modes by which the union of the two people was effected.

In many cases, too, the traveller may have opportunities of making useful observations respecting the general character and disposition of a people. Are they of a warlike or a peaceful disposition? Have they made any progress in the arts of civilization or of commerce? Do they possess any and what extent of literature? Are they remarkable for their honesty, or for contrary propensities? Are they open and frank towards strangers, or the reverse? Do they make any distinction in their dealings between natives and foreigners? How do they dress and live? What are their domestic habits and relations? Do they encourage or prohibit polygamy, and are women treated with respect and consideration? Without going profoundly into the study of these questions, the attentive observer cannot fail to pick up many interesting details and facts on these subjects, all of which may hereafter be of use to himself or to others.

2. *Language.*—The traveller will have many opportunities of collecting much interesting information respecting the languages of those countries which he visits, by taking notes of all the peculiarities he may have an opportunity of observing respecting them, when he feels confidence in

the accuracy of his information. These observations do not of course apply to the languages of Europe, and to those of the more civilized nations of the East, viz. the Arabic, the Persian, or Mahratta, &c., but are rather intended for the guidance of those who visit the islands of the Pacific, or the Indian Archipelago, Australia, Africa, and other lands, of which the languages are still unknown.

In this respect there will in all probability be great analogy with the previous subject. Where a nation has sprung up from the mixture of two races, it will generally, if not universally, be found that the language bears traces of the same admixture. Analogous elements of combination will have produced an analogous result in a language partaking of the essential characters of those of which it was composed. Any information, therefore, showing how far the grammatical construction of the resulting language or particular words are derived from one or the other of the parent tongues, will be important. Nor should these observations be confined to mere words, and their affinities in different languages. It is equally desirable to obtain information respecting the genius and character of languages; to remark how far the idioms of one correspond with those of another; and whether the resemblances observed between the languages of various nations can in any way be traced to any original connexion between the nations themselves, or to political or commercial relations existing between them at a former period.

But it is not alone with reference to the comparisons to be made between different languages that it is desirable to obtain correct information. Even when the traveller has no opportunity of comparing several languages, he may collect much valuable matter by attention to any one in particular. Above all things let him endeavour to make as complete a vocabulary as possible of all those words, of which he can depend on obtaining the true and precise meaning. Nor are words alone to be attended to: all peculiarities of diction, all idiomatical expressions and phrases ought to be remarked and carefully written down. With respect to the languages of many barbarous yet interesting people, it is only by the repeated observations of successive travellers, and by the comparison of such observations with those of others in different regions, that we can at last obtain any idea of their nature, their genius, and their origin. It may also be useful to ascertain how far foreign words have been introduced into the language, and to what extent they are used, whether confined to one or more classes of the population—whether they are more particularly used by the military, the commercial, or the manufacturing classes.

3. *Government.*—It is hardly to be expected that those for whom these remarks are principally intended will have the time or opportunity to make many inquiries, or to collect much correct information on the details of government in its various branches, in the countries they may visit. Many of these details, even if they could be obtained, would be more appropriately noticed under the head of Statistics. There are, however, several points connected with this subject on which an intelligent traveller can hardly fail to make useful and interesting observations. Amongst these we may mention all kinds of forms, ceremonies, and processions, whether of a religious

or civil nature; the observance of religious rites, where strangers are not superstitiously excluded; the ceremonies and processions which are generally a part of such rites, and which for the most part take place in the open air, afford many opportunities for remarks. Royal pageants and processions, military manœuvres and encampments, the dress and bearing of the troops, are all worthy of notice. Many municipal institutions necessarily come under the observation of travellers, as matters of police and surveillance, passports and other documents required by the authorities, as well as any other regulations necessary, or supposed to be so, for the maintenance of peace and order. What are the principal taxes, and how are they levied, and on what articles are they imposed? What is the principle of taxation—direct or indirect? Public institutions, also, in those countries where the state of society warrants their existence, and can secure their continuance, whether maintained by the liberality of the state or supported by the zeal and resources of individuals, may well deserve a passing notice, even if more detailed information is not accessible. These, too, may be of very different characters, and may have various objects in view: they may be intended for the promotion of literature amongst the old, or of education amongst the young; they may tend to the furtherance of trade and commerce, or they may only look to affording amusement and relaxation. Something at least on all these subjects will not escape the eye or ear of the most casual observer.

4. *Buildings.*—In considering the buildings of a people, they present themselves to our notice under several points of view. We may, in the first place,

consider them as public or private. Amongst the former we shall find such as belong to the nation generally, either as the residence of the sovereign, or as belonging to the different departments of the executive government, or to the legislature, or as devoted to the alleviation of suffering or to the maintenance of health, as poorhouses, hospitals, and infirmaries of various kinds. They may be devoted to the service of God, or to the deities worshipped by uncivilized nations, as churches, temples, mosques, and other similar edifices; or they may be intended for the advancement of literature and science, such as colleges and university buildings, museums, picture galleries, &c.; or erected for the amusement and recreation of the people, or for the furtherance of public business, as market-places, town-halls, theatres, &c. With regard to private residences, the different purposes are not so numerous; but even here we may distinguish the habitations of the rich and of the poor, and those intended for town or country residences; the different styles of villages in the country, and the character of streets and houses in the towns; villas, farm-houses, &c., and, in some cases, the different dwellings of different tribes. This, in the case of those nomadic people who still dwell in tents, is very remarkable.

Again, we may consider the buildings of a people either with regard to the degree of civilization of which they may be considered as the evidence, or to the progress in art and architecture which they may be held to indicate. For this purpose, not only is it desirable to point out the style in which they are erected, but also the materials which have been used, and the mechanical contrivances by which they have been assisted. In this case slight sketches will often convey a clearer idea of the object than long and minute description. Nor should we neglect altogether another class of buildings, partly private and partly public in their nature, which often convey much information with respect to the character and progress of a people: I mean their tombs and other sepulchral monuments erected to the memory of the dead, or for the purpose of preserving their bodies. It may be observed that few things indicate more directly the progress of a people through different stages and degrees of civilization than the successive changes which have taken place in the style and character of their buildings, and of the arts by which they have been embellished, from the first rude attempts of Druidical and Cyclopean structure to the more elaborate and symmetrical proportions of what may be called the Palladian style. Any information of this description which falls under the notice even of the most hurried traveller cannot fail to be productive of great interest.

5. *Agriculture.*—The geographer will have numerous opportunities, in his examination of a new country, of obtaining much valuable information on this and its collateral subjects, by a little attentive observation and a few concise inquiries. Amongst the chief points to which his attention should be directed, we may mention the use of tools and agricultural implements, for the purpose either of cultivating the soil or of transporting its produce from one locality to another, the mode of ploughing and preparing the land for different crops, the manner of raising the crops themselves, of sowing, planting, and transplanting, of reaping and gathering in the crops, of

threshing, and other similar occupations, the rotation of crops, and whether, and under what circumstances, more than one crop is raised in the year.

Other inquiries may be usefully directed towards the animals used for agricultural purposes or domestic economy, in the field or in the farm-yard; whether they are indigenous, or brought from distant or neighbouring countries; to what uses they are applied, whether for draught, for food, or for clothing. How are they fed? Are they of a hardy or delicate constitution? Have any changes taken place of late years in the state of agriculture and tillage? Is it in a course of progress or decay? What is the feeling of the inhabitants towards it? Is it practised by the majority, or only a small portion of the population? What buildings form a part of agricultural capital? farm-houses, barns, and cottages? All these depend on the social state of the inhabitants. Is the pursuit of agriculture esteemed or despised? What are the usual prices of provisions—animal and vegetable? To which do the inhabitants give a preference? What is the principal produce of the country—vegetables, fruits, cerealia, meat, or poultry? What is the tenure of land? Is it distributed in large estates, or subdivided into small properties? Is it chiefly in fee, or held on long or short leases from year to year? What is its chief feature—arable, meadow grass, or woodland? What are the respective quantities of each? What is the nature of the soil, and what distinctions are there in it? Is one kind more adapted for one species of cultivation than another, and whence is this difference derived, and by what natural causes has it been occasioned or modified?

6. *Trade and Commerce.*—Our information respecting a country cannot be complete without some knowledge of its trade and commerce, and the manner and the means by which they are carried on. In this respect, also, without stopping to inquire very minutely into the statistical details of the resources and means of a country, the travellers for whom we write can add much to our information by the mere recording of the facts which come under their own observation. What is the nature of the trades chiefly exercised by the different classes of the population, and by different tribes, when such exist? Are they principally employed in working up the raw materials produced in their own country, or those imported from other quarters? Are they workers in metal, and whence are the metals obtained? Or are they workers in leather and similar materials? Or do they spin and weave, and what are the materials worked up in their looms—whether wool, cotton, flax, or silk—and which, if any, of them are raised in their own country, and from what other districts do they draw their supplies when requisite? Is their commerce chiefly domestic, foreign, or transit, and by whom is it carried on? What are the principal articles of import and export? Where do they come from, and whither are they sent? What is the medium of exchange? What progress have they made beyond the mere principle of barter? Is money used as a medium of exchange? What coins are known? Have the natives any knowledge of bullion, paper, or bills of exchange? Have they any system of credit or bill-discounting? How is commerce conducted? What are the means of communication—water or land? If by water, what is the nature of their

ships and vessels? Are they employed at sea, or on rivers or canals? What is the character of their sailors? If by land, have they yet learnt the use of railroads? What is the nature of the roads and other tracks? Are they available for carts and waggons, or only for beasts of burthen? What beasts are used—horses, mules, asses, bullocks, or camels? Which are most useful? How are the roads kept up? Are they in good or bad condition? Are the bridges well built and well kept up? What is the ordinary rate of travelling, and the expense of carrying goods? What are the weights and measures used in the country? Are they the same in trade or commerce as in private life? Many of these questions are easily answered, and all will be found useful for one purpose or another.

There remains one subject on which it may not be irrelevant to make a few remarks, although there may be some question as to whether this is the proper place for its introduction. Our information respecting distant lands and their inhabitants cannot be said to be complete without some knowledge of their past history and their antiquities; and we therefore propose briefly pointing out to the traveller a few of the points to which his attention may be advantageously directed. In the first place, in his excursions in the country, let him carefully examine the sites and remains of ancient buildings. This identifying of ancient positions, and fixing the names of ancient cities, has not unfrequently been called comparative geography, as establishing a comparison between the ancient and modern state of things. Where the remains appear to indicate the site of a ruined city, let him carefully trace the line of the ancient walls, ascertain the position of the gates, describe or sketch the style of architecture, and state the materials of which they have been built. If the fallen fragments indicate the site of a temple or analogous building, let the traveller endeavour to obtain precise measurements of its different component parts, the length and diameter of the columns, the details of architraves, capitals, and cornices, and whatever other features may attract his attention. Above all things, let him diligently search for inscriptions, and then carefully copy *all* that he may find, endeavouring as much as possible to preserve the precise form of the characters in which they are written.

Two or three other evidences of ancient art or history remain to be noticed—coins and manuscripts, and works of art. With respect to the former, he cannot be too industrious in collecting all that his means allow him to procure of those which come in his way—taking care, of course, in those countries where such practices obtain, that he is not imposed upon by forgeries. Manuscripts are of more rare occurrence, but even these may safely be collected when possible, and there is less danger of deceit than in the case of coins. With regard to works of art it is more difficult to lay down any precise rule, on account of their greater variety, and a certain degree of vagueness attaching to the term, and also on account of their bulk and cost. Two classes, however, may be mentioned which particularly deserve attention—statues and gems. Of the former of these, the traveller will generally be enabled only to make drawings: their size will in most cases prevent their being moved. Gems, on the other hand, whether cameos or intaglios, are amongst the most valuable and portable works of art which a traveller can collect. But let him beware of imposition: nowhere is it more frequently and more notoriously practised. With due attention to these hints, the traveller whose fate or duty may lead him to the shores of classic land, cannot fail to obtain much information which will prove not only a source of interest to himself, but will be received with satisfaction and delight by every cultivated mind on his return to his native shores.

SECTION VI.

GEOLOGY.

BY CHARLES DARWIN, ESQ., F.R.S., F.G.S.

A PERSON embarked on a naval expedition, who wishes to attend to Geology, is placed in a position in some respects highly advantageous, and in others as much to the contrary. He can hardly expect during his comparatively short visits at one place, to map out the area and sequence of widely extended formations; and the most important deductions in geology must ever depend on this having been carefully executed; he must generally confine himself to isolated sections and small areas, in which, however, there can be no doubt many interesting facts may be collected. On the other hand, he is admirably situated for studying the still active causes of those changes, which, accumulated during long-continued ages, it is the object of geology to record and explain. He is borne on the ocean, from which most sedimentary formations have been deposited. During the soundings which are so frequently carried on, he is excellently placed for studying the nature of the bottom, and the distribution of the living organisms and dead remains strewed over it. Again, on sea-shores, he can watch the breakers slowly eating into the coast-cliffs, and he can examine their action under various circumstances: he here sees that going on in an infinitesimally small scale, which has planed down whole continents, levelled mountain-ranges, hollowed out great valleys, and exposed over wide areas rocks, which must have been formed or modified whilst heated under an enormous pressure. Again, as almost every active volcano is situated close to, or within a few leagues of the sea, he is admirably situated for investigating volcanic phenomena, which in their striking aspect and simplicity, are well adapted to encourage him in his studies

In the present state of the science, it may be doubted whether the mere collection of fragments of rock without some detailed observations on the district whence they are brought, is worthy of the time consumed and the carriage of the specimens. The simple statement that one part of a coast consists of granite, and another of sandstone or clay-slate, can hardly be considered of any service to geology; and the labour thus thrown away might have been more profitably spent, and thus saved the collector much ultimate disappointment. It is now generally recognized that both the sedimentary rocks, and those which have come from below in a softened state, are nearly the same over the whole world. A mere fragment, with no other information than the name of the place where collected, tells little more than this fact. These remarks do not at all apply to the collection of fossil remains, on which subject some remarks will presently be made; nor do they apply to an observer collecting suites of rock-specimens, with the intention of himself subsequently drawing up an account of the structure and succession of the strata in the countries visited. For this end, he can hardly collect too copiously, for errors in the naming of the rocks may thus be corrected, and the careful comparison of such specimens will often reveal to him curious relations which at the time he did not suspect.

In order to make observations of value, some reading and much careful thought are necessary; but perhaps no science requires so little preparatory study as geology, and none so readily yields, especially in foreign countries, new and striking points of interest Some of the highest problems in geology wait on the observer in distant regions for explanation; such as, whether the successive formations, as judged of by the character of their fossil remains, correspond in distant parts of the world to those of Europe and North America, or whether some of them may not correspond to blank epochs of the north, when sedimentary beds either were not there accumulated, or have been subsequently destroyed. Again, whether the lowest formation everywhere is the same with that in which living beings are first present in the countries best known to geologists. These and many other such wide views in the history of the world are open to any one, who, applying thought and labour to his subject, has the good fortune to geologise in little frequented countries.

A person wishing to commence geology, is often deterred by not knowing the names of the rocks; but this is a knowledge, he may rely on it, easily acquired. With half a dozen named crystalline rocks, or even by patiently familiarizing his eye (aided by a lens) to the aspect of the feldspar and quartz in granite, he will know the two most essential ingredients in most igneous rocks; and in granite he will often find the glittering scales of mica replaced by a dark green mineral, less hard than the feldspar and quartz; and then he will know the third most important mineral, hornblende. The sedimentary rocks can hardly be described, except by the terms in common use: impure limestone, which cannot be readily recognized by the eye, can be distinguished by its effervescence with acids. By the repeated comparison of freshly fractured sedimentary and igneous rocks, such as sandstone and clay-slate on the one hand, and granite and lava on the other, he will learn the difference between crystalline and mechanical structure; and this is a very necessary point. Let no one be deterred from geology by the want of mineralogical knowledge; many excellent geologists have known but little; and from this reason its value has perhaps sometimes been underrated, for many of the obscurer points in geology, such as the nature of the metamorphic changes in rocks, and all the phenomena of metallic and other veins, almost require such knowledge. The appearances presented by the different forms of stratification (that is, the original planes of deposition) may be soon learnt in the field; though no doubt the beginner would be aided by the diagrams given in many elementary works.

The two most useful works which the geologist can carry with him, are without doubt the 'Principles' and the 'Elements of Geology,' by Sir Charles Lyell. He should procure a treatise on mineralogy, for instance, 'Phillips's Mineralogy,' by Allan. If he has the opportunity to procure others, Sir H. Delabeche's 'Researches in Theoretical Geology' would be particularly desirable from discussing many of the questions which ought especially to engage the attention of a sea voyager. As he will probably visit many volcanic regions, Dr. Daubeny's 'Treatise on Volcanos' would be extremely useful; and a list is there given of special treatises on the volcanic countries likely to be visited by him. The 'Description Physique des Isles Canaries,' by Von Buch, may be cited as a model of descriptive powers. The voyager in the Temperate and Polar regions ought to have Agassiz' work on Glaciers.

The geologist fortunately requires but little apparatus; a heavy hammer, with its two ends wedge-formed and truncated; a light hammer for trimming specimens; some chisels and a pickaxe for fossils; a pocket-lens with three glasses (to be incessantly used); a compass and a clinometer, compose his essential tools. One of the simplest clinometers is that constructed by the Rev. Prof. Henslow: it consists of a compass and spirit-level, fitted in a small square box; in the lid there is a brass plate, graduated in a quadrant of 90 degrees, with a little plumb-line to be suspended from a milled head at the apex of the quadrant. The line of intersection of the edge of the clinometer, when held horizontally, with the plane of the stratum, gives its strike, range, or direction; and its dip or inclination, taken at right angles to the strike, can be measured by the plumb-line. In an uneven country, it is not easy without the clinometer to judge which is the line of greatest inclination of a stratum; and it is always more satisfactory to be certain of the angle than to estimate it. A flat piece of rock representing the general slope can usually be found, and by placing a note-book on it, the measurement can be made very accurately. In studying the cleavage or slaty structure of rocks, accurate observations are indispensable. A mouth blow-pipe with its apparatus, and a book with instructions for its use (Phillips's Mineralogy contains brief directions), teaches a little mineralogy in a pleasant manner. Besides the above instruments, a mountain barometer is often very necessary: a portable level would, in the case of raised sea-beaches and terraces, be useful. Messrs. Adie and Son, of Edinburgh, sell a hand-level, a foot in length, which is fitted with a little mirror on a hinge, so that the observer, whilst looking along the level, can see when the bubble of air is central, and thus instantly find his level in the surrounding district. This is a very valuable instrument. Mr. R. Chambers, moreover, and others have found, that an observer having previously ascertained the exact height of his eye when standing upright, can measure the altitude of any point with surprising accuracy; he has only to mark by the level some recognizable stone or plant, and then to walk to it, repeat the process, and keep an account how many times the levelling has been repeated in ascending to the point, the height of which he wishes to ascertain.

A few cautions may be here inserted on the method of collecting. Every single specimen ought to be numbered with a printed number (*those which can be read upside down having a stop after them*) and a book kept exclusively for their entry. As the value of many specimens entirely depends on the stratum or locality whence they were procured being known, it is highly necessary that every specimen should be ticketed on the same day when collected. If this be not done, in after years the collector will never feel an absolute certainty that his tickets and references are correct. It is very troublesome ticketing every separate fossil from the same stratum, yet it is particularly desirable that this should be done; for when the species are subsequently compared by naturalists, mistakes are extremely liable to occur; and it should always be borne in mind, that misplaced fossils are far worse than none at all. Pill-boxes are very useful for packing fossils. Masses of clay or any soft rock may be brought home, if small fossil shells are abundant in them. Rock-specimens should be about two or three inches square, and half an inch thick; they should be folded up in paper. To save subsequent trouble, it will be found convenient to pack up and mark outside, sets of specimens from different localities. These details may appear trifling; but few are aware of the labour of opening and arranging a large collection, and such have seldom been brought home without some errors and confusion having crept in.

To a person not familiar with geological inquiry, on first landing on a new coast, probably the simplest way of setting to work, is for him to imagine a great trench cut across the country in a straight line, and that he has to describe the position (that is, the angle of the dip and direction) and nature of the different strata or masses of rock on either side. As, however, he has not this trench or section, he must observe the dip and nature of the rocks on the surface, and take advantage of every river-bank or cliff where the land is broken, and of every quarry or well, always carrying the beds and masses in his mind's eye to his imaginary section. In every case this section ought to be laid down on paper, in as nearly as possible the real proportional scale, copious notes should be made, and a large suite of specimens collected for *his own* future examination. The value of sections, with their horizontal and vertical scales true to nature, cannot be exaggerated, and their importance has only lately been appreciated to the full extent. The habit of making even in the rudest manner sectional diagrams is of great importance, and ought never to be omitted: it often shows the observer palpably and before it is too late (a grief to which every sea-voyager is particularly liable), where his knowledge is defective. Partly for the same reason, and partly from never knowing, when first examining a district, what points will turn out the most important, he ought to acquire the habit of writing very copious notes, not all for publication, but as a guide for himself. He ought to remember Bacon's aphorism, that "Reading maketh a full man, conference a ready man, and *writing an exact man;*" and no follower of science has greater need of taking precautions to attain accuracy; for the imagination is apt to run riot when dealing with masses of vast dimensions and with time during almost infinity. After the observer has made a few traverses of the country and drawn his sections (and the coast-cliffs often afford him an invaluable one), he will be himself astonished how, in the most troubled country, over which the surface has been broken up and re-cemented, almost like the fragments of ice on a great river, how all the parts fall into intelligible order. He will in his mind see the beds first horizontally stretched out one over the other in a fixed order, and he will then perceive that all the disturbance has arisen from a few nearly straight cracks, on the edges of which the beds have been upturned, and between which he will sometimes find great wedges of once heat-softened, but now crystalline rocks. He will find that large masses of strata have been removed and denuded, that is ground down into pebbles and mud, and long ago drifted away to form in some other area newer strata. He will now have a good idea of the physical structure of his district; and this much can be acquired with much greater facility than he will at first readily anticipate.

In examining a district to make a section, many minor points of detail will occur for observation, which can hardly be specified; such as the nature and cause of the transitions and alterations of the different strata, the source of the sediment and pebbles, the alterations in chemical nature, either of the whole mass, or of parts, as in concretions; the presence, and grouping and state of the fossil remains; the depth and condition of the old sea-bottom, when the beds were deposited, and an infinity of similar points. Probably the best method of obtaining this power of observation, is to acquire the habit of always seeking an explanation of every geological point met with; for one mental query leads on to another, and this will at the same time give interest to his researches, and will lead him to compare what is before his eyes, with all that he has read of or seen. With his increasing knowledge he will daily find his powers of observation, his very vision, become deeper and clearer. No one, however, must expect to solve the many difficulties which will be encountered, and which for a long time will remain to perplex geologists; but a ray of light will occasionally be his reward, and the reward is ample.

Organic Remains.—In the sectional diagram which we have supposed to be made, the simple superposition of the beds gives their relative antiquity: but the best section which a sea-voyager can hope to make, will seldom include but a small portion of the long sequence of known geological formations. And as the voyager seldom passes over large districts, he will rarely succeed in placing in proper order, by the aid of superposition alone, the formations which he successively meets with even in the same country. Hence he must, more than any other geologist, rely on the characters of the embedded organic remains, and must sedulously collect every specimen and fragment of a specimen. By the means of fossil remains, not only will he be enabled to arrange (with the help of naturalists on his return home) the formations in the same country according to their age, but their contemporaneity with the deposits of the most distant parts of the world can thus and by no other method be ascertained; for it is now known that at each geological epoch the marine animals partook in the most distant quarters of a general similarity, even when none of the species were identically the same: thus beds have been recognized in North and South America, and in India, which must have been deposited when the chalk in Europe was accumulating beneath the sea.

It is highly necessary most carefully to keep the fossils found in different strata separate; it will often occur in passing upwards from one bed to another, and occasionally even without any great change in the character of the rock, that the fossils will be wholly different; and if such distinct sets of fossils are mingled together, as if found together, undoubtedly it would have been better for the progress of science that they had never been collected. As there is some inconvenience in keeping the fossils collected on the same day separate, this caution is the more requisite. The collector, if he be not an experienced naturalist, should be very cautious in rejecting specimens, from thinking them the same with what he has already got; for it requires years of practice to perceive at once the small, but constant, distinctions which often separate species: the same species, moreover, if collected in different localities, or in beds one placed far above the other, are generally more valuable to the geologist than new species.

In formations from a few hundred to a thousand feet and upwards in thickness, the whole of which does actually belong to the same geological age, and is therefore characterized by the same fossils, most curious and important results may be sometimes deduced, if the position or relative heights at which the groups of fossils are embedded be noted; and this is a point usually neglected. For, thanks to the researches of Professor E. Forbes, the depth of water under which a collection of shells lived can now be approximately told; and thus the movement of the crust of the earth, whilst the strata including the shells were accumulating, can be inferred. For instance, if at the bottom of a cliff, say 800 feet in height, a set of shells are buried, which must have lived under water only 50 or 100 feet in depth; it is clear that the bottom of the sea must have sunk to have allowed of the deposition of the 700 feet of superincumbent submarine strata; subsequently the whole 800 feet must have been upraised. For this same purpose, and for other ends, it is desirable that it should be noted which species are the most numerous, and whether layers are composed exclusively of single kinds. It should be also remarked, whether the more delicate bivalve shells retain their two valves united, and whether the burrowing kinds are embedded in their natural positions, as these facts show that the shells have not been drifted from afar. Where there are fossil corals, it should be observed whether the greater number of specimens are upright, in the positions in which they grew. The remark formerly made that the collection of mere fragments of rock is of little or no use to geology, is far from applicable to fossil remains. Every single fossil species, bones, shells, crustacea, corals, impressions of leaves, petrified wood, &c, should be collected, and it is scarcely possible to collect too many specimens. Even a single species without any information of any kind, if it prove a quite new form, will be valuable to the zoologist; if it prove identical with, or closely allied to a known species, it may interest the geologist. A set of fossils, however, and still more several sets, with their superposition known, cannot fail to be of the highest value; they will tell the age of the deposit, and perhaps give the key to the whole geology of the country: some of the highest problems in this science wait for solution on large collections of species carefully made in distant regions.

A collection of recent shells (both those living on the coast and those to be procured by the dredge off it) from the same country or island at which a collection of tertiary fossil shells is made, is generally of very great service to the palæontologist, who undertakes the description of the fossils. The collecting recent shells will, moreover, with the aid of a little study, teach the geologist some conchology, and this is an acquirement yearly becoming more necessary: the geologist should exert himself to learn some general zoology.

The bones of vertebrated animals are much more rarely found than the remains of the lower marine animals, and they are almost in proportion more valuable. A person not acquainted with the science will hardly be able to imagine the deep interest which the discovery of a skeleton, if of higher organization than a fish, in any of the oldest formations would most justly create. The age of such a formation would have to be judged of by the co-embedded shells, and therefore, if possible, part of the slab containing the bones should include one or two shells to demonstrate their contemporaneity. Bones, however, from any formation are sure to be valuable; even a single tooth, in the hands of a Cuvier or Owen, will unfold a whole history; the heads, jaws, and articular surfaces are the most valuable; but every fragment should be brought home. Where bones are found close together, and especially if some of the parts lie in their natural positions, they should be packed together. Every bone, if found even six inches beneath the black vegetable mould, should be collected; there can be no doubt that many most valuable relics have been neglected, from the supposition that they belonged to still living animals. Low cliffs of mud, gravel, and clay on the banks of streams and on sea-shores (as well as in bared reefs extending from them), are the most likely places for the discovery of the remains of quadrupeds. Gravel beds under streams of lava; fissures in volcanic rocks; peat beds, and the clay or marl underlying peat, are all favourable places. Fishes' bones are found occasionally in all sedimentary strata, and are highly interesting.

Caverns.—These most frequently occur in limestone rocks, and they have yielded a truly wonderful harvest of remains in Europe, South America, and Australia. The bones generally occur in mud, under a stalagmitic crust produced by the dripping of the lime-charged water, which requires being broken up by a pickaxe. As caverns have often been used by wild races of man as places of habitation and burial, a most careful examination should be made to detect any signs of the surface having been anciently broken up near where the bones are found. Even small islands, not now inhabited by any land quadruped, if not very distant from a continent, are almost as likely to contain osseous remains as larger tracts of land. The interest of the discovery of the remains of land quadrupeds in an oceanic island would be extreme: for instance, it has been stated that the tooth of a mastodon has been found in one of the Azores; if this were confirmed, few geologists would doubt that these islands had once been united to Europe, thus enlarging wonderfully our ideas of the ancient geography of the Atlantic: so also the remains of a mastodon are said to have been brought from Timor, thus perhaps indicating the road by which this great quadruped formerly reached Australia.

Fossil Footsteps.—As allied to organic remains, fossil footsteps may be here referred to. They have been observed in Europe and North America, but hitherto in no other part of the world. These curious vestiges not only proclaim the former existence of reptiles and birds at very remote periods, and in rocks often not containing a fragment of bone, but they generally prove that the level of the land subsided after the animal had left its impress on the ancient sea-beach, thus allowing thousands of feet of strata to be thrown down over them. The best place for searching for footsteps is in quarries of sandstone, in which the strata are separated by seams of shale. The best indication of their probable occurrence is the rock being "rippled," that is marked with narrow little wavy ridges, such as occur on most sandy shores when the tide is down, and which indicate that the now rocky surface was once either a tidal beach or a shallow surface, over which the ancient animals walked. In the case of fossil footsteps being found, the largest slab which could possibly be removed ought to be brought away, and accurate drawings, or still better, casts, made of several of the footsteps. A plan from accurate measurement ought to be taken of any row of steps. The value of such fossil footsteps would be in a manifold degree increased, if the age of the deposit could be determined by shells found in the same stratum, or above it.

Coal Deposits.—The origin of coal presents a most curious and difficult problem in geology, and though a vast amount of information has been accumulated on the subject, yet good observations in distant countries would be of the highest value. A very brief statement of the most prominent difficulties in the theory of its origin will, perhaps, be the best guide for further inquiries. If we look first to the coal itself, the frequency with which, both in Europe and North America, upright vegetables have been found in and on the coal, and the curious relation between the presence of coal and the nature of the clayey bed (abounding with roots) on which it rests, can leave no doubt that in these so frequent instances the vegetation, whence the coal has been derived, grew on the spot where now embedded. The regularity and wide extent of the beds of coal, and especially of certain subordinate seams in them, the stratification and fineness of the deposits alternating with the coal, and the rarity of channels (such as would have been formed by a stream or river) cutting through the associated strata, all seem pretty clearly to indicate that the coal was not formed on the surface, like a mass of peat, but under water. What, then, was the nature of those vast expanses of shallow water under which the coal was accumulated? The character of the upright fossil plants, according to our present knowledge, absolutely contradicts the idea of their having lived in the sea; yet occasionally strata, containing undoubted marine remains, are associated with the carboniferous series.

On the other hand, how can we believe that lakes, allowing of course their beds slowly to sink, could contain the enormous thickness, amounting in some instances to several thousand yards, of the coal-bearing strata? From these few remarks it will be seen how many points deserve careful examination in any new coal district; the chief points being, the presence of upright vegetables and trunks of trees (of the position of which careful drawings should be made), and whether furnished with roots,—the nature of the beds on which the coal rests, and generally of all the strata; the continuousness and form of the strata, and whether ripple-marked; the existence of marine animal remains, and whether such lived on the spot, or were drifted into their present positions, and many other similar points. It is superfluous to observe that all fossil plants should be collected; those found upright should be carefully distinguished from those embedded horizontally. The contents of any upright stems and of the roots should be examined; as it appears they have generally first become hollow from decay, and then been filled up with mud, which in some instances is charged with seeds and leaves.

Salt Deposits.—Information is much required on this subject; and this is a case in which good suites of specimens, illustrating the nature of the rocks beneath and above the salt, would possess much interest. Do they contain any organic remains? Did such live on the spot where now buried? Do the rocks show signs of having undergone in any degree the action of heat? Are the strata regular, or are they crossed by oblique layers, showing the probable action of currents? Are there ripple-marks, or beds of coarse pebbles, or other indications of the strata having been deposited in shallow water? What is the thickness, form, and dimensions of the beds of salt? Specimens of the salt, and of any associated saline substances, ought to be brought home in bottles for analysis. The origin of beds of salt, found in formations of very different ages in different parts of the world, is at present quite obscure; some authors attribute it to the sinking of superficial sea-water, rendered more saline by evaporation; others to the evaporation of sea-water periodically overflowing extensive low sandy tracts, like parts of the Run of Cutch; others suspect that its deposition is in some unknown way connected with the sea's bottom having been heated by volcanic action. In some countries there are large lakes of brine, often covering thick beds of salt; these deserve examination: on what does such salt or brine rest, whether on the bared underlying strata, or on sand or gravel, such as cover the surrounding country? Does the salt contain the remains of animals or plants? Specimens of the salt ought to be brought home in bottles, and attention paid, whether beneath it there is any thin layer of other saline substances.

Cleavage.—The slaty structure of rocks will at first perplex the young geologist; for in proportion as it becomes well developed, the planes of stratification or of original deposition become obscure, and are often quite obliterated. As the sea-voyager, and especially the surveyor, often visits numerous points on the same line of coast, he possesses some great advantages for studying this subject, and numerous observations made with care would probably give striking results. The range or strike of the cleavage is uniform over surprisingly large areas; whereas both the angle and point of dip varies much; but there is reason to believe that the planes of inclination, examined across a wide tract transversely to the range, will fall into order and show that they are the truncated edges of a few great curves or domes. The relation of the cleavage-planes to those of the stratification, or axes of elevation, should be carefully noted, and likewise to the general outline of the whole country. Long sections at right angles to the strike of the cleavage, with the dip carefully protracted on paper, would be highly interesting. When two chains of hills, each having its independent cleavage, cross each other, careful observations should be made. In all cases, any mineralogical difference, however slight, in the parallel cleavage-layers, deserves attention; but observations on this head would be hardly trustworthy, without the planes of stratification were so distinct that there could be no possibility of confounding (as has often happened) cleavage and stratification. Where a stratum of sandstone, or of any other rock without cleavage, is interstratified with a slaty rock, the surface of junction ought to be minutely examined, to see if the slate has slipped along the planes of cleavage, or whether again the mass has not been either stretched or compressed at right angles to these same planes. Fossil shells have been found by Mr. Sharpe in slaty rocks, which have had their shapes greatly altered, and all in the same direction; here then we have a guide to judge of the amount and direction of the mechanical displacement which the surrounding slate-rocks have undergone.* Observations on cleavage, to be useful, must be numerous and very accurately made.

The foliation of the metamorphic schists, that is, the origin of the layers of quartz, mica, feldspar, and other minerals, of which gneiss, micaceous, chloritic, and hornblendic schists are composed, is intimately connected with the cleavage of homogeneous slaty rocks. Nearly all the proposed observations on cleavage are applicable to foliation. Wherever large districts of foliated and ordinary slaty rocks unite, observations would be most desirable. These foliated rocks have all undergone metamorphic action, that is, they have been mineralogically altered and rendered crystalline by chemical attraction, aided by heat; but this is a most obscure subject, one on which it would appear that much further light will not be thrown without the aid of a profound knowledge of mineralogy or chemistry. It is now known that granitic rocks, which have been fluidified (as may be told by their sending great veins into, and including fragments of, the overlying rocks), are foliated in a more or less perfect degree: in these cases the relation of the planes of foliation with those of the adjoining rocks, which have been metamorphosed but not fluidified, would be eminently curious.

Nature of the Sea-bottom.— As every sedimentary stratum has once existed as the bed of the sea or of a lake, the importance of observations on this head is obvious;

* With respect to further observations on this important point, Mr. Hopkins remarks, in his paper 'On the Internal Pressure of Rock Masses' (Cambridge Philosoph. Transact., vol. viii.), that "the observer should direct his attention especially to those cases in which the inclination of the cleavage planes to the bedding is either small, or nearly 45°."

and no one is so favourably circumstanced for making them as a naval officer on a surveying expedition. The limits of depth under different latitudes at which the various marine animals live or are found strewed dead, is perhaps the most important point for further investigation which can be suggested in the science of geology: scarcely any observations with the dredge have been made within the tropics. Not only the shells, corals, sea urchins, crabs, &c., brought up from different stated depths, should be preserved, but the proportionate numbers of each kind be carefully noted, as well as the nature of the sea-bottom. An observer could not labour too much in this line, and especially if he would subsequently himself undertake to tabulate and work out the results.*

There is another point of view under which the bed of the sea would amply repay long-continued observations. It is well known that the nature of the bottom often changes very regularly in approaching a coast; the pebbles, for instance, increasing in size in a surprisingly steady ratio with the decreasing depth. But the means by which the pebbles are thus sorted is not known: is it by the oscillation of the waves at ordinary periods, or only during gales; or is it by the action of currents? A chart, with the nature of the bottom carefully noted on it and the currents laid down, would by itself throw some light on this question. The nature of the pebbles being observed, perhaps a point would be found whence they radiated. Excellent observations have been made by engineers on the travelling of shingle beaches, but scarcely anything is

* The best kind of dredge, and the manner of using it, are described under the Zoological Section.

known of their movement under water. In what condition are the pebbles?—are they encrusted (as often happens) with delicate corallines—after a heavy gale are the spines of such corallines found broken? In narrow channels where there are rapid currents, and in the open sea in front of straits, where the water often suddenly deepens, what is the nature of the bottom? To what depth does the sea in a storm render the water muddy? How far from the beach, and to what depth, does the recoil of the waves, or the "undertow," act, for instance, on light anchors? At what depth can the sea wear solid rock? This may sometimes be judged of by the nature of the bottom; thus, where soft mud overlies the rocky surface, we may infer that the sea can hardly now be a destroying agent, even if the inclination of the strata on the adjoining coast shows that rocky strata must once (probably, when the land stood at a different level) have extended much further. Is it at the line of high or low water, or between them, that the breakers most vigorously eat into coast-cliffs? Gigantic fragments of rock, much too large to be themselves rolled about, may be seen at the foot of almost every line of high cliffs; by what means in the course of time will these be removed, as must have happened with their innumerable predecessors? Are they slowly worn away or broken up? It may be well to recollect that in the tropics the powerful action of frost in splitting stones is entirely eliminated. Our observations, moreover, on the alluvial and sub-littoral deposits of these latitudes are not perplexed by the ancient effects of floating ice. The spray of salt-water, above the line of breakers, corrodes by chemical decomposition calcareous rocks; does this play any important part on other rocks? Most bold coasts are fronted by sharp promontories and even isolated pinnacles; are these *exclusively* due to the greater hardness of the rocks composing them, or do not the breakers act more efficiently when eddying round any slight projection?

Rocks rising steeply out of the open ocean, and exposed to the incessant wash of the heaviest surf, are often thickly coated over with various marine animals, and this would seem to indicate that pure water has not the power of gradually wearing away hard rocks, though the waves may occasionally tear off large fragments. Is the washing to and fro of pebbles, or of sand, a necessary element in the corroding power of waves on hard rocks? but how comes it that small land-locked harbours, where the waves can hardly have force to move the shingle, should ever be surrounded by cliffs, which, in most cases, clearly prove that considerable masses of rock have been worn down into mud and removed? Again, at a moderate depth, where the bottom is covered with shingle, does the rolling to and fro of the pebbles wear away solid rock? if so, the pebbles would be clean, and the submarine rocky surface probably worn into furrows or channels at right angles to the beach. Where there are violent currents and eddies, are deep round holes worn in the bottom, like those produced by eddies at the foot of cascades? This, perhaps, might be ascertained by a long pole at the turn of the tide: deep round holes have been observed on rocks formerly covered by the sea, and their origin has perplexed geologists.

Any person steadily attending to these subjects will occasionally be enabled to form an opinion on points at first appearing hopelessly obscure to him. The common deep-sea lead, especially if made a little bell-shaped and well armed, gives a surprisingly good picture of the bottom. There can be no doubt that whoever will for a long period collect and compare observations, made over wide areas and under different circumstances, will arrive at many curious, novel, and important results.

An observer occasionally may arrive at a district where lately some great aqueous catastrophe has occurred, such as the bursting of a lake temporarily formed by a slip, or the rush of a great earthquake-wave over low land. In such cases all the effects produced, such as the thickness and nature of any deposit left—whether stratified irregularly or continuously—whether any rocky surface, over which the debacle has passed, be scored or smooth; all such points should be minutely described, and measurements taken of any great blocks which may have been transported: the great desideratum is accuracy and minuteness.

Ice Action.—The voyager in the Polar Seas would render an excellent service to geology by observing all the effects which icebergs produce in rounding, polishing, scoring, and shattering solid rocks, and likewise in transporting gravel and boulders. Floating ice under two forms is known to transport fragments; namely, coast-ice, in which the stranded boulders are frozen, and icebergs formed by glaciers entering the sea, on the surface of which masses of rock had previously fallen from the surrounding precipices. It is obvious that in the latter case the fragments would generally be quite angular, and they could not be landed in water shallower than the thickness of the submerged ice, requisite to float the berg. On the other hand, the boulders frozen in coast-ice would generally be previously water-worn, and they could be landed on an ordinary beach, and might be driven by the force of the pack high and dry, and perhaps left piled in strange positions. All facts illustrating the difference in the results produced by coast-ice and true icebergs would be very valuable. Do the boulders fixed on coast-ice, when driven over rocky shoals, become themselves scored? Wherever there was reason to believe that a surface had been scored by recent ice-action, a minute description and drawings ought to be made of the depth, length, width, and direction of the grooves; and even large slabs brought home. On true icebergs are the fragments of rock generally fixed or loose; when icebergs turn over, are fragments frequently seen embedded in that part which was under water; and how were they fixed there? The nature, number, size, form, and frequency of occurrence of all fragments of rock seen on floating ice ought to be recorded, and the distance from their probable source. A polar shore, known from upraised organic remains to have been lately elevated, would be eminently instructive. Do great icebergs force up the mud and gravel at the bottom of the sea in ridges like the moraines of glaciers? Can shells, or other marine animals, live in a shallow sea, often ploughed up and rendered turbid by the stranding of icebergs? The dredge alone could answer this. The means to distinguish the effects of ancient floating ice from those produced by ancient glaciers is, at present, a great desideratum in geology. M. Agassiz' work on Glaciers, with its admirable plates, ought to be procured by any one going to the colder regions of the north or south.

Erratic boulders occur in Europe, N. America, and in the southern parts of S. America, which, it is believed by most geologists, were transported by ice; those near mountains, by ancient glaciers; and those on the low-lands, by floating ice. Erratic boulders, when not of gigantic size, may be confounded with rounded stones, transported by occasional great floods or by the coast-action of the surf during slow changes of level of the land. Masses of granite, from often disintegrating into large, apparently water-worn boulders, and then rolling downwards, have several times been erroneously described as belonging to the erratic class. Where the nature of all the rocks in the vicinity is not perfectly known, great size and the angularity of the fragments (though by no means a constant concomitant) are the most obvious distinctive characters; but even when the surrounding country is not at all known, the composition of a single isolated hill or small island may easily be ascertained, and if large fragments of foreign rock lie strewed on its surface, these may be assumed almost certainly to be erratic boulders. Here, however, a caution has been found necessary; for in the case of fragments of *sedimentary* rocks, they may be the last remnant of a denuded overlying formation. Wherever erratic boulders are found, their composition, form—especially attending to whether they are angular, water-worn, or scored, and their size, from actual though rude measurements, should be given.

Both in the north and south a peculiar formation called "till" has been found connected with erratic boulders; it consists generally of mud, containing angular and rounded stones of all sizes up to the largest boulders, mingled in utter confusion, and generally without any stratification. Such deposits should be examined. Sometimes when they are stratified, the upper beds have been found violently contorted, whilst the lower ones are undisturbed, showing that the violence has not proceeded from below, as in ordinary geological cases. Sir C. Lyell has suggested that this effect has been produced by the stranding of great icebergs.

As far as our present knowledge goes, the above enumerated phenomena—such as scored, mamillated, and polished rocks, moraines, erratic boulders, and beds of till, though occurring in latitudes where glaciers do not now occur, where the sea is never frozen, and where icebergs are never drifted, yet have not been observed in either hemisphere higher than about latitude 40°. Hence, on whatever coast ancient ice-action might be discovered, the limit of latitude towards the tropics at which it ceases ought to be carefully investigated. Observations are much wanted on the west coast of N. America and the east coast of Asia; and again in New Zealand and other islands of the Southern Ocean. The period of the ice-action is pretty well ascertained in Europe and North America, and a very great service would be rendered to geology if the same point could be clearly made out in the southern hemisphere; for it might greatly influence our ideas on the climate of the world during the late tertiary periods. Any shells embedded in "till" (though, unfortunately, of very rare occurrence) would decide this point, and it might probably be closely judged of, if till or boulders were found resting on, or covered by, shell deposits.

Distribution of Organic Beings.—As geology includes the history of the organic inhabitants, as well as of the inorganic materials, of the world, facts on distribution come under its scope. Earth has been observed on icebergs in the open ocean; portions of such earth ought to be collected, washed with fresh-water, filtered, gently dried, wrapped up in brown paper, and sent home by the first opportunity to be tried, with due precautions, whether any seeds still alive are included in it. Again, the roots of any tree cast up on an island in the open ocean should be split open, to see if any earth or stones are included (as often happens), and this earth ought to be treated like that from icebergs: it is truly surprising how many seeds are often contained in extremely small portions of earth. Any graminivorous bird, caught far out at sea, ought to have the contents of its intestines dried for the same object. The zoologist who, with a towing-net, fishes for floating minute animals, ought to observe whether seeds are thus taken. These experiments, though troublesome, undoubtedly, would be well worth trying. All facts or traditional statements by the inhabitants of any island or coral-reef, on the first arrival of any bird, reptile, insect, or remarkable plant, ought to be collected. In those rare cases in which showers of fish, reptiles, shells, earth, seeds, confervæ, &c., have fallen from the sky, every fact should be recorded, and specimens collected.

Volcanic Phenomena.—The voyager will probably have

ample opportunities of examining volcanic islands, and perhaps volcanoes in eruption. With respect to the latter, he ought to record all that he sees: should the exact position of the orifice be known, he might, perhaps, by observing some point in a cloud, measure with a sextant to what height the fragments were shot forth, and the height of the often flat-topped column of ashes. Having surveying instruments, he ought to map, as carefully as time will permit, any crater remarkable for its size, depth, or peculiar form. M. Elie de Beaumont has found that, owing to the fluidity of lava, streams never consolidate into a thick, moderately-compact mass, except on a surprisingly gentle inclination. On a slope of above 2 or 3°, the stream consists of extremely irregular masses, often forming a hollow vault within. Fresh observations on this point are much wanted in regard to lavas of different composition. The measurements can easily be made by a sextant and artificial horizon.* In

* M. Élie de Beaumont gives the following directions (Mémoires pour servir, &c., tom. iv. p. 173):—

" The method I am in the habit of employing for these kinds of measurements is simple and easy, and a description of it may save useless trouble to others. I place on the edge of the sextant, and behind the fixed mirror, a small piece of white paper, in which there is a narrow opening (ouverture étroite) corresponding to the axis of the telescope. On the exterior surface of the paper a black line is drawn, perpendicular to the plan of the graduated circle, and passing through the centre of the opening above mentioned. A quantity of mercury is poured into a vessel sufficient to form a plane horizontal surface of a certain extent. The telescope of the sextant is then directed vertically over the mercury, and the image of the black line sought for. When this is found, I am certain that the visual ray from the image in the mercury can only deviate from the perpendicular, in so far as the line is not without breadth, and the opening has a sensible size. These two sources of error can be diminished so that the maximum of error shall not exceed a minute. Being once certain of the verticality of the visual ray from the making such observations, comparatively recent streams must be chosen, so that there can be no doubt that the whole consists of a single stream: this cannot be judged of without examining the whole line between the two points of measurement, for some liquid lavas thin out to a very fine edge; and two streams, one over the other, may be thus very easily mistaken for a single one. The composition, thickness, and degree of cellularity of any lava-stream, of which the slope is measured, ought to be described as seen on the sides of fissures, and wherever its internal structure can be made out.

Round many active and extinct volcanoes, both on continents and on islands, there is a circle of mountains, steep on their inner, and gently inclined on their outer flanks. The volcanic strata, of which they are composed, everywhere dip away from the central space, but at a considerably higher angle than it is believed lava can consolidate into such thick and compact masses. These mountains form the so-called "craters of elevation," the origin of which has excited much controversy, and which demand further examination. There is a grand range of mountains of this class at the Mauritius and at St. Jago in the Cape de Verdes, parts only of which have been described. The chief points to attend to are, the *inclination* of the streams by actual measurement, their thickness, compactness, and composition; the form and height of the mountains, whether traversed by very many dikes,

image of the black line, I have only to make the image of any object reflected from the moveable mirror coincide with that of the black line, to have the angle between the vertical, and the line drawn from the centre of the instrument to the object in question, which may be any distant point on the surface of a bed of lava, a glacier, a road, a river, &c."

of which the common direction ought to be recorded; how far the mountains stand apart, and the diameter and outline of the rude circle which they together form. In fact, a most useful service would be rendered by mapping any of these "craters of elevation," or, what would be more feasible, drawing from actual measurements two sections at right angles to each other, across the circle.

Some streams of lava, especially those belonging to the trachytic series (harsh, generally rather pale-coloured lavas, with crystals of glassy feldspar), are laminated. The course of the layers with respect to the course of the stream ought to be minutely studied, both on the surface, at the termination, and flanks of the stream; and, if by a most fortunate chance there should have been formed a transverse section, throughout its entire thickness: this would be a very interesting subject for investigation. A series of specimens ought to be brought away to illustrate the nature of the lamination.

Aerial Dust.—Fine brown-coloured dust has often fallen on vessels far out at sea, more especially in the middle of the Atlantic. This should be collected; the direction and force of the wind (and the course of any upper current, as shown by the movement of the clouds) on the same day, and for some previous days, ought to be recorded, as well as the date, and the position of the ship. Such dust has been shown by Ehrenberg to consist, in many cases, almost entirely of the siliceous envelopes of infusoria. The distance to which real volcanic dust is blown is, likewise, in some respects well worth determining.

Elevation of the Land.—The changes of level, often

accompanying earthquakes, will be treated of by Mr. Mallet, but a few remarks on the nature of the evidence to be sought, on changes of level not actually witnessed by man, may be here inserted. Many appearances, such as lines of inland cliffs, of sand-hillocks, eroded rocks, and banks of shingle, often indicate the former effects of the sea on the land when the latter stood at a lower level. But the best evidence, and the only kind by which the period can be ascertained (for the appearances above enumerated, though well preserved, may sometimes be of considerable antiquity), is the presence of upraised recent marine remains. On land which has been elevated within a geologically recent time, sea shells are often found, either embedded in thin layers of sand and mould, or scattered on the bare surface. In these cases, and especially in the latter case, great caution is requisite in testing the evidence; for man, birds, and hermit-crabs often transport, in the course of ages, an extraordinary number of shells. In the case of man, the shells generally occur in heaps, and there is reason to believe that this character is long preserved. To distinguish the shells transported by animals from those uplifted by the movement of the earth, the following characters may be used:—Whether the shells had long lain dead under water, as indicated by barnacles, serpulæ, corallines adhering to their *insides:* whether the shells, either from not being full grown or from their kind, are too small for food; remembering that certain shells, as mussels, may be unintentionally transported by man or other animals in their young state adhering to larger shells; and lastly, whether all the specimens have the same appearance of antiquity. Some shells, which have been exposed for many ages, yet retain their colours in a surprising manner. The very best evidence is afforded by barnacles and boring shells being found attached to or buried in the rock, in the same posi tions in which they had lived; these may be sometimes found by removing the earth or birds' dung covering points of rock. Where shells are embedded in a superficial layer of soil, though it may appear exactly like vegetable mould, specimens of it should be preserved, for the microscope will sometimes reveal minute fragments of marine animals. In all these cases, specimens of the shells, though broken and weathered, and having a wretched appearance, must carefully be preserved; for a mere statement that such upraised shells resembled those still living on the beach is absolutely of no value. It should be noticed whether the proportional numbers between the different kinds appear to be nearly the same in the upraised shells and in those now cast on the beach. The height at which the marine remains occur above the level of the sea should be measured. In confined situations where the change of level appears to have been small, much caution must be exercised in receiving any evidence; as a change in the direction of the currents (resulting from alterations in neighbouring submarine banks) may cause the tide to flow to a somewhat less height, and thus give the appearance of the land having been upraised.

Wherever a tract of country can be proved to have been recently elevated, its surface, as exhibiting the late action of the sea, is a fertile field for observation. On such coasts, terraces rising like steps, one above another, often occur. Their outline and composition should be studied, diagrams made of them, and their height measured at many and distant parts of the coast. There is reason to believe that in some instances such terraces range for surprisingly long distances at the same height. Where several occur on opposite sides of a valley a spirit level is almost indispensable, in order to recognize the corresponding stages. Where ranges of cliffs exist, the marks of the erosion of the waves may sometimes be expected to occur, and as these generally present a defined line, it is particularly desirable that their horizontality should be ascertained by good levelling instruments, and if not horizontal, that their inclination should be measured. Where more than one zone of erosion can be detected all should be levelled, for it does not necessarily follow that the several lines are parallel. Along extensive coasts, and round islands which have been uplifted to a considerable height, and where we now walk over what was, within a late geological period, the bed of the sea, it would be well to observe whether extensive sedimentary deposits have been upraised; for it has often been tacitly assumed that sedimentary deposits are in process of formation on all coasts.

Subsidence of the Land.—This movement is more difficult to detect than elevation, for it tends to hide under water the surface thus affected. Evidence, therefore, of subsidence is very valuable; and this movement, moreover, has probably played a more important part in the history of the world than elevation, for there is reason to believe that most great formations have been accumulated whilst the bed of the sea was sinking. Subsidence may sometimes be inferred from the form of the coast-land;

for instance, where a line of cliffs, too irregular to have been formed by elevation alone, plunges precipitously into a sea so profoundly deep that it cannot be supposed that the now deeply submerged portions of the cliff have been simply worn away by the currents. The direct evidence of subsidence, if not witnessed by man, is almost confined to the presence of stumps of trees, peat-beds, and ruins of ancient buildings, partly submerged on tidal beaches. Ancient buildings may sometimes afford such evidence in unlikely situations: it has been asserted, that in one of the volcanic islands in the Caroline archipelago there are ruins with the steps covered by the sea. Again, at Terceira, at the Azores, there is an old church or monastery said to be similarly circumstanced.

*Coral Reefs.**—The most important point with respect to coral reefs, which can be investigated, is, the depth at which the bottom of the sea, *outside the reef*, ceases to be covered with a continuous bed of living corals. This can be ascertained by repeated soundings with a heavy and very broad bell-shaped lead, armed with tallow, which will break off minute portions of the corals or take an exact impression of them: it can thus also instantly be seen how soon the bottom becomes covered with sand. This limit of depth ought to be ascertained in different seas, under different latitudes, and under different exposures. For collecting specimens of the corals, it is to be feared that the dredge would become entangled, but chains and hooks may be lowered for this purpose. There is reason to suspect that different species of corals

* The only work specially written on this subject is 'The Structure and Distribution of Coral Reefs,' by Mr. Darwin.

grow in different zones of depth; so that in collecting specimens, the depth at which each kind is found, and at which it is most abundant, should be carefully noted. It ought always to be recorded whether the specimen came from the tranquil waters of a lagoon or protected channel, or from the exposed outside of the reef. The small reefs within the lagoons of certain atolls (or lagoon-islands) in the Indian Ocean all rise to the surface; whereas in other atolls not a single reef rises within several fathoms of the same level. It would be a curious point to ascertain whether the corals in these cases consisted of the same species; and if so, on what possible circumstance this singular difference in the amount of their upward growth has depended.

Any facts which can elucidate the rate at which corals can grow under favourable circumstances, will ever be interesting: nor should negative facts, showing that within a given period reefs have not increased either laterally or vertically upwards, be neglected. In a full-grown forest, to judge of its rate of growth, a part must be first cut down; so is it probably with reefs of corals. The aborigines of some of the many coral islands in the great oceans might perhaps adduce positive facts on this head; for instance, the date might be known when a channel had been cut to float out a large canoe, and which had since grown up.

For the classification of coral reefs, the most important point to be attended to, is the inclination of the bed of the adjoining sea; and secondly, the depth of the interior lagoon in the case of atolls, and of the channel between the land and the reef, in Encircling or Barrier, and in

Fringing reefs. Whenever it is practicable, soundings ought to be taken at short ascertained distances, from *close* to the breakers in a straight line out to sea, so that a sectional outline might be protracted on paper. In those cases in which the bottom descends by a set of ledges or steps, their form ought to be particularly attended to; and whether they are covered with sand or by dead or living coral; and whether the corals differ on the different ledges: the same points should be attended to within the lagoon, wherever its bed or shore is step-formed: the origin of these steps or ledges is at present obscure. In the Indian and Pacific oceans there are entire reefs, having the outline of atolls or lagoon-islands lying several fathoms submerged; there are likewise defined portions of reefs both in atolls and in encircling reefs similarly submerged. It would be particularly desirable to ascertain what is the nature of these submerged surfaces, whether formed of sand or rock or living or dead corals. In some cases two or more atolls are united by a linear reef; the form of the bottom on each side of this connecting line ought to be examined. Where two atolls or reef-encircled islands stand very near each other, the depth between them might be attempted by deep soundings: the bottom has been struck between some of the Maldiva atolls. Generally the form and nature of the reefs encircling islands ought to be compared in every respect with the annular reefs forming atolls.

On the shores of every kind of reef, especially of atolls and of land encircled by barrier reefs, evidence of the slow sinking of the land should be particularly sought for;

for instance, by stumps of trees, the foundation-posts of sheds, by wells or graves or other works of art, now standing beneath the level of high-water mark, and which there was good reason to believe must have once stood above its level. The observer must bear in mind that cocoa-nut trees and mangroves will grow in salt-water. If such evidence be found, inquiry ought to be made whether earthquakes have been felt. On the other hand, all masses of coral standing so much above the level of the sea that they could not have been thrown up by the breakers during gales of wind, at a period when the reef had not grown so far out seaward, should be investigated and their height measured. There is reason to believe that some coral-reefs have been thought to have been upraised, owing to the effect of the lateral or horizontal extension of the reefs having been overlooked; for the necessary result of this outward growth is gradually to break the force of the waves, so that the rocks, now further removed from the outer breakers, become worn to a less height than formerly, and the more inland corals not being any longer constantly washed by the surf, cease to live at a level at which they once flourished. It is indispensable that specimens of all upraised corals, and especially of the shells generally associated with them, should be collected; for there can be no doubt that ancient strata containing corals, have in some instances been confounded with recent coral-rock. The importance of ascertaining whether coral-reefs have undergone, or are undergoing, any change of level, depends on the belief that all the characteristic differences between Atolls and Encircling reefs on the one hand, and

Fringing reefs on the other, depend on the effect produced on the upwardly-growing corals by the slow sinking or rising of their foundations.

A thick and widely-extended mass of upraised recent coral-rock has never yet been accurately examined, and a careful description of such a mass—especially if the area included a central depression, showing that it originally existed as an atoll—is a great desideratum. Of what nature is the coral-rock; is it regularly stratified or crossed by oblique layers; does it consist of consolidated fine detritus or of coarse fragments, or is it formed of upright corals embedded as they grew? Are many shells or the bones of fish and turtle included in the mass, and are the boring kinds still in their proper positions? The thickness of the entire mass and of the principal strata should be measured, and a large suite of specimens collected.

In conclusion, it may be re-urged that the young geologist must bear in mind, that to collect specimens is the least part of his labour. If he collect fossils, he cannot go wrong; if he be so fortunate as to find the bones of any of the higher animals, he will, in all probability, make an important discovery. Let him, however, remember that he will add greatly to the value of his fossils by labelling every single specimen, by never mingling those from two formations, and by describing the succession of the strata whence they are disinterred. But let his aim be higher: by making sectional diagrams as accurately as possible of every district which he visits (nor let him suppose that accuracy is a quality to be acquired at will), by collecting for his own use, and carefully examining numerous rock-specimens, and by acquiring the habit of patiently seeking the cause of everything which meets his eye, and by comparing it with all that he has himself seen or read of, he will, even if without any previous knowledge, in a short time infallibly become a good geologist, and as certainly will he enjoy the high satisfaction of contributing to the perfection of the history of this wonderful world.

SECTION VII.

ON OBSERVATION OF

EARTHQUAKE PHENOMENA.

BY R. MALLET, A.B., Mem. Ins. C.E., M.R.I.A.

WHENEVER a blow or pressure of any sort is suddenly applied, or the passive force of a previously steady or slowly variable pressure is suddenly either increased or diminished, as these affect material substances, all of which, whether solid, liquid, or gaseous, are more or less elastic, then a *pulse* or *wave* of force, originated by such an *impulse*, is transferred, through the materials acted on, in all directions from the *centre of impulse*, or in such directions as the limits of the materials permit. The transfer of such an *elastic wave* is merely the continuous forward movement, of a change in the relative positions, of the integrant molecules or particles, of a determinate volume, of the whole mass of material.

Ordinary sounds are waves of this sort in air. The shaking of the ground felt at the passage of a neighbouring railway-train is an instance of such waves in solid ground or rock. A sound heard by a person under water, or the shock felt in a boat lying near a blast exploded under water, are examples of an elastic wave in a liquid.

The velocity with which such a wave traverses, varies in different materials, and depends principally upon the degree of elasticity, and upon the density in any given one. This *transit period* is constant for the same material, and is irrespective of the amount or kind of original impulse: for example, in air its velocity is about 1140—in water about 4700—and in iron about 11,100 feet per second—all in round numbers.

Thus, if one stand upon a line of railway near the rail, and a heavy blow be delivered at a few hundred feet distant upon the iron rail, he will almost instantly hear the wave through the iron rail—directly after he will feel another wave through the ground on which he stands—and, lastly, he will again hear another wave through the air; and if there were a deep side-drain to the railway, a person immersed in the water would hear a wave of sound through it, the rate of transit of which would be different from any of the others—all these starting from the same point at the same moment.

The size of such a wave—that is, the volume of the displaced particles of the material in motion at once, depends upon the elastic limits of the given substance, and upon the amount or power of the original impulse. By the *elustic limit* in solids is meant the extent to which the particles may be relatively displaced without fracture or other permanent alteration: thus glass, although much more perfectly elastic than India rubber, has a much smaller elastic limit.

Nearly all such elastic waves as we can usually observe originate in impulses so comparatively small that we are only conscious of them by sounds or vibrations of various sorts, the advancing forms of whose waves are imperceptible to the eye; but when the originating impulse is very violent, and the mass of material suddenly acted on very great, as in an earthquake, the size of the wave becomes so great as to produce a perceptible undulation of the surface of the ground, often visible to the eye; and by whose transit bodies upon the earth are disturbed (chiefly through their own inertia), thrown down, &c.

There is every reason to believe that an earthquake is simply "*the transit of a wave of elastic compression in any direction, from vertically upwards to horizontally in any azimuth, through the surface and crust of the earth, from any centre of impulse or from more than one, and which may be attended with tidal and sound waves, dependent upon the impulse and upon circumstances of position as to sea and land.*" This truth has not yet been fully and experimentally demonstrated. It is of the highest importance to a wide region of science that observations should be made, enabling it to be so.

Observers in earthquake countries should make themselves familiar with the usual features, and with the succession of events, and concomitants which with a certain sort of regularity apply to all earthquakes. For this Sir C. Lyell's 'Geology,' in loco, will be sufficient. The greatest shocks are not the most instructive, except as to secondary effects; but every great shock is usually followed by several smaller—the first should therefore be viewed as a "notice to observe" the latter carefully.

The phenomena of every earthquake may be divided into—1st, *Primary*, or those which properly belong to the transit of the wave or waves through the solid or watery crust of the earth, the air, &c.; 2, *Secondary*, or the effects produced by this transit—and both must be kept distinct from co-existent forces, such as those of volcanic eruption, permanent elevation or depression of land, &c., which, however closely they may be connected with the originating impulse of the earthquake, form no true part of it—they merely complicate its phenomena.

The *centre of impulse*, or *origin* of earthquakes, is generally conceived to be a sudden volcanic outburst, or sudden upheaval or depression of a limited area, or sudden fracture of bent and strained strata. This origin should be carefully sought for, as to its nature and position.

An earthquake may have its *origin* either *inland or at sea;* and as this may be, a different set of phenomena will present themselves. In the former case we may expect, in the following order—1st, The *Great Earth-wave*, or *true shock*, a real roll or undulation of the surface travelling with immense velocity outwards in every direction from the centre of impulse. If this be at a small depth below the surface, the shock will be felt principally horizontally; but if the origin be profound, the shock which is propagated from it in every direction in spherical shells will be felt more or less vertically; and in this case also we *may* be able to notice two distinct waves, a greater and a less, following each other almost instantaneously; the first due to the originating normal wave, the second to the wave vibrating at right angles to it. If we can find the point of the surface vertically over the origin, and the direction of emergence of the shock at a distant point or at several, we can find the depth of the origin from the surface.

There are, therefore, certain lines at points in which the shock, in passing outwards from the origin, is simultaneously felt on the earth's surface. These may be called *Coseismal lines.*

An erroneous notion, of the dimensions of the great earth wave must not be formed from its being called an undulation—its *velocity* of translation appears to be frequently as much as thirty miles per minute, and the wave or shock moving at this rate often takes ten or twenty seconds to pass a given point; hence its *length* or *amplitude* is often several miles.

During the passage of the great earth wave or main undulation, a continuous violent tremor or short quick undulation (like a short chopping sea) is often felt. This arises from secondary elastic waves, upon the surface of the great earth wave, like the small curling or capillary waves on the surface of the ocean swell, but which within its mass are analogous to the dispersion of light. The physical cause of these is not here in point; but it is very desirable that the interval in time between these minor oscillations should be observed by a seconds watch.

2nd. When the roll or undulation of the earth wave, coming from inland, reaches the shores of the sea (unless these be precipitous, with deep water), it may lift the water of the sea up, and carry it along on its back, as it were, as it goes out into deep water; for the rate of transit of the shock is so immense that the elongated heap of water lifted up has not time to subside laterally. This may be called the *forced sea-wave;* its elevation will be comparatively small, and nearly the same as that of the earth-wave, when close to the shore on a sloping beach; and where the water is still, any observations that can be made as to the height of this fluid ridge will afford rude indications of the height of the earth-wave or shock.

Earthquakes, whether at sea or on land, seem to be only accompanied with subterranean noises when strata are fractured or masses of matter blown away at volcanic origins. Where such is not the case, the two preceding are the only waves to be expected from an earthquake of inland origin; but when fracture occurs, then at the moment of the shock, or very slightly before or after it, we shall hear, 3rd, the *Sound wave, through the earth;* and at an interval longer or shorter after this, 4th, the *Sound wave through the air.*

Again, when the origin of the earthquake is under the sea (and such seems to be the case with most great earthquakes), we may expect in the following order—1. The great earth-wave or shock; 2. The forced sea-wave, which is formed as soon as the true shock or undulation of the bottom of the sea gets into shallow water, and forces up a ridge of water directly above itself, which it brings in to shore, and which seems to be the cause of that slight disturbance of the margin of the sea often noticed as occurring at the moment of the shock being felt; 3. The sound-wave through the earth (as in the former case); 4. The sound-wave through the sea, which arrives after that through the earth, but prior to 5. The sound-wave through the air. Where the original impulse is not a single impulse, but a quick succession of these, or a single impulse extending along a considerable line of operation, passing away from the observer: the sound-waves will be rumbling noises, and may be confounded in each medium more or less: and where no fractures or explosions occur, the sound-waves may be wholly wanting.

Lastly, and usually a considerable time after the shock, the *great sea-wave* rolls in to land. This is a wave of translation: a heap of sea-water is thrown up at or over the origin of the earthquake by the disturbance of the sea-bottom, and begins to move off in waves like the circles on a pond into which a pebble is dropped; and its phenomena depend upon laws different from any of the other (elastic) waves of earthquakes.

The original altitude (above the plane of repose of the fluid) and volume of this wave depend upon the suddenness and extent of the originating disturbance, and upon the depth of water at its origin. Its velocity of translation on the surface of the sea varies with the depth of the sea at any given point, and its form and dimensions depend upon this also, as well as upon the sort of sea-room it has to move in. In deep-ocean water one of these waves may be so long and low as to pass under a ship without being observed; but as it approaches a sloping shore its advancing slope becomes steeper, and when the depth of water becomes less than the altitude of the wave, it topples over, and comes ashore as a great breaker. Sometimes, however, its volume, height, and velocity, are so great that it comes ashore bodily and breaks far inland. The direction from which it arrives, at any given point of land does not necessarily infer that in which the origin may be; as this wave may change its direction of motion greatly, or become broken up into several minor waves in passing over water varying much and suddenly in depth, or in following the lines of a highly-indented or island-girt shore.

Observations of each of these classes of waves which we have thus briefly described may be made either directly by the aid of instruments, specially provided or extemporaneously formed, or indirectly by proper notice of certain effects which they produce, on objects upon the earth's surface.

Direct observations by complete self-registering Seismometers do not come within our present scope. We, therefore, proceed to direct observations with extemporaneous instruments on the earth-wave or shock. The elements necessary to be recorded are such as will enable us to calculate—1. The direction in azimuth of the wave's motion; and (if it have an upward motion) also its direction of emergence at the points of observation. 2. Its velocity of transit. 3. Its form—*i. e.* its amplitude and altitude.

If a common barometer be moved a few inches up and down by the hand, the column of mercury will be found to oscillate up and down in the tube with the motions of the instrument and in opposite directions, the range of the mercury depending upon the velocity and range of motion of the whole instrument. A barometer fixed to the earth, therefore, if we could unceasingly watch it, would give the means of measuring the vertical element of the shock-wave; and if we could lay it down horizontally, it would do the same by the amplitude or horizontal element. This we cannot do; but the same principle may be put into use by having a few pounds of mercury, and some glass tubes bent into the form of L, sealed close at one end, and open at the other; the bore being about three-eighths of an inch in diameter, and each limb about eighteen inches long. We shall also require some common barometer tubes of the same calibre: the open end being turned up like an inverted syphon, and equal in bore to the rest of the tube, (See *Fig.* 4.) The **L** tubes are used for the horizontal, the others for the vertical elements.

To fit the **L** tubes for use, fill each partly with mercury, and so adjust it that a column of six inches in length shall be in each limb of each tube, when held as in *Fig.* 1;

Sealed end.

Fig. 1.

a b

the limb *a b* horizontal, and the vertical column being supported as in a barometer. Tie four of these tubes so prepared together, back to back, so that if one horizontal limb face the north, the others shall face east, south, and west respectively, as in *Fig.* 2. In this position secure them all down upon a broad stout board, that can be itself fixed to a surface of rock, or other flat fixed surface of the earth.

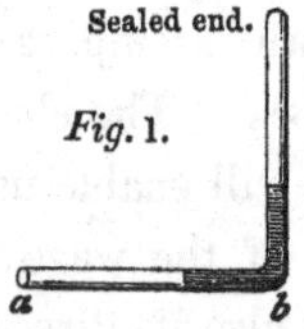

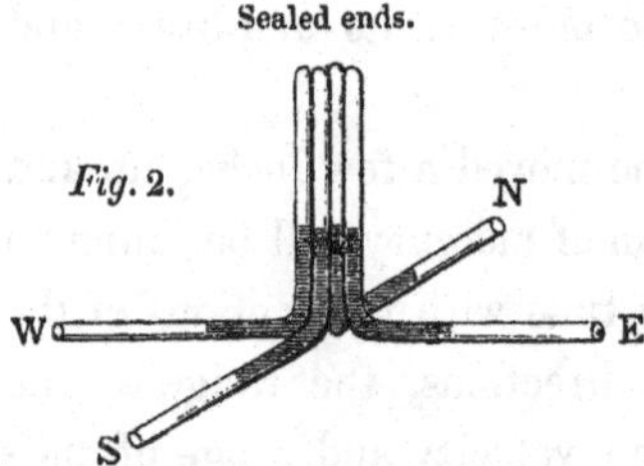

Fig. 2.

An index or marker must now be prepared for each tube; for one of these cut a common piece of card two inches long by rather more than five-eighths of an inch wide, and double it down the long way, so that the two segments shall stand at rather less than right angles to each other; cut a cylindrical slice of cork one-eighth of an inch thick, of five-sixteenths of an inch diameter, so that it will go easily into the tubes (these being all three-eighths of an inch in diameter): attach the bit of cork with glue or sealing-wax to the end of *one* wing or segment of the folded card, leaving the other free, and thrust the whole into the horizontal limb of the tube until the cork just touches the mercury, and so for the others. This marker is shown at full size (about) in

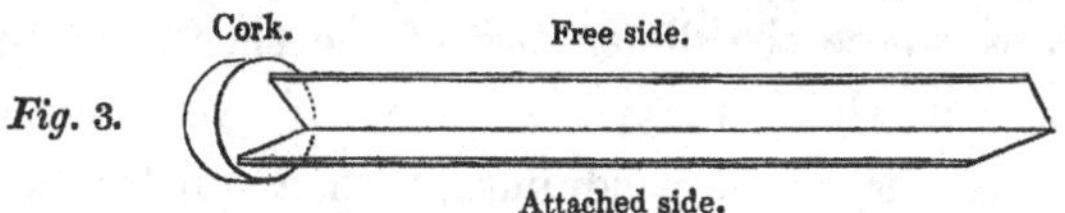

Fig. 3.

The edges of the card having a certain amount of elastic extension, must slightly grip the inside of the tube.

It will now be found that if horizontal motion be given to the system of four tubes—say, from south to north,—that the marker in the southern tube will be pushed southwards a certain space by the movement of the mercury, and will remain to point out the space when the mercury has returned to rest. If the motion be in some direction between two adjacent tubes—say, from south-east to north-west,—the markers in the south and east tubes will both show a certain motion, equal in this case, but in others with a certain ratio to each other, by which the direction between the cardinal points may be calculated.

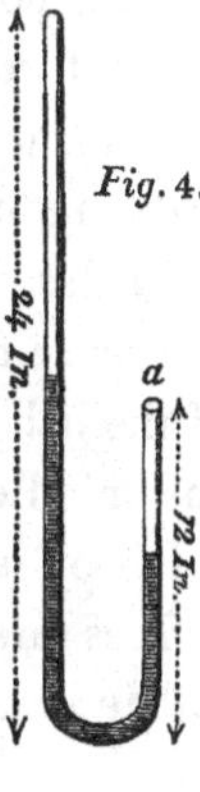

For the vertical element: let the barometer-tube *Fig.* 4, be filled with mercury, so that about six inches shall stand in the open end *a*, into which thrust a marker, as in *Fig.* 3, and about twelve inches in the sealed limb; place this vertically, and secure it to a fixed mass of rock, a heavy low building, or large tree; the amount to which the marker is found moved up in the tube will give the altitude of the wave; and it is obvious, that by the conjoint indications of

Fig. 4.
24 In.
a
12 In.

the four horizontal tube-markers and this vertical one, the direction of emergence of the wave is determinable.

These instruments are of the nature of fluid pendulums. They are much superior to common solid pendulums, for these uses, where the dimensions of the shocks are moderate; but where these are great and very violent, heavy solid suspended pendulums, with a quick time of vibration, will be found alone applicable: the seconds pendulum for lat. Greenwich will always be desirable. Where fluid pendulums are not attainable, a solid pendulum to answer some of the purposes may be thus prepared:—Fix a heavy ball, such as a four-pound shot, at one end of an elastic stick, whose direction passes through its centre of gravity: a stout rattan will do. Fix the stick vertically in a socket in a heavy block of wood or stone, and adjust the length above the block as near as may be to that of the seconds pendulum for Greenwich. Prepare a hoop of wood, or other convenient material, of about eight inches diameter; bore four smooth holes through the hoop in the plane of its circle, and at points ninety degrees distant from each other; adjust through each of these a smooth round rod of wood (an uncut pencil will do well), and make them, by greasing, &c., slide freely, but with slight friction through the holes.

Secure the hoop horizontally at the level of the centre of the ball by struts from the block, and the ball being in the middle of the hoop, slide in the four rods through the hoop until just in contact with the ball.

It is now obvious that a shock, causing the ball to oscillate in any direction, will move one or more of the rods through the holes in the hoop, and that they will remain to mark the amount of oscillation.

A similar apparatus, with the pendulum-rod secured horizontally (wedged into the face of a stout low wall, for example), will give the vertical element of the wave. Two of these should be arranged, one north and south: the other east and west.

It will be manifest that the observer must record minutely the dimensions and other conditions of such apparatus, where it is not permanently kept, to enable calculations of scientific value as to the wave to be made from his observations of the range of either fluid or solid pendulums.

A common bowl partly filled with a viscid fluid, such as molasses, which, on being thrown by oscillation up the side of the bowl, shall leave a trace of the outline of its surface, has been often proposed as a Seismometer. This method has many objections: it can only give a rude approximation to the direction of the horizontal element; but as it is easily used, should never be neglected as a check on other instruments. A common wooden tub, with the sides rubbed with dry chalk and then carefully half filled with water or dye stuff, would probably be the best modification.

Another extemporaneous instrument for measurement of vertical motion in the wave may be sometimes useful. Make a spiral spring of eighteen inches or so in length by twisting an iron wire of one-eighth of an inch diameter round a rod of about 1¼ inch diameter (the staff of a boarding-pike); suspend it by one end vertically from a fixed point, and fix a weight (a twelve-pound shot will do) to the lower end, and below and in a line passing vertically through the centre of gravity of the weight fix the stem of a common tobacco-pipe; let the lower end of this stem just dip into a deep cup filled with pretty thick common ink or other coloured fluid: the action of this needs no description.

The preceding instruments suffice at once to give the direction of transit of the earth-wave and its dimensions; its rate of progress or transit over the shaken country remains to be observed.

Several distant observers, with chronometers, will of course best observe this, but such observations cannot be very numerous or extend over a large tract of country; yet it is most desirable that a network of such observing points should be stretched over the shaken country. For this purpose common house-clocks, situated at several distant points, may be easily arranged, so that the pendulum shall be brought to rest and the clock stopped at the moment that the shock passes.

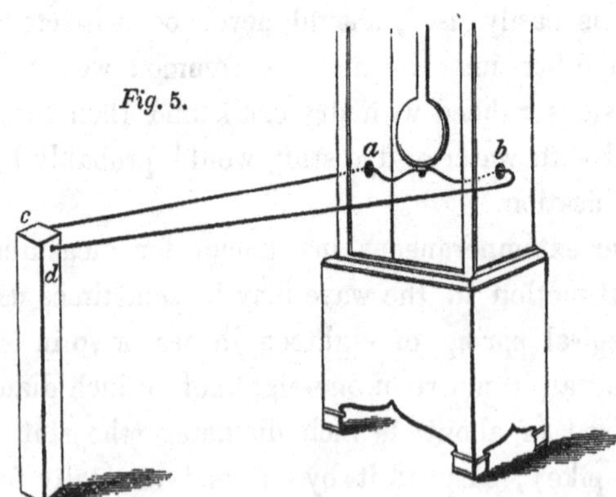

Fig. 5 shows part of the case and pendulum of a common clock. To fit it for this purpose bore two holes of a quarter of an inch diameter, one through either side of the clock-case, at *a b*, at the level of the lowest point of the pendulum-bob and in the plane of its vibration ; round off the edges of these holes, and grease them.

In the centre of a piece of fishing-line or stretched whip-cord, make a loop and pass it round the screw or other lower projection of the pendulum bob : pass the two free ends of the cord out, one through each of the holes in the sides of the clock-case ; provide a squared log of heavy wood of about five or six inches thick each way, and from four to five feet in height ; cut both ends off square, and stand the log upright on one end directly opposite the dial of the clock.

Measure off equal lengths of the cord at each side of the pendulum, and make fast their extremities to the two opposite sides of the upright log, *c d*, close to the top ; bring the log backwards from the clock now, until the pendulum being at rest, both cords are drawn tight ; and then advance the log two or three inches towards the clock, so that the cords may be slacked down into a fes-toon or bend at each side of the pendulum, and *within the clock-case*, so that the pendulum may have room to swing freely ; and *very slightly* wedge the cord to keep it so, through the holes in the clock-case, and from the out-side ; see that the log rests firmly and upright upon a firm floor ; and now set the clock a-going. The length of the cords, or the distance of the log from the clock in re-lation to its height, must be such that if it fall *towards* the clock it shall bring the cords up tight before the upper part of the log touches the ground. It is now obvious that in whatever direction the log may fall, it will arrest the motion of the pendulum and stop the clock within less than a second of the true time of transit of the wave at the spot.

If the adjustments are similar for all the clocks this error will be constant for them all; and if the true time be noted at the principal station it can be got for the rest.

Clocks with seconds pendulums only should be chosen for this use. They should be all set by one chronometer, and their errors afterwards taken.

Where convenient, the pendulums should be all placed to swing north and south, or east and west; and in this case the sides of the logs will face the cardinal points, and the directions of their fall (where not entangled) be a rude index of that of the wave. It will be also desirable to place a bowl of fluid to mark direction with each clock.

The positions chosen for the clocks must vary with circumstances, but they should, as far as possible, surround the principal station; their distances apart must be considerable, as the speed of the wave or shock is immense —probably five miles is the ordinary minimum, and thirty to fifty miles a convenient maximum. Such arrangements should be made as rapidly as possible after the first shock has given the expectation of others in succession.

When practicable, the following method of fitting common clocks may be adopted advantageously.—Let *a*, *Fig.* 6, be the pendulum-bob; fix a pin of stout wire into a hole in the centre of it, *b*, at right angles to the plane of vibration; cut two small mortices through the

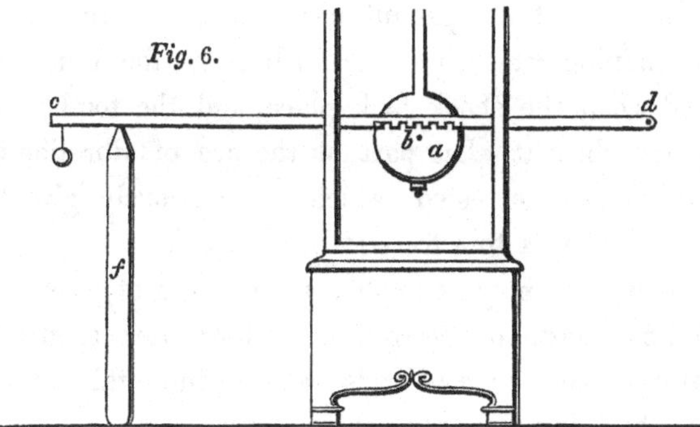

Fig. 6.

sides of the clock-case, so that a lath of deal or other light wood, of about an inch and a half wide by a quarter of an inch thick, may be passed through from *c* to *d*, just in front of the bob and clear of it.

Mark the length of the arc of vibration on the lower edge of the lath, and cut this length into nicks or teeth like a rack, of about three-eighths of an inch in depth and breadth each. Place the lower edge of the lath horizontally, and just above and clear of the pin *b*; secure the end of the lath *d* by a wire pin or stud, as a fixed point, so that the end *c* is free to move in an arc of a few inches up and down round *d* as a centre. Prepare a vertical log of wood *f*, of the size and form already described, but cut its upper end to a square pyramid, the flat surface at the top being reduced to about a quarter of an inch square; adjust the length and position of the log, so that it shall form a support for the end of the lath *c*, as in the figure.

It is obvious that the moment the log *f* is overthrown by a shock the lath will drop at the end *c* (which should be slightly weighted), and the teeth or rack nicks catching the pin *b* of the pendulum-bob will stop the clock; on examining which, the dial will show the time to a second when the shock took place, and the tooth in the rack will show at what part of the arc of vibration the pendulum was arrested, which will obviously give the time of the shock to a fraction of a second.

This method may be applied to any form of clock, and with any length of pendulum. Observation should be accurately made by a seconds watch of the total duration of the shock in passing the observer's station.

Returning now to observations to be made indirectly upon the earth-wave, or by its effects, consisting principally of—1. Observations on buildings and other objects thrown down; 2. On bodies projected, displaced, or inverted; 3. Bodies twisted on a vertical axis, with more or less displacement.

The observer must bear in mind that all these motions are due to the inertia of the bodies at the moment of the wave transit. The first tendency, therefore, of every body is to fall in a direction contrary to that of the wave's motion; but this is often perplexed by mutually-supporting bodies, as cross walls—by the direction of the wave being one, in which a fall is impossible, as when passing very diagonally through a long line of wall—by disintegration from the first wave, though so altering the conditions of the bodies (walls, towers, &c.) short of producing a fall, as that the dislocation and fall produced by a succeeding one is not contrary, but in the same direction as the wave motion. Long walls, in or nearly in the line of wave motion, are often split vertically, but not overthrown. When the shock emerges at a large angle to the horizon bodies are often projected, as stones out of or from the coping of walls: the size, weight, form, cement, sort of stone, distance thrown, and all other conditions of projection should then be carefully noticed. Bodies twisted on a vertical axis (such as the Calabrian obelisks, see Lyell, 'Geology') were formerly supposed due to a vorticose motion of the earth. This motion arises from the centre of gravity of the body lying to one side of a vertical plane in the line of shock, which passes through that point in the base on which the body rests, in which the whole adherence, by friction or cement of the body to its support, may be supposed to unite, and which may be called the *centre of adhesion.* The observer who fully masters these mechanical conditions of motion will see what elements he must collect, so that the motion impressed on bodies thus twisted may be used to calculate the velocity, &c. of the wave. All observations of this class, to be of scientific value, must comprise the materials, size, form, weight, sort of cement, base or foundation of the bodies disturbed, and measurements of the amount, &c. of disturbance, with any other special conditions which occur; and these will always be very numerous, and demand the utmost alertness and scrutiny of the observer.

Amongst the doubtful phenomena on record of this class are *inversions* of bodies, such as pavements turned upside down (see Lyell, 'Calabrian Earthquake'): any such cases deserve special attention.

In traversing an extensive city, or thickly built-over country, to observe the shattered buildings—having first known the general line of motion of the wave—the observer should remark if its direction of motion has appeared to change as it passed along, and obtain decisive evidence of its actual transit, for *sometimes the wave seems to emerge* all but simultaneously over a vast tract of country, where the origin is deep-seated, and nearly vertically below. Changes in the rate of transit horizontally, or in the force of the wave, should be noted by its effects on similar objects at distant spots. These changes may be expected at the lines of junction of different rocks or other formations. Evidence should also, if possible, be got of any breaking up of the primary wave into secondary waves, as of several shocks being felt where only one has occurred further back.

All evidence should, as far as possible, be *circumstantial.* Nature rightly questioned never lies; but men are prone to exaggerate, at the very least, where novel and startling events are in question.

Various *local conditions* must be recorded:—the nature of the geological formations at the spot, not merely the underlying rock, with the directions of its bedding, lamination, joints, &c.; but *the character* of surface, the depth and description, of its loose materials, their variations and extent, the geological formations of the surrounding districts from whence and towards which the earth-wave travels especially. The deeper a knowledge can be got by exposed sections, &c. of the rocks of the shaken district the better, the proximity or otherwise to volcanic vents, active or passive, the character of surface of the country shaken, mountain or plain, even or broken, solid or fissured; if the latter, their general directions, dip, &c., whether dry or flooded, and the effects on the elements of the wave, of changes in any or all of these conditions. Places least and most affected by the shock, and those free from any, and their local conditions, to be particularly noted.

Referring now to *secondary phenomena*, or effects resulting from the transit of the earth-wave (other than merely measures of it), we should observe falls of rock, or land-slips, to which most of the conditions of shattered buildings apply. Land-slips change their directions frequently, in consequence of moving over curved or twisted surfaces of rock: thus the previously straight furrows of a field may be found twisted after an earthquake. Scratches or furrows engraven on rocky surfaces by such land-slips should be looked for.

Sometimes great sea-waves are produced by the fall into the sea of rock or land-slips, which need to be carefully distinguished from the true great sea-wave produced by an original impulse of the sea-bottom. Land-slips often dam rivers, fill up lakes; and various changes of surface again produce basins for new lakes, to be filled by the changed river-courses. The circumstances, as far as possible, should be accurately observed, and the chain of events unwound, and all such phenomena cautiously separated from actual ejections of water (temperature to be ascertained), which are said sometimes to have happened on an immense scale.

Fissures containing water often spout it up at the moment of shock. Wells alter their water-level, and sometimes the nature of their contents. The directions of the fissures and any changes in the temperature of wells should be noted. Ejections from holes or fissures of strange liquid or solid matters, sometimes of dry ashes or dust, are recorded, and occasionally fiery eructations have occurred, especially near volcanic centres, and blasts of steam-vapours, or gases, whose chemical characters should be in all the above cases observed, as far as possible. The dust of overthrown buildings must not be confounded with these. Fissures, often of profound depth, open and remain so, or close again: their directions, dimensions, time and order of production, and closing up, and the formations in which they occur, to be noted; bodies engulphed to be detailed, as future organic remains. Fissures most probably arise from the range of displacement of molecules by the passing wave, exceeding the elastic limit of the materials disturbed. Permanent elevations and depressions of the land usually accompany earthquakes, and are of much importance to science. The modes of observing these rather belong to geology proper. The half-tide level must in all cases be taken as the datum-line, and opportunities along beaches, quays, wharfs, or inland, along mill-streams or irrigating channels, &c., diligently sought for, where evidence may be trustworthily collected of changes of depth of water. Occasionally local but widely-extended permanent elevations or depressions accompany earthquakes, which seem to result from lateral compression, and not from direct elevatory forces. These should be distinguished from the preceding.

Rivers are stated to have sometimes run dry during earthquakes, and again begun to flow after the shock. This is presumed to arise either from the transit of an earth-wave along their courses up stream, thus damming off their sources, or from sudden elevation of the land, and as sudden depression. This demands careful observation.

Observations of the *forced sea-wave*, whether produced by the earth-wave going out to sea or coming in from it, will be nearly the same. It is desirable to find its height above the surface of repose referred to half-tide level, and its length or amplitude; but from the extreme rapidity of its production and cessation, or conversion into small oscillatory waves lapping on the beach, and its generally small altitude, observations are extremely difficult—they are only possible when the surface of the sea is perfectly calm, and then must be left to the skill of the observer in taking advantage of local circumstances, and of evidence as to the visible circumstances of this wave, which occurs at the instant the shock is felt.

Observations of the *waves of sound* through the earth, the sea, or fresh water and the air, are indicated pretty fully by the description of these waves already given. The sound-wave through the earth travels at the same rate as the shock, or earth-wave; it is in fact *the shock (or its fractures) heard.* Notice if the sound is heard before, along with, or after the shock is felt. An observer, putting one ear in close contact with the earth, and closing the other, will hear the sound-wave through the earth separate from that through the air. So also an observer immersed in the sea will hear the sound-wave through it sometimes without any complication of that through the earth.

The character and loudness of the sound through each medium, and the places in an extensive district where each was heard loudest and faintest, should be noted.

The duration of the sound from first to last, through either medium, accompanying each shock, is important.

Observations on the *great sea-wave* should embrace, for each wave, its height, its amplitude or length, its velocity and direction of translation. The height to be taken above the plane of repose of the fluid, and referred to half-tide level. These waves, when on their grandest scale, defy any methods of admeasurement (unless by soundings where they break before coming to the beach), but chance observations of their results, such as the height to which they have reached on mural faces of rock, or on such buildings, &c. as may withstand them, or eye-sight observations made at the moment of transit of the crest of the wave, cutting distant objects; but when of a manageable size, the height of the crest may be pretty closely obtained by the traces on wharfs, buildings, &c., or on posts, or piles driven into the littoral bottom. This may be taken from any convenient fixed points of level, and all ultimately referred to half-tide level as the datum for all earthquake observations as to level.

The sextant may be occasionally used to get the elevation of the crest of the passing wave, several observers making a simultaneous observation of an expected wave. The velocity of the wave may be got by noticing from a suitable position, by a seconds watch, the time of its transit inwards between two distant points which should have an interval of water whose depth is or may be known. Islands off the land are advantageous posts for this purpose.

The length of the wave (while entire) should be sought

for by a similar method; a knowledge of its length and of the depth of water infers its height.

The point approaching the coast at which the wave is first observed to break, when capable of being accurately found, gives the height of the wave, which is here equal to the depth of the soundings. This breaking point and depth should always be anxiously tried for. Besides the dimensions of the wave, observations should be made on the interval of time after the great earth-wave, or shock, before the great sea-wave comes in, reckoning from the *commencement* of the shock; how many such waves in succession; what is the period between each; what are their relative dimensions; what changes are observable in the directions whence they come at the same point of coast, and what are the several in-coming directions at various points along a great stretch of coast (the latter must be had usually from collected testimony); what reflux from the beach before or after the coming in of the wave; after the wave has come in and broken, what oscillatory waves are produced, their character and dimensions; is the level of the surface of the sea in repose, the same before and after the subsidence of the great sea-wave and its secondary or oscillatory waves; what the state of tide above or below half-tide level at the moment of shock and of great sea-wave.

As accurate a section as possible of the form of the littoral bottom, beach, offing, and out to deep water, should be got by soundings in the line of the coming in of the wave, and laid down on paper; and, where possible, a cruise should be made out to sea in the direction whence the waves came, to look for pumice, dead fish, volcanic ashes, or other indications of the distant origin or centre of disturbance.

The *secondary effects* of the great sea-wave most worthy of remark are the materials, if any, carried in from deep sea, such as loose mineral matter, new animal or vegetable forms. Unless the water is very deep close to land this is unlikely to occur, as the range of transferring power of a great sea-wave (wave of translation) is only equal to the length of the wave itself.

If fish or testacea are thrown inland into fresh water, the effects on them should be noticed.

In recording the transporting power of the wave (*i. e.* its absolute transferring power, without reference to distance), the size, form, specific gravity, and lithological character of rocks or boulders moved, the distance moved, and height lifted are to be given. The base on which moved, if rock, the scratches or furrows produced. The mode of motion; if swept or rolled along. Obstacles overcome in their progress. Where gravel or loose materials are moved, an estimate of the mass moved, and to what distance; the character, external and internal, of its deposition; the mutual relations of its fine and coarse parts. The effects on buildings variously exposed; on vertical and sloping sea-walls; on steep faces of cliffs. The *denuding* effects of the wave in sweeping off sand, gravel, trees, animals, &c. The disruption and abrasion of stratified rocks, especially of nearly level and nearly vertical beds. Effects of vertical sea-walls or cliffs in the reflection or extinction of the wave.

Specimens should be taken of the rock of which very remarkable boulders or architectural fragments moved by the wave consist; of any new or strange matters cast up from the sea, or ejected from fissures, cavities, wells, &c. on land; of mineralized or fouled water suddenly made so; of gases evolved from fissures. Where possible, immediate chemical qualitative examination should be made.

Specimens in particular should be brought home of the rocks or other mineral masses through which the speed of transit of the earth-wave has been carefully observed, such as will enable the *modulus of elasticity of the mass* to be determined. Where this is rock, three specimens should be taken of maximum, minimum, and average hardness, density, and compactness, as representatives of the whole shaken district, noticing especially in stratified rock the depth from surface of ground and from top of the formation at which taken: each specimen to be of a size enabling a block to be sawn out of it of at least three feet in length by four inches square. Where convenient, this operation may be done on the spot. Where the district is a deep detrital or alluvial one, the depth and characters of the loose materials should be carefully observed, and illustrative specimens, as far as possible, brought home.

Collateral conditions to be observed are—barometer before, during, and after the earthquake; thermometer, rain-gauge; hydrometer and electrical state of the air during the phenomena; magnetometrical observations where these are practicable; all unusual meteorological appearances to be noted; the state of tide at time of each shock; the rise and fall of tide at the place; and any tidal anomalies produced after the earthquake. Active volcanic phenomena occurring before, during, or after the earthquake, in adjacent or distant regions, to be recorded.

Records or trustworthy traditions are to be sought for, in volcanic countries or those neighbouring to them, as to the state of activity or repose of these vents for a long period prior to and during the earthquake; also as to their state before and during any previous earthquakes—all remarkable facts as to which should be collected. Where meteorological or tidal tables exist, they should be transcribed for the times, correlative to the above records.

Any changes of permanent level of sea and land that accompanied former earthquakes that are on record should be obtained, with their particulars; whether the same points have been affected in successive earthquakes, and by successive upheavals; whether the same or different volcanoes were in action during successive earthquakes.

Maps may be advantageously made of the lines of direction through the country in which the shock was simultaneously felt; *coseismal lines:* also of the incoming directions or *cotidal lines* of the great sea-waves on a long coast-line, showing origin where possible. Maps of fissures formed in relation to the coseismal lines, and generally sketches of all visible remarkable effects of the earthquake on natural or artificial objects. The effects of earthquakes on the lives of men and animals; statistics of mortality; modes of entombment by the convulsion, as bearing on future organic remains; burying of objects of human art; production of presumed epidemics or pestilences, &c. are all worthy of notice.

Observers will do well to consult at least the following

works, to enable them the better to grasp beforehand the subject of earthquakes:—Lyell's 'Geology,' *passim;* Dolomieu's 'Account of the Great Earthquake in Calabria,' neglecting his theoretic views; Herschel's Art. 'Sound,' Airy's Art. 'Tides' ('Encyclopædia Metropolitana'); Russel's 'Report on Waves,' Brit. Asso., 1844; Mallet's 'Dynamics of Earthquakes,' Trans. Royal Irish Acad., 1846; Hopkins's 'Report,' Trans. Brit. Ass., 1847-48, and the several narratives of earthquakes.

It sometimes happens that a shock of *earthquake is felt at sea,* at great distances from land, and over profound depths; a sudden blow is felt, as if the ship had struck a rock.

The earth-wave *emerging* from an origin closely beneath, is here transferred to the ocean, through which it passes upwards as an elastic wave, with the same speed as the sound-wave through the sea. When such an event occurs, and circumstances are favourable, we should look out for the passage almost immediately in form of a single, low swell, of the great sea-wave, which may be formed directly over the origin, at no very great distance off. A cruise about and soundings may be made in search of indications (pumice, &c.) of the origin, the direction of which will be indicated by the degree of lateral stroke felt at the moment of the shock and by the direction of the great sea-wave, if any.

Section VIII.

MINERALOGY.

By SIR HENRY DE LA BECHE, C.B., F.R.S., &c.

A glance at the best treatises on mineralogy, even those wherein the matter is most condensed, is sufficient to show that a profound acquaintance with this science can only be acquired by long-continued study, and by means of a competent knowledge of certain other sciences, the aid of which must be obtained properly to comprehend the internal and external structure and chemical composition of minerals. The naval man may nevertheless accomplish much, more especially respecting the mode of occurrence and probable origin of minerals under certain conditions, and he may also add by his researches to the catalogue of known substances of this class.

When we see a diamond, we consider that we have before us an arrangement of the particles of carbon in the most perfect manner, that is, these particles have been enabled freely to adjust themselves, so that they have finally been aggregated in a definite form. So also a ruby or a sapphire presents us with the particles of alumina (with usually some slight admixture of other substances, such as oxide of iron, silica, &c.) arranged in a definite and most perfect manner, the conditions having been such that they also could freely adjust themselves;

with this difference, however, in the case of alumina, that it is not one of the simple substances which chemists consider carbon to be, but a compound of a metal (aluminium) and of oxygen. The ruby and sapphire are well-known transparent minerals, but it is not necessary that the particles of even a simple substance should be arranged in what, in common language, may be termed a perfect manner to make the mineral transparent. We may take as familiar examples of the contrary crystals of gold, silver, and copper.

As the knowledge of mineralogy advanced, it was discovered that there existed an intimate connexion between the chemical composition and physical structure of minerals when their constituent substances could form those arrangements of their particles known to us as crystals. This led to the view that when minerals possess the same chemical composition, they also always present the same crystalline system.

This is now known not to be strictly true. The same bodies have been found to occur under two different and incompatible forms, and to this the term *dimorphism* has been applied. Certain substances have also been discovered to replace others, without altering the form of a mineral, and to this the name *isomorphism* has been given. The known *dimorphous* bodies are very few, not more than about 10 in 350 crystallized minerals. The substances which are *isomorphous* being ascertained, no very great difficulty is experienced on this head. M. Dufrénoy has well remarked that "it is not necessary, in order to present the same composition, that minerals should exactly contain the same weight of their simple constituent substances; it is sufficient that there is an exact relation between the bases and the acids they contain, or between their isomorphous substances."*

The external geometric forms of minerals were, as far back as 1784, discovered by Bergman and Haüy not always to be those which might be considered fundamental or primary, since many were found capable of being split or divided into other forms, representing the solid arising from the free adjustment of the component particles, the body of the crystal and external form being made up of an aggregation of many primary crystals, or of some modification of a primary crystal. Of the aggregation of primary crystals the common mineral, calcareous spar, affords a familiar example. The fundamental crystal of carbonate of lime is a given rhombohedron, yet the external forms of this mineral are very varied, so much so that the Comte de Bournon was enabled to describe nearly 800 modifications of them. The primary form alone, as external, is much more rare.

The crystallization of carbonate of lime also well illustrates *dimorphism.* The common kind, as we have seen, has a fundamental crystal of a given rhombohedral form, but there is another kind, named arragonite, wherein,

* Dufrenoy, 'Traité de Minéralogie,' tome i. p. 19. This is a most excellent work, and should be in the hands of those who desire an extended knowledge of mineralogy. We may also mention Phillips's 'Elementary Treatise on Mineralogy,' an English edition by Allan, and an American edition by Alger; Dana's 'System of Mineralogy' (New York and London); Beudant's 'Traité Elémentaire de Minéralogie;' Rammelsberg's 'Handwörterbuch des Chemischen Theils des Mineralogie;' Blum's 'Die Pseudomorpheu des Mineralreichs;' Rose's 'Elements der Krystallographie;' Dr. Karsten's 'System der Mineralogie' (Berlin); and others.

though the proportions of carbonic acid and lime are precisely the same, the crystals are hexagonal prisms. At one time this very different crystalline structure was attributed to the presence of a small per centage of carbonate of strontia (from 0·7 to 4·1); but as arragonite has been found solely composed of carbonate of lime, this opinion seems abandoned. It has been lately stated that when carbonate of lime is crystallized from a warm solution it takes the form of arragonite, and when from a cold one that of common calcareous spar. It should be added, that the packing of the particles of arragonite is such that the specific gravity of this mineral is greater than that of common carbonate of lime.

At one time, though crystals of definite forms, constant internal structure, and chemical composition, allowing for isomorphous substitutions, were being obtained in a multitude of chemical processes carried on upon the great scale, as well as in the laboratory, much stress was laid upon distinctions between artificial and natural products. Now, however, that bodies, once only discovered in various positions among rocks, have been formed artificially, sometimes by accident, at others by design, there appears a disposition to look at inorganic matter more generally, however convenient it may be to describe those bodies by themselves which have been found in some natural position.

Among the researches which have tended to break down the barriers once thought to exist between natural and artificial minerals, the recent labours of M. Ebelmen may be mentioned as the most remarkable, since among the minerals produced are some commonly considered as insoluble by our processes and infusible in our furnaces, and some of them moreover belong to the class of gems. M. Ebelmen inferred that, inasmuch as many substances in solution in water crystallize when the water is evaporated, he would obtain certain minerals if he dissolved their elementary substances in some body (in a state of igneous fusion) capable of so doing, and which at a still higher temperature would evaporate and leave the elementary substances to adjust themselves in a crystalline form. Most perfect success attended the labours founded on this view, and in this manner he obtained crystals identical with rubies, spinels of various colours, chrysoberyl, chrysolite, and others. Crystals of emerald were also formed from pounded emerald. The crystals of chrysoberyl were of sufficient size to have their optical properties tried, and were found identical with those of the natural mineral.

To classify the natural substances described under the head of mineralogy, very various methods have been adopted, chiefly, however, divisible into those based upon their external characters or chemical composition. The following is that adopted by M. Dufrénoy in 1845, founded on chemical composition.

First Class.—*Simple substances, each being one of the essential principles of compound minerals.*

Electro-negative bodies; never acting as a base with the bodies of other classes, and always forming a constituent part of binary compounds.

Genus.	Genus.	Genus.
I. Oxygen.	X. Silicium.	XIX. Tellurium.
II. Hydrogen.	XI. Titanium.	XX. Mercury.
III. Nitrogen.	XII. Columbium.	XXI. Molybdenum.
IV. Chlorine.	XIII. Sulphur.	XXII. Tungsten.
V. Bromine.	XIV. Selenium.	XXIII. Chromium.
VI. Iodine.	XV. Arsenic.	XXIV. Osmium.
VII. Fluorine.	XVI. Phosphorus.	XXV. Rhodium.
VIII. Carbon.	XVII. Vanadium.	
IX. Boron.	XVIII. Antimony.	

Second Class.—*Alkaline Salts.*

The different salts composing this class are soluble in water, and possess a marked taste.

Genus.	Genus.	Genus.
XXVI. Ammonia.	XXVII. Potash.	XXVIII. Soda.

Third Class.—*Alkaline Earths and Earths.*

The substances composing this class have a stony aspect; pure, they are without colour or of a milky white; they are not generally hard. With the exception of corundum, none scratch glass; their specific gravity is between 2·7 and 4·6; tungstate of lime alone forms an exception to this general rule.

Genus.	Genus.	Genus.
XXIX. Baryta.	XXXI. Lime.	XXXIII. Yttria.
XXX. Strontia.	XXXII. Magnesia.	XXXIV. Alumina.

Fourth Class.—*Metals.*

This class comprises two divisions, each distinct in aspect:—

1. Native metals, and the combination of many metals with each other in a metallic state.
2. Combinations of metals with oxygen or with acids.

The minerals of the first division have generally a metallic lustre, which gives them a remarkable external character, distinguishing them from other minerals.

The combinations of the metals with oxygen or with acids rarely present this lustre; in this respect they range among the minerals of the class silicates. They nevertheless, for the most part, possess a peculiar colour, guiding us in their study; their specific gravity is generally high, and almost all upon assay immediately give a regulus or metallic scoria.

Genus.	Genus.	Genus.
XXXV. Cerium.	XLI. Cadmium.	XLVII. Silver.
XXXVI. Manganese.	XLII. Lead.	XLVIII. Gold.
XXXVII. Iron.	XLIII. Tin.	XLIX. Platinum.
XXXVIII. Cobalt.	XLIV. Bismuth.	L. Iridium.
XXXIX. Nickel.	XLV. Uranium.	LI. Palladium.
XL. Zinc.	XLVI. Copper.	

Fifth Class.—*Silicates.*

The minerals of this class have all a stony aspect, whence they were long known especially as *stones.* They form two distinct groups—the hydrous and the anhydrous silicates: the first are soft, and easily dissolve in acids; the second are hard; a portion with difficulty soluble in acids; the greater part insoluble in them.

The specific gravity of the silicates is between 2·5 and 3·6; a small number only approaching the latter limit.

Genus.	Genus.
LII. Aluminous silicates.	and their isomorphic substances.
LIII. Hydrated aluminous silicates.	LVII. Non-aluminous silicates.
LIV. Silicates of alumina, of lime, or their isomorphic substances.	LVIII. Silico-aluminates.
	LIX. Silico-fluates.
	LX. Silico-borates.
LV. Aluminous and alkaline silicates, and their isomorphic substances.	LXI. Silico-titanates.
	LXII. Silico-sulphurets.
	LXIII. Aluminates.
LVI. Aluminous hydrated silicates, with alkalies, lime,	LXIV. Substances of unknown composition.

Sixth Class.—*Combustibles.*

The minerals constituting this class for the most part still present traces of their organic origin; when crystallization has, as in *mellite*, effaced this essential character, we are reminded of it by the nature of the elements which enter into the composition of the mineral.

The combustibles of organic origin generally burn with flame at a moderate temperature, giving out a marked odour. They are soft; their specific gravity, generally very low, does not exceed 1·6.

They may be divided into the following:—

1. Resins. 2. Bitumens. 3. Fossil combustibles, comprising *anthracite, coal, lignite,* and *peat.*

Genus.	Genus.	Genus.
LXV. Resins.	LXVI. Bitumens.	LXVII. Fossil combustibles.

Under these 67 heads are now classed more than 500 minerals, supposed really to differ sufficiently to be so separated, independently of many merely considered as varieties, or accidental.

It will be obvious that a voyager, especially when his general time may be occupied with other duties, (only a portion of it applicable to mineralogy, and that irregularly,) cannot expect to make himself familiar with all these substances. With many of those more commonly found he will have little difficulty, and by practice he will readily detect them when presented to his attention. Those which form the constituents of rocks it is especially necessary to learn and distinguish, since so much of geological importance often turns upon their proper determination. Those which are referable to the useful class should engage his attention, since while, on the one hand, valuable ores of the useful metals and other important substances are often neglected (even in our mining districts unusual though valuable ores have been thrown away at no remote times), on the other, many a mineral, commercially worthless, is treasured up, often even to the neglect of those of high value, some particular brilliancy of appearance or fancied resemblance to precious or metallic substances having misled the collector.

However desirable it may be to consider inorganic matter as a whole, the conditions under which its parts have been found to combine either naturally or artificially being only regarded with reference to the general subject, so that the natural bodies, commonly termed minerals, merely constitute a portion of this whole, it is important that the voyager be enabled to distinguish natural minerals, both as respects science and its applications. The foregoing classification being founded on chemical composition, if he possessed no other means of distinguishing minerals from each other than chemistry afforded him, he would in many instances, from the want of the needful space and appliances on board ship, have the extent of his mineralogical labours greatly abridged. At the same time, with a box containing certain chemical substances, a slight apparatus, and a blowpipe, he will, after a little practice, find his power to distinguish minerals chemically far greater than he might at first anticipate.

Postponing for the present the assistance which these means of distinguishing minerals chemically, by the wet or dry methods, may afford, it may be convenient to refer to the mode of aggregation of the particles of minerals internally, as also to their external characters, without entering upon those refinements which require the application of a high order of investigation. For this reason we pass by those properties of minerals which are termed optical, beautiful and important as they are, and so valuable with reference to a knowledge of the arrangement of the component particles of minerals. So also with regard to their electrical properties and the effects of heat upon them. Should the voyager eventually find himself sufficiently interested in the study of minerals, he must consult works and memoirs dedicated especially to these researches.

With respect to the characters of minerals, they have been arranged under the following heads by M. Dufrénoy :—

1. *State of Aggregation.*—While minerals are commonly solid, some, like native mercury and certain bitumens, are liquid; so that they may be distinguished as *liquid, friable,* and *solid.*

2. *Colour.*—Colours are either constant or accidental: when the former, and connected with chemical composition, they are important; thus peroxide of iron is red, sulphuret of lead a peculiar blue-grey, and so on. Accidental colours are chiefly due to the mixtures of mineral substances. The peculiar appearance known as *chatoyant* depends upon the structure, and is referred to the cleavage-planes, the reflected light from which changes according to their position. Labradorite is a good example of this property.

3. *Form.*—This term is not intended to include the geometric form of a mineral, which is considered under the head of its crystallographic characters, but comprises only common, imitative, pseudo-morphous, and pseudo-regular forms. The first term is applied to the mode of occurrence of the mineral in mass, fragments, plates, or in an amorphous condition. The second to its occurrence in grains, nodules, &c. The third, when a mineral takes the form of a pre-existing body, whether organic or inorganic. The term *pseudo-regular* is applied to such arrangements of parts as are presented by basaltic columns and other prismatic forms of igneous rocks, apparently alse extending to the parallelopipeds arising from the intersection of the divisional planes, commonly termed the *joints* and *cleavage* of rocks.

4. *Lustre.*—Such as vitreous, waxy, silky, nacreous, adamantine, semi-metallic, and metallic.

5. *Transparency.*—Varying from diaphanous through semi-diaphanous, translucent, and translucent at the edges, to opaque. Rock crystal is diaphanous, chalcedony translucent—both different aggregations of the particles of silica.

6. *Fracture.*—This is distinguished as lamellar, granular, fibrous, radiated-fibrous, schistose, and compact.

7. *Hardness.*—This character is relative. The following is a scale of hardness proposed by Mohs, and somewhat commonly adopted :—

1. Lamellar Talc. 2. Selenite (crystallized sulphate of lime). 3. Iceland spar (carbonate of lime). 4. Fluor spar (fluate of

lime). 5. Phosphate of lime. 6. Lamellar felspar. 7. Rock crystal. 8. Topaz. 9. Ruby or Sapphire. 10. Diamond.

8. *Toughness.*—This character consists in the resistance which a substance offers to be broken or torn. A soft mineral may be very tough, such as sulphate of lime; a hard one readily fractured, as flint; and some are both hard and tough, as jade.

9. *The Scratch.*—Trials for hardness give a scratch and powder, which are useful in the determination of minerals. Thus the ores of iron, named hematites, give a red or yellow ochre powder, which at once distinguishes this mineral from the concretionary ores of manganese, the powder of which is black.

10. *The Stain.*—This character is only applicable to a few minerals, and those soft. It consists in marking paper or linen with the mineral—chalk and plumbago thus leave marks. Plumbago may be thus distinguished from sulphuret of molybdenum, which it otherwise much resembles.

11. *Unctuosity.*—Many minerals are soft and soapy to the touch, such as talc and serpentine, magnesian minerals.

12. *Flexibility.*—Several are flexible, such as native silver and copper. Some are both flexible and elastic, as mica.

13. *Ductility.*—Principally applicable to native metals. Though sulphuret of silver and halloysite cannot be lengthened under the hammer, they are nevertheless termed ductile by the mineralogist.

14. *Taste.*—Only applicable to certain substances, distinguished as bitter, sweet, salt, &c.

15. *Adhesion to the tongue.*—Generally sufficient for distinguishing argillaceous from pure limestones.

16. *Odour.*—Such as of the bitumens and other similar substances, or by means of breathing on or rubbing a mineral, when a peculiar smell is perceived.

17. *Cold.*—The feeling of cold when a mineral is placed in the hand. In this manner rock crystals and gems can be distinguished from glass and enamel, which otherwise may be made closely to imitate them.

18. *Sound.*—This property must be taken in its ordinary acceptation, and not with regard to the motion given to the molecules by percussion. Some substances are very sonorous; phonolite is so named from this property.

19. *Weight.*—This property also to be taken in its common acceptation, the mineral being only supposed to be weighed roughly in the hand. In this manner carbonate of lime, sulphate of baryta, and carbonate of lead may be easily distinguished.

These characters may be regarded as aids towards the approximate knowledge of a mineral, and as such may be useful, more especially to one who may in the first instance be desirous of availing himself, with as little loss of time as possible, of the differences or resemblances of any minerals he may have collected, endeavouring to refer them to known substances, so that while opportunities may continue to be afforded he might institute still further search for any minerals respecting which it may be desirable to collect additional information.

Supposing the voyager in possession of minerals collected on some excursion, he will find it useful to try their specific gravity when in harbour, or the weather may be sufficiently calm for the purpose. This is no difficult task, and the use of the needful apparatus for the purpose is soon acquired.* As the specific gravity of minerals,

* Much may be done with the common balance made for the purpose, whereby, after weighing the body in the air, it is weighed in the water, which should be distilled water, and enough can be got by dexterously condensing the steam coming out of a common tea-kettle.

For very exact determinations a flask of the kind herewith represented is used. *c d* is a ground-glass stopper, made so exactly to fit, that its bottom coincides with the line *a b*, so that the volume of the interior of the flask is constant. A capillary tube traverses the stopper through its length, so that when the flask is filled any excess of water escapes through this tube. The flask full of water is first weighed, and then the weight of the mineral is taken in the air. The mineral is then placed in the flask, so that it remains full, after the volume of water equal to that of the substance inserted has escaped. If now the flask, with the mineral in it, be weighed, the difference of weight gives that of the volume of water displaced, and by finding the relation of the weight of the mineral to that of the water, the specific gravity is obtained. If the substance tried

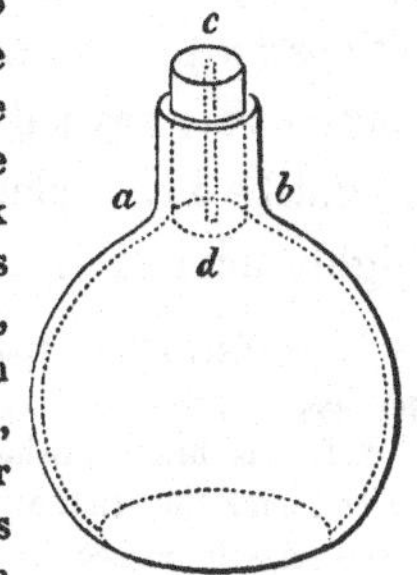

when these are pure, is considered to be intimately connected with their chemical composition, a knowledge of it becomes a useful preliminary inquiry. It may be almost unnecessary to remark that the specific gravity of a mineral is obtained by first weighing it in air, then in water, that the loss it sustains in water gives the weight of the displaced water, and that, by dividing the weight in air by the loss, or the weight of an equal bulk of water, the specific gravity is obtained. Thus if W be the weight of a substance in the air, W^1 its weight in water, $W-W^1$ will be that of the displaced volume of water, and $\frac{W}{W - W_1}$ the specific gravity.

It has been above stated that with the exceptions arising from *dimorphism* and *isomorphism*, the chemical composition and crystallization of mineral substances bear a marked relation to each other. Whatever the forms of the ultimate atoms of matter or those of the integrant molecules* of the substances of the minerals may be, very definite and constant fundamental polyhedral solids are generally found accompanying definite chemical compositions (with the exceptions mentioned), and the study of these forms has led to the science of *crystallography*. It would be out of place here to attempt to enter upon this highly interesting branch of knowledge; it will be found

be in powder, it is important that all air-bubbles be removed, and if porous, that the air in the pores be replaced by water.

* It has been considered that all crystals may be derived from a prism, and thus that all minerals might be composed of prismatic particles closely joined to each other without void spaces. On the other hand polyhedral forms have been thought most probable, void spaces being interposed between the molecules, and this the researches of M. Biot on lamellar polarization are considered to have proved.

sketched in most treatises on mineralogy, and works have been dedicated solely to it.

Experiments in the laboratory have pointed out, what the mode of occurrence of natural minerals would have led us to expect, that the multiplied modifications of some primary or fundamental form observed, much depend upon the conditions under which inorganic substances may have crystallized. So long ago as 1788 the experiments of Leblanc showed this, and the every-day experience in laboratories and chemical manufactories proves it. If by accident or design a solution of some given substance be added to another, the crystals of that which would be otherwise formed from the first solution become modified in shape: all the crystals so produced being generally similar. Thus also in nature, all collectors of minerals know that certain localities, in other words, given conditions arising from the combination of rocks and other circumstances where the minerals are found, produce crystals of some substance with a marked crystalline exterior, so that not unfrequently it is not difficult for a mineralogist, when differently modified crystals of the same substance are before him, to point out from whence each may have been obtained. Again, mineralogists are well aware that some mine, or in other words, some mineral vein, or part of a mineral vein, will afford a modification of a known mineral most abundantly for a time, and no similar modification be afterwards discovered in it. This is but the result of certain conditions, which have obtained in the particular cavity of the crack in the enclosing rocks, and which otherwise variously filled constitutes the vein.

Although thus liable to be modified by the conditions under which they may be formed, it is found that many substances will bear an admixture, sometimes considerable, of other substances, without having their power to crystallize in certain fundamental forms prevented, the particles of the one substance, as it were, compelling those of the other to adjust themselves in a subordinate manner to them. This is more especially the case where the admixture is clearly mechanical, as regards the substance so compelled to adjust its particles that they do not interfere with the fundamental or primary form of crystal of the more powerful body. Of this one of the most marked is the well-known crystallized sandstone, as it is often termed, of Fontainebleau, where grains of siliceous sand are gathered up by crystals of carbonate of lime, and do not injure the form of the latter. It is not improbable that the matter found in many minerals, and considered extraneous, may often thus be mechanically mixed. However this may be, the proportions in which one substance may be sometimes mingled with another, the crystallization obtained being that proper to the minor quantity (as if though less in that respect it was greater in crystalline power), are very remarkable. Thus M. Beudant was enabled to produce crystals of the form of sulphate of iron which contained 85 per cent. of sulphate of zinc, the remainder only (15 per cent.) being composed of the substance which gave the form to the crystals.*

With regard to *dimorphism*, or the crystallization of the same chemically composed substance in two distinct forms, considered fundamentally different, we have

* Beudant, 'Annales des Mines' (1817), tome ii. p. 10.

already remarked that carbonate of lime has this property, and that a difference in the heat of the solution whence it may be formed is found to accompany this variation of shape. This apparent influence of difference in temperature upon the arrangement of the component particles becomes the more interesting when we couple it with the experiments of Mitscherlich.* Right rhomboidal crystals of sulphate of nickel exposed in a vase to the sun, were found changed in the interior, without passing through the liquid state, into octahedrons with a square base, the exterior crust of the original crystals retaining its first form. As previously observed, the natural minerals hitherto ascertained to have two different primitive forms amount to so few as scarcely to interfere materially with the determination of minerals by their crystalline form.†

With *isomorphism*, or the replacement of one substance by another, so that the resulting crystals are not altered, as before stated, the voyager will not be embarrassed when he becomes familiar with such substances. Magnesia, lime, protoxide of iron, and protoxide of manganese thus replace each other in any proportion. With the crystals termed *pseudomorphous*, or those which do not belong to the chemical character of the substance thus presenting them, the case is different. There would appear two classes of these bodies; at least it is convenient

* Mitscherlich, 'Annales de Chimie et de Physique,' tome xxxvii.

† M. Dufrénoy points out that the two forms of carbonate of lime, carbonate of iron, and carbonate of lead are of the same nature, all occurring both as rhombohedrons and as right rhomboidal prisms, and adverts to the probability of the carbonates of baryta and of strontia being dimorphous also. 'Traité de Minéralogie,' tome i. p. 204.

in our present state of knowledge so to divide them. The first class shows the mere filling of a mould, left by the disappearance of one mineral, with the matter of another of dissimilar chemical composition, and which, if it could have crystallized out freely in a cavity, or amid yielding particles of matter, would also have had a dissimilar external form. The study of the minerals forming the substances which have filled up the cracks or dislocations in rocks, commonly known as mineral veins, or faults, as they may or may not contain ores of the useful metals, abundantly shows us that, after the formation in them of some crystallized minerals, and the envelopment of these by another mineral substance, a change of conditions took place in the crack or fissure, in such cases only partially filled with solid matter, so that the original crystals were removed. It is again obvious that while the cavities left by these crystals continued empty, or formed little hollows in which minerals of various kinds have merely covered the sides, partly filling these hollows, at others some given mineral has completely occupied the cavities, so that while internally keeping the structure peculiar to it, the exterior form corresponds exactly with that of the original and removed crystal. Further observation shows us that while in the greater number of instances the mineral substance forming the mould has remained firm, frequently covered over by other mineral substances, deposited from variable solutions in the crack or fissure until this be finally and completely filled with solid matter; sometimes the mould has been dissolved and removed, the pseudomorphous or externally false-shaped crystals appearing uncovered and by themselves. At least they so remain, if not again covered in consequence of any of those changes which may have taken place in the mineral vein, and by which new mineral matter may be thrown down, being of course liable, like any other of the pre-existing minerals, to be covered up by deposits of this kind. As might be expected, the minerals thus, as it were, cast in a mould, vary considerably; silica, often as chalcedony, being in some districts frequent.

When we find the hollows that have been left in rocks by the disappearance of the original substances of shells, and other organic remains which have been entombed in them, filled by various mineral substances, some totally different from the original matter of the organic remain, such as silica, and the sulphurets of iron, copper, and lead, we are not surprised at finding pseudomorphous crystals of minerals in the body of a rock itself. In spring waters, those clearly derived from rain percolating through rocks, and thrown out on the sides of hills by some bed called impervious, a term which should be only regarded as comparative, we find abundant evidence of the chemical solution of some parts of the beds which the water has passed through. In all kinds of springs, including those which in volcanic countries may be regarded as water formed by the condensation of steam, we still find the same thing, so that we are prepared for the filling up of hollows and cavities, no matter how formed, by matter brought in solution into them, and partly or wholly left there, according to circumstances. Of this kind of filling up, the vesicles and gas, vapour, or air-cavities of igneous rocks of different geological ages afford us excellent examples, more particularly when it is seen, as it often can be, that at one time the elements of some mineral substances have succeeded to others in percolating into, and being deposited in these hollows or cavities. If a crystal of any substance, such as felspar, disappears in the body of a rock, leaving its mould, any mineral substance entering the hollow may take its form, as is the case with the well-known pseudomorphous crystals of peroxide of tin from St. Agnes, Cornwall, the peroxide of tin having thus filled cavities left by felspar crystals. The other class of pseudomorphous minerals would appear to have been formed in a different manner, there being little reason to suppose that like the pseudomorphous minerals before mentioned they have merely filled up moulds left by the disappearance of the original minerals. On the contrary, the elements of the new mineral seem gradually, molecule by molecule, to have replaced the old mineral, so that the original form is always retained. At first sight, perhaps, this kind of replacement may be difficult to conceive, but when we learn that a plate of steel was found in part replaced by silver, having been left eight years in a case at the mint in Paris, in contact, by one of its ends, with a solution of nitrate of silver, which reached it slowly from a fissure in the vessel containing the solution, this difficulty vanishes. Now and then specimens are found wherein the parts of the original crystals still occur, the remainder replaced by another substance.

By a careful consideration of these exceptions to the agreement between the external forms of minerals with their composition, which are by no means so formidable when we regard the subject as a whole, the voyager may derive most important aid in the determination of the crystallized minerals he may collect from their crystalline forms.

To cleave minerals, so as to acquire a knowledge of any form made thereby apparent, thus exposing a structure aiding in the determination of the mineral, requires both dexterity and a fair knowledge of the probable planes of cleavage. While some minerals, like selenite, cleave most readily, in others it is only by a smart blow upon a chisel, placed in the supposed lines of cleavage, that this is effected. In others, again, cleavage is more to be traced by lines observed on the faces of the crystals than to be obtained mechanically. In some again the structure can only be traced by means of optical researches. It is found that in the same mineral the cleavages are always disposed in the same manner, forming constant angles with each other, and with the faces of the crystal. When there are three directions of the cleavage planes, the resulting solid always presents the same angles for the same kind of mineral. When the planes of cleavage are in more than three directions, one set is termed *principal*, the other *supplementary cleavages.*

The crystalline types, as they have been termed, those under which the different forms can be classified, have been variously treated by crystallographers. The following is the arrangement adopted by Dufrénoy, founded on that of Haüy :—

With perpendicular axes :—

I. *Cube.*—The modifications of which are the octahedron, the regular rhomboidal dodecahedron, the hexatetrahedron, the trapezoihedron, the octotriahedron, and some other forms.

II. *Right prism with a square base.*—The modifications of which are the octahedron with a square base, the dioctahedron, and others.

III. *Right rectangular prism.*—The modifications of which are the right rhomboidal prism, the rectangular octahedron, the rhomboidal octahedron, &c.

With oblique axes:—

IV. *Rhombohedron.*—Including equiaxial rhombohedrons, scalene triangular dodecahedrons, two regular prisms with six faces, and isosceles triangular dodecahedrons.

V. *Oblique rhomboidal prism.*—With its modifications.

VI. *Non-symmetrical oblique prism.*—With its modifications.

Instruments, named *goniometers*, have been invented to measure the angles made by the different planes of the crystals, as well those considered primary as their modifications. The most simple is that of Haüy, consisting of a graduated semicircle, with metallic adjustments, so that by applying these adjustments the angle sought is read off upon the graduated semicircle. This was found a somewhat rough process in many cases, and Wollaston invented his reflective goniometer, by means of which adjustments are applied so that the bright surfaces of the crystalline planes are made to coincide with them, and the angles are thus closely read off. This goniometer being inapplicable to crystals, the surfaces of which are not sufficiently bright, a goniometer has been constructed by M. Adelmann, by which both methods can be employed, as the cases for their use in preference may arise. Wollaston's goniometer is a very valuable instrument, and its use is not difficult to acquire.* A

* There are boxes containing the models of crystals, one of which (for they occupy little space, and could readily be put away in a berth or cabin) might be found very convenient for the voyager desirous of availing himself of the external characters of minerals for their determination. The number of models in these boxes varies, but most of them contain the crystalline types according to some system, their chief modifications not unfrequently having the names of some characteristic minerals upon them.

very simple goniometer has also been constructed by Dr. Leeson.*

It is probable that to chemical composition the voyager will chiefly look for aid, more especially if he be a medical officer, and therefore likely to have become sufficiently acquainted with chemistry for the purpose. The modes of investigation will readily present themselves to one so qualified,† and we would suggest that no surveying voyage should be sent, more particularly to distant countries, without one of those little chests of needful things for chemical research which are prepared for the purpose.‡ For those not sufficiently versed in chemistry it might be

* 'Memoirs of the Chemical Society,' vol. iii. p. 486.

† The following works will be found useful:—Will's 'Outlines of Qualitative Analysis,' Fresenius's 'Qualitative and Quantitative Analysis,' Parnell's 'Qualitative and Quantitative Analysis,' and Rose's 'Analysis,' translated by Normandy.

‡ Griffin (of Glasgow, and of Baker Street, London) and others fit up very compact and useful chests of this kind. They necessarily vary in price according to their contents. For about 8*l*., a chest of about 1½ cubic feet, not a cumbrous size for a cabin, may be obtained. It would contain apparatus and substances sufficient for discriminating all well-known ores and minerals, including a blowpipe apparatus with the necessary fluxes and reagents, as also a selection of the most useful instruments for testing in the wet way, with a collection of tests in the dry state, and stoppered bottles to contain solutions; also a set of bottles with pure acids.

More complete chests may be obtained for about 15*l*. or 16*l*.—far more valuable for long voyages, during which deficiencies cannot be expected to be supplied. These are divided into two chests, one containing the things needful for more constant, the other large articles for occasional use, as well as duplicates of apparatus liable to be broken, with an extra stock of chemicals. These chests usually occupy about 4 cubic feet, and contain apparatus and chemicals sufficient for the complete quantitative analyses of minerals, or the separation of the component parts of a mineral, in quantities sufficient for an accurate analysis. They include platinum crucibles, Bohemian test tubes, Berlin porcelain crucibles and capsules, complete blowpipe apparatus, &c. &c.

hazardous to attempt the wet method of investigation, but by a little practice, much knowledge may be acquired by the blowpipe, or what may be termed the dry method. Works are especially dedicated to this mode of investigation. As these may not be at hand, Mr. Warington Smyth, Mining Geologist of the Geological Survey of the United Kingdom, and who has long employed the blowpipe in his researches, has prepared the following short notice of the mode in which this instrument may be rendered useful :—

The ordinary blowpipe is so well known as scarcely to need description. Various forms have been recommended by their inventors, but for common purposes it is only important that the orifice be not too large, and that the tube be provided with a reservoir for the reception of the moisture which is carried into it with the breath. The flame of a neatly trimmed lamp is undoubtedly the most convenient, but that of a common candle is quite applicable to the qualitative tests with which we shall have occasion to deal.

In looking at the flame of a candle, we may observe two principal divisions, which it is necessary by the assistance of the blowpipe to use separately, since their action on the same substances is so different, as on the one hand greatly to facilitate certain processes of analysis, and on the other to cause much perplexity unless clearly understood.

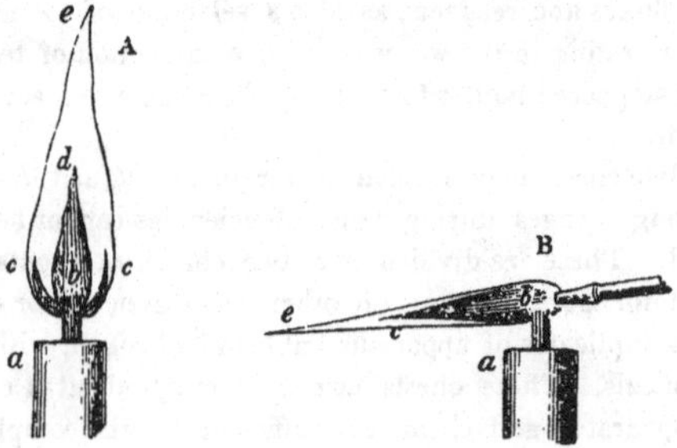

The outer and larger part of the flame *c, d, e*, which is the source of its light, is caused by the full combustion of the gases derived from the oil, wax, or tallow which rises into the wick, and is called the *reducing*

flame, because when concentrated upon the substance to be tested, it tends to abstract oxygen from it and thus to reduce it. In the lower part of the flame a narrow stripe of deep blue may be observed, *b, c*, which when acted on by the current of air from the blowpipe forms a cone, *b, c* (B). This is technically called the *oxidizing flame*, from its property of imparting oxygen to the substance upon which it is directed. To produce the latter, the point or jet of the blowpipe should be inserted into about a third of the flame, and the assay is then to be held at the extremity of the cone of blue flame. For reduction the point of the tube should scarcely penetrate the flame, and the assay should be so placed as to be completely enveloped in it, and thus prevented from receiving oxygen.

A little practice is sufficient to overcome the slight difficulty which at first is felt in keeping up a continual and even stream of air. The tyro may begin by accustoming himself to breathe through the nostrils whilst his cheeks are inflated, and will soon find it easy to maintain an uninterrupted supply for several minutes.

Of the instruments used in experimenting by the blowpipe, the following are the most necessary:—1st. A pair of fine-pointed forceps, tipped with platinum. 2nd. A small spoon of platinum. 3rd. An agate pestle and mortar. 4th. Thin platinum wire and holder. 5th. A magnet. 6th. A few small tubes of thin glass. 7th. Some small porcelain capsules or saucers. Charcoal is required as a support in many cases, particularly in the reduction of ores; and the following re-agents are also indispensable, the three first being fluxes applicable under different circumstances:—

1st. Soda, or carbonate of soda.

2nd. Borax, or borate of soda.

3rd. Microcosmic salt, or phosphate of soda and ammonia.

4th. Saltpetre, to increase the degree of oxidation of certain metallic oxides.

5th. Borax-glass, for the determination of phosphoric acid, and of small quantities of lead in copper.

6th. Nitrate of cobalt, in solution, to distinguish alumina, magnesia, and oxide of zinc.

7th. Oxide of copper, for determining small quantities of chlorine in compounds.

8th. Fluor-spar, for the recognition of lithia, boracic acid, and gypsum.

9th. Lead in a pure metallic state.

10th. Bone-ashes (9th and 10th are used for separating the silver from certain argentiferous ores).

11th, 12th, and 13th. Hydrochloric, sulphuric, and nitric acids.

14th. Litmus-paper, blue and red, for detecting the presence of acids and alkalies.

The experiments on an unknown mineral must be made systematically, and referred for comparison to some list or table of minerals in which their behaviour before the blowpipe is described, as Von Kobell's tables.*

The first point to examine is, whether it be fusible; and if so, in what degree. The various grades of fusibility may be conveniently divided into six; as representatives of which it is convenient to take the following minerals, species which are everywhere easy to obtain, and which may therefore be often practised upon:—

1. Antimony-glance, or sulphuret of antimony, which melts at the candle.
2. Natrolite, or mesotype, fine splinters of which may be rounded by the candle-flame.
3. Almandine, or precious garnet, which fuses in large pieces before the blowpipe.
4. Actinolite (hornblende), fusible only in smaller portions.
5. Orthoclase (felspar) offers some difficulty: and
6. Bronzite can only be rounded by the flame in the finest splinters.

According to this scale, the mineral in question may be referred to either of the above numbers, or placed half-way between any two of them; as for instance, apophyllite, being more easily fused than natrolite, and yet more refractory than antimony-glance, will have its comparative fusibility represented by 1·5.

The fragment to be experimented upon is generally held in the platinum forceps, but it is necessary to guard against the melting of the test upon the points, since the platinum, though infusible, is by that means rendered brittle.

In other cases the mineral may be supported upon charcoal; but whatever be the means of holding it, the phenomena exhibited by the action of the flame must be noted, as

1st. The manner in which it fuses, whether quietly, or with decrepitation, exfoliation, intumescence, or phosphorescence; whether it loses or retains colour and transparency.

2nd. The appearance of the product, whether a *glass*, an *enamel*, or a *slag;* or, as in the case of ores reduced upon charcoal, a metallic bead or *regulus*.

3rd. The separation of volatile substances, and the colour of the deposit on the charcoal, by which we may recognise

* Von Kobell, 'Tafeln zur Bestimmung der Mineralien, München,' and the same translated into English by R. Campbell.

a. Lead, giving a greenish yellow deposit.

b. Zinc, having a white crust, which when heated becomes yellowish and difficult to volatilize.

c. Antimony, a white deposit, easy to volatilize.

d. Bismuth, a crust partly white, partly orange-yellow, without colouring the flame.

e. Sulphur, with the well-known odour of sulphurous acid.

f. Selenium, in an open glass tube, gives a red deposit of selenium.

g. Tellurium, in a similar glass tube, gives a grayish-white crust of its oxide.

h. Arsenic gives off a grayish-white vapour, which smells like garlic.

i. Quicksilver, in a glass tube, will be precipitated in minute metallic globules.

k. Water, from hydrous minerals, deposited by condensation in the same manner.

4th. The colour of the flame when the tip of the blue part is neatly directed upon the mineral; whence may be distinguished

a. Red tint, given by several minerals containing strontia and (?) lithium.

b. Green, produced by some phosphates and borates, sulphate of baryta, some copper ores and tellurium ores.

c. Blue, given by chloride of copper, chloride of lead, &c.

5th. The development of magnetic properties after treatment in the reducing-flame, as in ores of iron, nickel, and cobalt.

So far the assay has been considered by itself, but it is frequently necessary to mix it with fluxes, either to render it fusible or to produce a glassy compound of a characteristic colour.

Thus if borax or microcosmic salt be fused into a glass at the end of a platinum wire bent into an eye, and a little powder of the unknown mineral be added to it, we shall obtain by the use of the oxidizing flame the following results:—

Manganese, in all its compounds, gives a beautiful violet or amethyst colour.

Cobalt causes a sapphire-blue colour; chromium an emerald-green.

Oxide of iron produces a yellowish-red glass, which becomes paler as it cools, and at length grows yellow or disappears.

Oxide of cerium gives a red or dark-yellow colour, which also grows paler as it cools.

Oxide of nickel renders the glass a brown or violet tint, which after cooling becomes reddish-brown.

Oxide of copper in very small quantity gives a green tint, which grows blue in cooling.

Oxide of uranium renders the glass bright-yellow, which in cooling takes a greenish tint.

Oxide of antimony gives a pale yellow colour, which soon disappears.

When soda is used as a flux, it is generally upon charcoal, and by this aid the metals may be obtained from most of their combinations in a pure state. For this purpose the powdered ore is either mixed with the moistened soda in a paste, or is enveloped in a piece of thin paper which has been dipped in a solution of soda. After fusion, that portion of the charcoal which has absorbed any of the fluid substance is to be cut off and ground down with it in the mortar, when the metal, if malleable, will at once be recognised. If several metals are combined, of which one is more easily oxidized than another, as for instance lead when combined with silver or copper, the latter may be separated by adding metallic lead or boracic acid, according to circumstances, and maintaining a continued oxidizing flame, till the whole of the lead has passed into the state of litharge.*

We will now suppose the voyager landing upon some coast, and desirous, among other things, either of adding to our knowledge of minerals or their localities, or of discovering ores of the useful metals or coal. With respect to many minerals and the ores of the metals, it fortunately so happens that precisely the same places may be searched, and these are cracks and fissures, or those dislocations of rocks known as faults, either partially or wholly filled with mineral matter. Should he see before him such veins as *a* and *b* traversing the rocks of a cliff, he should not neglect to land there. If any hollow spaces present themselves, let him there search for the crystalline minerals.

* By means of more complete apparatus and extended operations, the most exact assays may be undertaken with the blowpipe; and those who desire a further insight into the subject may consult Plattner's 'Art of Assaying by the Blowpipe;' Berzelius 'On the Blowpipe;' and the above-mentioned work by Von Kobell of Münich,—all of which are translated into English.

The vein *a* is represented as filling a fault, the dislocation having brought different rocks into contact; and we may

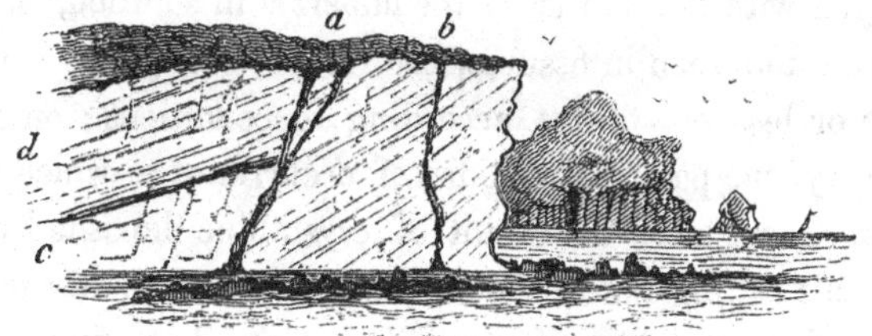

suppose, for illustration, that *c* is porphyry, and *d* some schistose rock. The fissure *b* is intended to be a mere crack. Often when dissimilar rocks are brought into contact, mineral substances are found in the fissures, and this is a point which the voyager should not neglect. In certain countries the occurrence of the ores of the useful metals is not unfrequent under such conditions. On tidal coasts, should a vein of this kind be found productive, it may be desirable to wait for low water to trace the direction of the vein among any ledges or rocks which may be then laid bare. This may give the run of the vein inland, but not with certainty; for though fissures or faults may take general lines on the large scale, they, as would be expected, are very irregular for minor distances.

Should crystals be found in any such vein, it is often desirable to ascertain how they occur relatively to other bodies, crystalline or otherwise. Whole groups of crystals are thus frequently seen placed on certain projecting surfaces, facing one direction, and this as well on surfaces of crystals of other substances, as upon the sides of the vein, or *walls*, as they are technically termed. Such modes of occurrence are found as well in what, in common terms, may be called a horizontal as a vertical manner. They are not due to the drippings of water charged with the matter of the minerals in solution, such as are often seen in fissures, the resulting deposits being more or less crystalline according to conditions; on the contrary, the particular modes of occurrence to which we allude, seem more the result of crystalline deposits from solutions (filling the whole or a large part of the fissure), so acted upon, that projections in a given line, more especially if composed of certain substances, received these deposits—in fact much in the same manner as substances in solution may be thrown down by well-known methods when galvanic action is employed. We have before us an illustrative specimen from the Consols Mine, Gwennap, Cornwall, in which large crystals of quartz are on the one side covered by crystals of sulphuret of iron, and on the other by crystals of copper pyrites. Cases where crystals of one substance abundantly occur on one side of prior-formed crystals of another substance, and not on the reverse or opposite sides, are sufficiently common, and best seen in the fissures or mineral veins themselves.

Although when exposed to the action of weather, the minerals which may be found in veins or fissures, open on the faces of cliffs, are not very often (except when of substances not easily injured) in a good state of preservation, they show that such minerals are found in the vein, so that if time and opportunity permit, some unexposed part of the vein may be broken into. Success may not, certainly, always attend such a search, for it is curious to observe how very local, even in the same vein, the occurrence of a particular mineral may be.

In collecting minerals in a vein, should a boat be at hand, so that they may be readily taken to the ship, it is better not to limit the specimen to some mere crystal itself, but to break off some of the body (either part of the vein or of the rock, as the case may be) upon which it has been formed, so that when more leisure may be obtained, any illustration the whole specimen may afford of the manner in which the mineral may have been formed, should be preserved. By such specimens we often learn the history, as it were, of the mineral accumulations which, taken together, may, wholly or in part, have filled up a fissure. In this way it may often be seen that crystalline coatings of many substances have successively covered each other up towards the centre of the fissures.

The contents of veins are far often from being definitely crystalline ; thus quartz and other mineral substances, such as the ores of many metals, have an amorphous appearance, their deposit having been effected under conditions which did not permit their particles to adjust themselves in definite crystalline forms. Again we find that, during the filling up of veins, fragments of rocks from the sides or upper parts of the fissure have dropped in ; by their want of contact and by their isolation in many parts of the vein showing that this happened when the mineral or minerals thrown down from solutions were accumulating. A mineral vein sometimes forms a complete breccia, and this as well from the cause just assigned as from the mere filling up of the chinks left by fragments from the adjoining rocks, accumulated in the fissure before any deposit of mineral matter from solutions was effected. As might be expected, both varieties are to be sometimes seen in the same vein. Ores of the useful metals, such as sulphuret of lead, copper pyrites, and peroxide of tin, may, and do, as well form the cementing matter of such fragments, as common quartz, carbonate of lime, or other minerals. In collecting some minerals which have covered others, it may be frequently desirable to obtain enough of the first to show how the latter may have occurred. Rock crystals are thus often seen investing other minerals, the most delicate threads of the latter being preserved in them. By a little care we may take out enough of the crystals to show completely how these threads may have radiated from a centre or have been otherwise disposed.

As with other mineral substances, we find that ores of the useful metals have been sometimes thrown down in a fissure at one time and not at another, the deposit of one ore sometimes repeated, at others not. Thus there may have been a coating of a zinc ore at one time, of copper ore at another, and a covering of tin ore upon these, sometimes separated by other mineral substances, at others in deposits one above the other. Again, we find, in the successive dislocations which are sometimes seen to have effected the lines of fissures, that while the lines of least resistance to the applied force have been chiefly through the contents of the original fissure, occasionally a new fissure has been made through portions of the adjoining rock; so that the minerals which may have been subsequently deposited in the new crack or fissure will be partly in the old line, and partly amid the newly-broken and adjacent rocks.

It would be out of place to attempt a general notice of

those veins which, because they contain the ores of the useful metals, are commonly termed *mineral:* it will be sufficient to observe that from decomposition the upper or exposed parts of many do not show the ores in the manner they occur beneath. Thus, above veins wherein the ore from which the largest amount of copper is produced, namely the compound of copper, iron, and sulphur known as *copper pyrites,* a mass of ferruginous matter is often found, known by many of our miners as *gossan,* and by the French miners as *chapeau de fer.* This is the result of a decomposition arising from exposure to atmospheric influences of various kinds, and occasionally from other influences. It is probable that the sulphur by a union with the needful oxygen became sulphuric acid, and that, this formed, the copper was attacked and removed, to be dealt with like any other solution of sulphate of copper. And beneath this gossan, or the *back of the lode,* as it is often termed, we observe appearances strongly reminding us of the common electrotype process for procuring copper from a similar solution. The pure metal is gathered together in chinks and cavities between the main mass of gossan and the body of the undecomposed copper pyrites, mingling, perhaps, occasionally with the lower part of the former. Sometimes this *native copper,* as it is called, may retain its metallic character, but at others it becomes converted into an oxide, and this again into a carbonate by the percolation of waters containing common air and carbonic acid. The iron seems in a great measure to have been left behind, and this forms the rusty substance above mentioned. It will be readily understood that, the needful conditions obtaining, other parts of a mineral vein than the mere upper portion may become decomposed in the same manner. In fact, the changes which have been effected in the fissures containing mineral veins, the mode of throwing down a mineral substance, its subsequent removal, its reappearance, or apparent transport elsewhere, the pseudomorphous filling up of crystalline cavities, the substitution of particles of one substance for another, the evident alterations produced by new fissures, particularly when these have traversed the original fissures at right angles, the differences of contents of fissures when they take different directions traversing the same country and association of rocks, are objects of high interest; and though no doubt best studied in mining countries, where opportunities are so numerous, and veins are so extensively laid open, a voyager, with some little time on a favourable portion of coast, may often nevertheless acquire much information on these heads. To do so, and procure illustrative specimens and a highly valuable collection, interesting in many respects, it is not necessary that the vein should be one containing the ores of the useful metals—the contents of those fissures and dislocations, termed common faults, are often in a scientific point of view equally important.

The cavities of many igneous rocks, and indeed holes and cavities in all, afford good places wherein to search for minerals. They are often found in such situations well crystallized and in good condition, from not having been exposed to destructive influences until the containing rock be broken—always, it being understood, at distances or depths from the surface where the atmospheric action may not have been much felt, or matters have entered the cavities that have decomposed or injured the mineral substances in them. This has sometimes happened not only to the depth of a few inches, but to many feet: the vesicles of some igneous rocks, for example, having been completely emptied, near the surface, of the mineral matter which once filled the cavities, and which still fills those beneath, so that externally the rocks present much the same vesicular aspect as when they flowed in a molten and viscous state.

Minerals of the zeolite family are very common in the vesicular cavities of some igneous rocks; and at one time, before their mode of occurrence was properly understood, the quantity of water found in many of them was thought to militate against the igneous origin of the containing rock. They form an interesting class of minerals, and, opportunities offering, should always be collected. They come under the head of hydrated aluminous silicates, with potash, soda, lime, and their isomorphous substances. The great proportion of them contain from 8 to 18 per cent. of water in combination. In the same kind of vesicles, siliceous deposits in the form of agates are not uncommon. In these and in cavities of various rocks, even those of aqueous origin, such, for example, as the dolomitic rocks of the new red sandstone series, in Somerset and Gloucestershire, the agate linings of the cavities have continued only for certain distances, after which the elements of other minerals have entered the hollows, and various crystallized substances have been the result. Cavities, therefore, in all rocks may be searched. With respect to the successive siliceous coatings forming agates, while some kinds of coatings show an adjustment to the walls of the cavity, others have accumulated in flat layers, generally considered to have been formed horizontally. Sometimes part of a cavity has been filled in one way, and the remaining portion in the other. Occasionally, from cavities left after a part of the hollow has been filled horizontally, stalactites of the matter of the agate have descended from above, as in the annexed figure. It is desirable always to ascertain how far such flat layers correspond with the present horizon; and, if the vesicles or hollows are almond-shaped (elongated more in one direction than another), how far these are constant in the same direction, thus pointing out that in which the molten viscous rock moved.

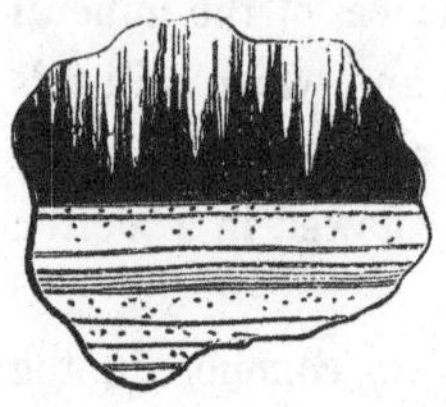

Many nodules in rocks, those which have clearly not been formed as gravel or boulders by attrition, afford examples of the aggregation of similar matter from a mass, such as one of clay, in which that matter has once been more generally diffused. In this way we have siliceous nodules, calcareous nodules, and those valuable nodules the clay ironstones. These last are fundamentally carbonates of iron, with a variable addition of the matter of the mud or silt amid which the carbonate of iron has once been more generally diffused. In many such nodules there has been a shrinking from the centre to the sides, causing cracks, that have been variously filled with mineral matter, as in the subjoined figure. Occasionally in the cracks so formed, and not quite filled

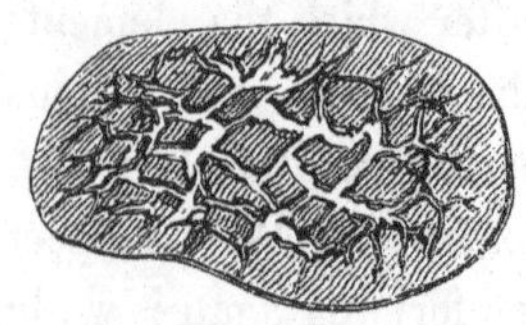

up, various mineral substances are obtained well crystallized. It may be here observed, as regards the metallic titanium frequently discovered in iron furnaces when blown out, that we have found the oxide of titanium crystallized in the cavities of clay ironstones. Taken as a whole, the observer will do well to look into any cavities or cracks he may discover in rocks, even in the hollows among organic remains, for various mineral substances. Many a crystallized body will thus be frequently found, and the replacement of one substance by another be well seen.

Not only in cracks or hollows, but in the body of the rocks themselves, minerals may be observed well crystallized, their component molecules having had free power to adjust themselves according to the affinities and forms proper to them. This is well seen in the class of igneous rocks known as porphyries—that is where a general paste or base, confusedly crystalline, compact or earthy, may happen to contain isolated and well-formed minerals of different kinds. From experiments in the laboratory, and the results of metallurgical and chemical operations carried on upon the large scale, we know that this isolation of crystals may readily be obtained, it being merely needful that the conditions for their production should be such that their component particles could freely adjust themselves first in the cooling mass. In the igneous *dykes*, as they are termed—that is, where igneous matter in fusion has been forced up, filling cracks formed in the rocks which they traverse—we sometimes see good illustrations of the mode in which isolated mineral crystals may be produced. Let us take as an example some of the granitic dykes, known as *elvans* by the miners of Cornwall, and let the annexed figure represent a section of one of them, *a a*

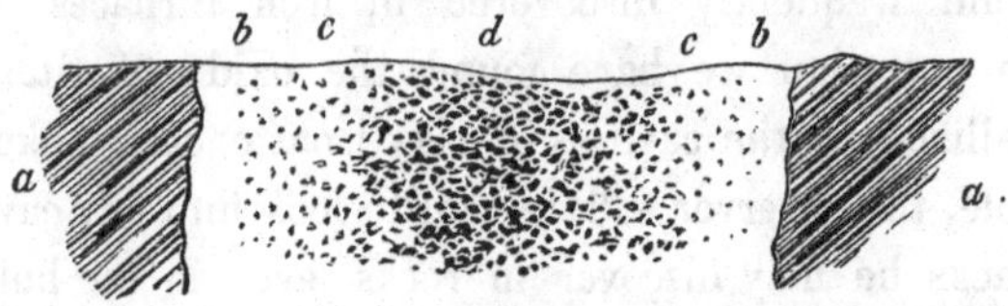

being some schistose rock broken through or fractured, (it may be any rock previously consolidated; granite is thus frequently fractured, and the fissure filled by an elvan.) We find that while the central portion *d* may be a granite, the parts *c c* are porphyritic, and *b b* some compact rock. Upon investigation, we see that all parts are chemically the same, and that these various characters are due to differences in cooling. The central portion retained its heat longest, while the portions adjoining the bounding and fractured rocks were more speedily cooled. In such porphyries various minerals are found, those of the felspar family being very common. Such results from differences of cooling can be imitated artificially with substances under our control. In this way crystals of silicate of lime may be beautifully obtained, isolated in transparent glass.

Whole mountain masses are occasionally composed of porphyritic rocks, including the porphyritic granites among them; and it is desirable to obtain specimens of these, selecting portions where the crystals may be well formed, and observing, should more than one kind of isolated mineral be present, how far when one kind becomes common another may disappear, and if different kinds continue mixed through the general mass, or only in patches. It is equally desirable to obtain good characteristic specimens of the base or paste, and from situations where they have been uninjured by exposure to the weather, and have lost little of the soluble substances which may have once been contained in the rock. The chemical study of the whole of such igneous rocks is every day becoming more interesting.

It is not only among the igneous rocks which have once been in a molten state that the observer should look for minerals, but also, in volcanic regions, for those evidently sublimed upon the faces of craters, or in cracks or chinks of their sides, or of lava currents. Many of the substances thus obtained are difficult of preservation; but by putting them away in bottles, well stopped, much may be accomplished.

The minerals often seen isolated in those rocks which have been termed metamorphic, or altered, in consequence of the upburst or protrusion of some rock in a state of igneous fusion near them, constitute a class of much interest. Here again we see conditions well fitted to the adjustment of the integrant molecules of minerals; but this case so far differs from that of the porphyries, that whereas in the latter the whole mass has evidently been in a fluid or viscous state, in the former the stratified character of the rocks of that class is preserved. The manner of observing this order of rocks belongs to geology. It is only necessary here to call attention to the kind of isolated minerals found. Among them staurolites, andalusites, and garnets are frequent under certain conditions, which it may be advisable to guard the observer from supposing merely those of temperature. The freedom with which the particles of the isolated minerals thus formed could often adjust themselves, the main mass retaining its general structure, is highly interesting. We have seen crystals of garnet, perfectly formed, amid the grains of a sandstone, close in contact with granite, the beds of the sandstone retaining their original shape, and the mechanically-produced grains well distinguishable. The component parts of andalusite have often been gathered together in such a manner to form that mineral in altered rocks, that the resulting crystals have pushed aside mica and other substances.

We have chiefly referred hitherto to minerals found crystallized, or produced under the conditions proper for the best arrangement of their constituent molecules, either alone or entangling some other substances. Many important mineral substances occur, so that, whatever may be the form of their ultimate molecules, they appear to us in mass, sometimes forming beds mingling with others, or occupying clefts in rocks; occasionally constituting portions which have separated out from the body of the rock, as in the instances of the nodules previously mentioned. In these various forms many minerals are found, some being ores of the useful metals, such as iron ores, including bog-iron among them, and iron pyrites, valuable for the sulphur in it. As filling clefts in rocks, mingled with other matter, many ores of lead, tin, copper, &c. Other substances, also employed for various purposes, are obtained in the massive state, such as rock-salt, gypsum, and coal. For these minerals qualitative chemical researches will be found valuable, it being desirable to test the composition of many substances, which may offer certain general resemblances in appearance, before some given locality may be quitted.

Among the minerals occurring in beds, we should more particularly notice coal and other substances of that class, which have of late become so important for the extension of steam navigation. Our shipping daily bring home specimens of coal or lignite from localities where they were not previously known to occur. And it may be here needful to state, as now well known to geologists, that good coal is not confined to rocks of a particular geological date, but that, the needful physical conditions having obtained, it has been produced from vegetable matter accumulated at different geological times. When we have a cliff before us, there is little difficulty in seeing that a coal-bed occurs among others of sandstone, shale, or other substances. Not unfrequently coal-beds are based upon clays, or argillaceous strata which have a clayey character from exposure, and then it sometimes happens, from the slipping and falling of the general mass, that the real thickness and importance of a coal-bed may not appear on a cliff. Thus it was at Labuan, where now a valuable coal-bed about nine feet thick is worked: when first seen on the coast it did not appear more than eighteen inches thick.

Of whatever geological age an accumulation of mud, silts, sands, and gravels, now more or less consolidated as shales, sandstones, and conglomerates, and containing interstratified coal, may happen to have been, it rarely occurs that the bed upon which the coal itself reposes has not some peculiar character, easily observed. In many cases we feel assured that this has arisen from these beds having formed the ground, often perhaps marshy or with a slight covering of water, on which the plants, now converted into coal, have grown. Should these marked deposits be found, they often form valuable aids in tracing coal-beds, where the outcrop of the latter may not be very apparent, and they are especially serviceable, as in many of the hilly coal-measure districts of the British Islands, where these beds throw out springs of water. Whole lines of such springs coinciding with the bottom of coal-beds can be traced in the hilly coal districts of South Wales and Monmouthshire, and often on a hill side faults traversing the general mass can be as well seen, where these lines are interrupted, as if a diagram section were before us.

In some of the beds immediately subjacent to the coal peculiar fossil plants are found. In the palæozoic coal of the British Islands a fossil plant, named *Stigmaria ficoides*, is very characteristic. Peculiar fossil plants, not the one mentioned, are discovered, it is thought, well marking the beds on which the coal-beds rest in the Burdwan coal district in India, and other instances of a similar kind are recorded. It will be obvious that, although the conditions for the production of marked accumulations may have preceded the growth of most of the coal-vegetables themselves, the latter may not have sometimes grown, so that no coal rests upon such beds, a fact observed, and according to conditions more in one part of a coal district than in another. Still these beds, when any such occur, are useful to trace, since while we find in one locality no vegetable accumulation to have taken place upon them, or if effected the vegetable matter to have been subsequently removed, upon an extension of the same beds we may often see good workable coal.

Though in cliffs, either on the shore or on the sides of rivers, hills, and mountains, we commonly find the most direct evidence of the existence of coal, it may be often traced to its beds, where such occur, by means of the detritus brought down by brooks and rivers. By following rolled pebbles up such water-courses they may be often seen to end near some bed or beds whence they have been derived. If these cross the stream, a good opportunity may be afforded for examining their quality and thickness. The pebbles may, however, come from the sides of some adjacent hills sloping towards the streams, the beds of coal not crossing them, fragments only of their outcrop being mingled with any others of associated beds. The thickness of such coal-beds may be thus concealed, as will be readily seen by the annexed section, in which *a* represents the river course, up which pebbles of coal may be traced; *b b* beds of coal, the outcrops of which, *c c*, may be much concealed by fragments

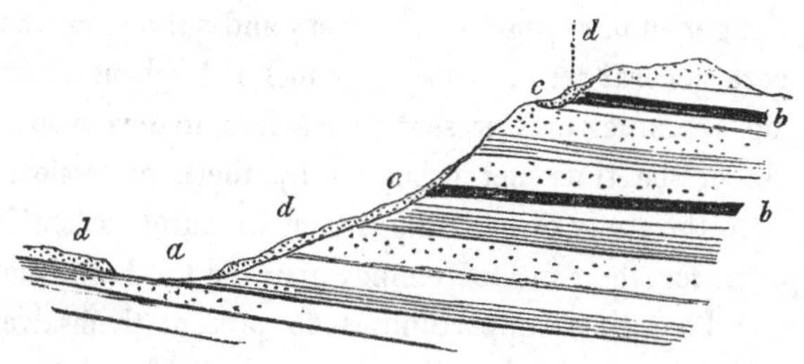

of rock descended from above and mingled in a fragmentary covering *d d*. The best should be done to obtain a knowledge of the associated beds by tracing up the rills of water descending the sides of the hills. Excellent evidence may thus be often obtained, and the true position of the coal-beds found. In selecting specimens of coal in such cases, it rarely happens that a portion of it can be procured fairly exhibiting its qualities, injury having arisen from atmospheric influences. If the outcrop of the coal can be attained, it is always desirable to penetrate as far as circumstances will permit into the body of the bed, thence selecting a fair specimen. When this cannot be done, and a voyager often has but little time for his researches, fragments lying about should be selected which may appear the least decomposed, and if these be different qualities, as if of portions of different beds, they also should receive attention. In all cases where fossil plants are mingled with the coal or associated beds, specimens as various as can be obtained should be secured. These have a geological bearing which may often turn out of great practical importance in some given region.

It scarcely requires remark that the foregoing observations are but hints which it is hoped may be useful to those engaged in voyages of discovery and survey, or who, on more general service, may feel inclined, whenever fitting opportunities may present themselves, to devote some portion of the time not occupied by their professional duties to the study of minerals, either for purely scientific purposes, for their useful employment, or for both combined. That these opportunities do present themselves we well know, or rather if sought will be found more frequently than might be imagined. Many a walk along a coast may thus be advantageously turned to account, and an interest be excited not at first thought probable. Not only may a naval man thus add to his own stock of knowledge, but he may most materially by his exertions promote the advance of science and its applications generally, minerals being objects of great interest, whether we regard them with reference to their importance to man, and the aid many of them afford to the spread of civilization, or as connected with several sciences, even those of the highest order.

Section IX.

METEOROLOGY.

By the EDITOR.

There is no branch of physical science which can be advanced more materially by observations made during sea voyages than meteorology, and that for several distinct reasons. 1stly. That the number and variety of the disturbing influences at sea are much less than on land, by reason of the uniform level and homogeneous nature of its surface. 2ndly. Because, owing to the penetrability of water by radiant heat, and the perpetual agitation and intermixture of its superficial strata, its changes of temperature are neither so extensive nor so sudden as those of the land. 3rdly. Because the area of the sea so far exceeds that of the land, and is so infinitely more accessible in every part, that a much wider field of observation is laid open, calculated thereby to afford a far more extensive basis for the deduction of general conclusions. 4thly. The sea being the origin from which all land waters are derived, in studying the hygrometrical conditions of the sea atmosphere, we approach the chief problems of hygrology in their least involved and complicated form, unmixed with those considerations which the perpetually varying state of the land (as the recipient at uncertain intervals of derivative moisture) forces on the notice of the meteorologist of the continents. Nor ought it to be left out of consideration that this, of all branches of physical knowledge, being that on which the success of voyages and the safety of voyagers are most immediately and unceasingly dependent, a personal interest of the most direct kind is infused into its pursuit at sea, greatly tending to relieve the irksomeness of continued observations, to insure precision in their registry, and to make their partial or complete reduction during the voyage an agreeable, as it always is a desirable object.

It happens fortunately, that almost every *datum* which the scientific meteorologist can require is furnished in its best and most available state by that definite, systematic process, known as the " *keeping a meteorological register*," which consists in noting at stated hours of every day the readings of all the meteorological instruments at command, as well as all such facts or indications of wind and weather as are susceptible of being definitely described and estimated without instrumental aid. Occasional observations apply to occasional and remarkable phenomena, and are by no means to be neglected; but *it is to the regular meteorological register, steadily and perseveringly kept throughout the whole of every voyage, that we must look for the development of the great laws of this science.*

The following general rules and precautions are necessary to be observed in keeping such a register :—

1. Interruptions in the continuity of observations by changes of the instruments themselves, or of their adjustments, places, exposure, mode of fixing, reading, and registering, &c., are exceedingly objectionable, and ought

to be sedulously avoided. Whenever an alteration in any of these particulars is indispensably necessary, it should be done as a thing of moment, with all deliberation, scrupulously noted in the register, and the exact amount of change thence arising in the reading of the instrument (whether by alteration in its zero point, or otherwise) ascertained.

2. As far as possible registers should be complete: but if, from unavoidable causes, blanks occur, no attempt to fill them up subsequently from general recollection, or (which is worse, and amounts to a falsification) from the apparent course of the numbers before and after, should ever be made. The entries in the register made at the time of observation should involve no reduction or correction of any kind, but should state the simple readings off of the several instruments, and other particulars, just as observed. This does not of course prevent that blank columns left for reduced and corrected observations should be filled up at any convenient time. On the contrary, it is very desirable that such should be the case —the sooner after the observation, consistent with due deliberation, the better, on every account, unless some datum be involved requiring subsequent discussion for its determination.

3. The observations of each kind should, if possible, all be made by one person; but as this is often impracticable, the deputy should be carefully instructed by his principal to observe in the same manner, and the latter should satisfy himself by comparative trials that they observe alike.

4. If copies be taken of registers, they should be care-

fully compared with the originals by two persons—one reading aloud from the original, and the other attending to the copy, and then exchanging parts—a process always advisable when great masses of figures are required to be correctly copied.

5. The registers should be regarded (if kept in pursuance of orders, or under official recommendation) as official documents, and dealt with accordingly. If otherwise, a verified copy, or the original (the latter being preferred), signed by the observer, should be transmitted to some public body interested in the progress of meteorological science, through some official channel, and under address "*To the Secretary of*, &c. &c." Circuitous transmission hazards loss or neglect, and entails expense on parties not interested.

6. The register of every instrument should be kept in parts of its own scale as read off; no reduction of foreign measures or degrees to British being made. But it should of course be stated what scale is used in each. British observers, however, will do well to use instruments graduated according to British units.

7. The regular meteorological hours are 3 A.M., 9 A.M., 3 P.M., and 9 P.M., mean time at the place. Irksome as it may be to landsmen to observe at 3 A.M., the habits of life on shipboard render it much less difficult to secure this hour in a trustworthy manner; and the value of a register in which it is deficient is so utterly crippled, that whatever care be bestowed on the other hours, it must on that account hold a secondary rank. The above hours, it must be borne in mind, are the fewest which any meteorological register pretending to completeness can embrace.

By any one, however, desirous of paying such particular attention to this branch of science as to entitle him to the name of a meteorologist, a three-hourly register—viz., for the hours 3, 6, 9, A.M., noon; 3, 6, 9, P.M., midnight—ought to be kept; and in voyages of discovery, where scientific observation is a prominent feature, the register ought to be enlarged, so as to take in every *odd* hour of the twenty-four; thus including, *without interpolation*, the six-hourly or standard series.

8. Hourly observations should be made throughout the twenty-four hours on the 21st of each month (except when that day falls on a Sunday, and then on the Monday following), commencing with 6 A.M., and ending at 6 A.M. on the subsequent day, so as to make a series of twenty-five observations. At all events, if this cannot be done monthly, it ought not to be omitted in March, June, September, and December. These are called "term observations." If any remarkable progressive rise or fall of the barometer be observed to pervade this series, it will be well to continue it until the maximum or minimum is clearly attained, with a view to comparison with other similar series elsewhere obtained, and thus to mark the progress of the aërial wave effective in producing the change. These term observations should be separately registered under that head.

9. Occasional *hourly* series of observations may be made with advantage under several circumstances, as, for instance—1stly. When becalmed for any length of time, especially when near the Equator, with a view to determining the laws and epochal hours of diurnal periodicity.

2ndly. When a party leaves the ship, furnished with a portable barometer or other instruments, for the measurement of heights of mountains, or with other objects. 3rdly. During threatening weather, and especially during the continuance of gales, and for some time after their subsidence, as will be more particularly specified under the head of "Storm Observations." 4thly. In certain specified localities mentioned in a subsequent article by Mr. Birt. 5thly. Whenever a *continued* rise or fall of the barometer has been noticed as at all remarkable, it should be pursued up to and past the turn, so as to secure the maximum elevation or depression, and the precise time of its occurrence; and a register of such maxima or minima should be kept distinct from the regular entries.

Of Meteorological Instruments; and first, of the Barometer and its attached Thermometer.

The barometer on shipboard should be suspended on a gimbal frame, which ought not to swing too freely, but rather so as to deaden oscillations by some degree of friction. Before suspending it, it should be carefully examined for air-bubbles in the tube and for air in the upper part above the mercury, by inspection, and by inclining the instrument from the vertical position rather suddenly till the mercury rises to the top with a slight jerk, when, if it do not *tap sharp*, the vacuum is imperfect; and if the sound be puffy and dead, or is not heard at all, air exists to an objectionable extent, and must be got rid of by inversion and gently striking with the hand to drive the bubble up into the cistern. The lower end of the tube, which plunges into the cistern in well-constructed marine barometers, is contracted so as to diminish the amount of oscillation produced by the ship's motion. The instrument should be suspended out of the reach of sunshine, but in a good light for reading, as near midships, and in a place as little liable to sudden changes of temperature and gusts of wind as possible. The light should have access to the back of the tube, so as to allow of setting the index to have its lower edge a tangent to the convex surface of the mercury. In well-constructed barometers the slider has its lower part tubular, embracing the tube, and can be made to descend by the rack-motion of the vernier till it becomes an upper tangent to the mercury: the eye being on its exact level, a reflected light by day, or white paper strongly illuminated from behind at night, will throw the light properly for setting the vernier correctly. The exact height of the cistern above the ship's water-line should be ascertained and entered on the register.

The attached thermometer ought to indicate a temperature the exact mean of that of the whole barometric column. Its bulb, therefore, ought to be (though it seldom is) so situated as to afford the best chance of its doing so, that is to say, fifteen inches above the cistern, enclosed within the wooden case of the barometer, nearly in contact with its tube, and with a stem so long as to be read off at the upper level. To ensure a fair average and steady temperature, it were well to enclose the whole instrument, thermometer and all, in an outer case of leather, over a wrapper of flannel, leaving only the setting and reading parts above and below accessible, and that no more than is absolutely necessary.*

In choosing a barometer, select one in preference in which the lower level (of the mercury in the cistern) is adjustable to contact with a steel or ivory fiducial point, and *that* not by altering the height of the mercurial surface, but by depressing the steel point *carrying down with it the whole divided scale*, the zero-point of which is of course the apex of the point itself. Care should be taken that air have free but safe access to the lower surface.

In transporting a compared barometer to its place of destination great care is necessary. Carry it upright, or considerably inclined, and *inverted;* and over all rough roads in the hand, to break the shocks it would otherwise receive. A "portable barometer" strapped obliquely across the shoulders of a horseman travels securely and well; and with common care in this mode of transport its zero runs no risk of change. If merely fastened to any kind of carriage, and abandoned to its fate, it is almost sure to be broken.

To make and reduce an Observation of the Barometer.—First read off and write down the reading of the attached thermometer. Then give a few gentle taps on the instrument to free the mercury from adhesion to the glass, avoiding to give it any violent oscillation. Adjust

* For a permanently suspended or fixed barometer, the best thermometer would be one with a tubular bulb of equal bore and thickness of glass with the barometer tube, and extending in length from the cistern to the exposed face of the instrument, and as close to the barometric column as is consistent with the structure of the upper works. Immersion of the ball of the attached thermometer in the cistern is the worst arrangement of any.

the lower level to the fiducial point, if such be the construction of the instrument. Then set the index to the upper surface of the mercurial column, placing the eye so as to bring its back and front lower edges to coincidence, and to form a tangent to the convexity of the quicksilver. If the instrument have no tubular or double-edged index, the eye must be carefully placed at the level of the upper surface to destroy parallax. Whatever mode of reading is adopted should be always adhered to. A magnifier should be used to make the contact and to read the vernier, and the reading immediately written down and carefully entered on the register.

As soon after the observations have been made as circumstances will permit, the reading of the barometer should be *corrected* for the relation existing between the capacities of the tube and cistern (if its construction be such as to require that correction), and for the capillary action of the tube; and then *reduced* to the standard temperature of 32° Fahr., and to the sea-level, if on shipboard. For the first correction the *neutral point* should be marked upon each instrument. It is that particular height which, in its construction, has been actually measured from the surface of the mercury in the cistern, and indicated by the scale. In general the mercury will stand either above or below the neutral point; if *above*, a portion of the mercury must have left the cistern, and consequently must have *lowered* the surface in the cistern: in this case the altitude as measured by the scale will be *too short—vice versâ*, if below. The relation of the capacities of the tube and cistern should be experimentally ascertained, and marked upon the instrument by the maker. Suppose the capacity to be $\frac{1}{50}$, marked thus on the instrument, "*Capacity* $\frac{1}{50}$:" this indicates that for every inch of variation of the mercury in the tube, that in the cistern will vary contrariwise $\frac{1}{50}$th of an inch. When the mercury in the tube is *above* the neutral point, the difference between it and the neutral point is to be reduced in the proportion expressed by the "capacity" (in the case supposed, divided by 50), and the quotient *added* to the observed height; if *below*, *subtracted* from it. In barometers furnished with a fiducial point for adjusting the lower level, this correction is superfluous, and must not be applied.

The second correction required is for the capillary action of the tube, the effect of which is always to depress the mercury in the tube by a certain quantity inversely proportioned to the diameter of the tube. This quantity should be experimentally determined during the construction of the instrument, and its amount marked upon it by the maker, and is always to be *added* to the height of the mercurial column, previously corrected as before. For the convenience of those who may have barometers, the capillary action of which has not been determined, a table of corrections for tubes of different diameters is placed in the Appendix, Table I.

The next correction, and in some respects the most important of all, is that due to the temperature of the mercury in the barometer tube at the time of observation, and to the expansion of the scale. Table II. of the Appendix gives for every degree of the thermometer and every half-inch of the barometer, the proper quantity to be added or subtracted for the reduction of the observed height to the standard temperature of the mercury at 32° Fahr.

After these the index correction should be applied. This is the amount of difference between the particular instrument and the readings of the Royal Society's flint-glass barometer when properly corrected, and is generally known as the *zero*. It is impossible to pay too much attention to the determination of this point. For this purpose, when practicable, the instrument should be immediately compared with the Royal Society's standard, and the difference of the readings of both instruments, when corrected as above, carefully noted and preserved. Where, however, this is impracticable, the comparison should be effected by means either of some other standard previously so compared, or of an intermediate portable barometer, the zero-point of which has been *well determined*. Suspend the portable barometer as near as convenient to the ship's barometer, and after at least an hour's quiet exposure, take as many readings of both instruments as may be necessary to reduce the probable error of the mean of the differences below 0·001 inch. Under these circumstances the mean difference of all the readings will be the *relative* zero or index error, whence, if that of the intermediate barometer be known, that of the other may be found. As such comparisons will always be made when the vessel is in port, sufficient time can be allowed for making the requisite number of observations: hourly readings would perhaps be best, and they would have the advantage of forming part of the system when in operation, and might be accordingly used as such.

It is not only desirable that the zero point of the barometer should be well determined in the first instance; it should also be carefully verified on every opportunity which presents itself. And in the first instance, previous to sailing, after suspending the barometer on shipboard, it should be re-compared with the standard on shore by the intervention of a portable barometer, and no opportunity should be lost of comparing it on the voyage by means of such an intermediate instrument with the standard barometers at St. Helena, the Cape of Good Hope, Bombay, Madras, Paramatta, Van Diemen's Island, and with any other instruments likely to be referred to as standards, or employed in research elsewhere. Any vessel having a portable barometer on board, the zero of which has been well determined, would do well on touching at any of the ports above-named to take comparative readings with the standards at those ports, and record the differences between the standard, the portable, and the ship-barometers. By such means the zero of one standard may be transported over the whole world, and those of others compared with it ascertained. To do so, however, with perfect effect, will require that the utmost care should be taken of the portable barometer; it should be guarded as much as possible from all accident, and should be kept safely in the "portable" state when not immediately used for comparison. To transport a well-authenticated zero from place to place is by no means a point of trifling importance. Neither should it be executed hurriedly nor negligently. Some of the greatest questions in meteorology depend on its due execution, and the objects for which these instructions have been prepared will be greatly advanced by the zero points of all barometers being referred to one common standard. Upon the arrival of the vessel in England, at the termination of the voyage, the ship's barometer should be again compared with the same standard with which it was compared previous to sailing; and should any difference be found, it should be most carefully recorded.

The correction for the height of the cistern *above* or *below* the water-line is *additive* in the former case, *subtractive* in the latter. Its amount may be taken, nearly enough, by allowing 0·001 in. of the barometer for each foot of difference of level.

An example of the application of these several corrections is subjoined :—

Attached Therm. 54°·3.

Barometer reading . .		29·409
Corr. for capacity .	—	·017
		29·392
Corr. for capillarity .	+	·032
		29·424
Corr. for temperature	—	·068
		29·356
Corr. for zero and water-line	+	·040
Aggregate = pressure at sea-level		29·396

Data for the correction of the Instrument.

Neutral point		30·123
Capacity $\frac{1}{42}$		
Capillary action . .	+	·032
Zero to Royal Society.	+	·036
Corr. for altitude above water-line . . .	+	·004

Thermometers.—The observer should be furnished with a delicate and accurate thermometer, most carefully compared with a perfectly authentic standard, at several temperatures, differing considerably, and of which the freezing point has been most scrupulously verified. This he should keep *solely* as a thermometer of reference, and every thermometer he employs should be compared with it, so as not to leave a doubt as to the amount of their constant difference exceeding a tenth of a degree. To make such comparisons, long rest, in contact, in a box stuffed with cotton, allowing only the portions of the scales where read off to be from time to time uncovered for that purpose, is the best mode of insuring their perfect identity of temperature. If, in any instance, the zeros differ in different parts of the scales, a table of reduction to the standard will require to be constructed. The comparisons should be repeated at not very distant intervals of time, especially in the case of self-registering thermometers, whose index-error is constantly changing, and requires great watchfulness. In registering thermometers *record* but do not *apply* their zeros.

In placing the *External Thermometer*, an exposure should be chosen perfectly shaded both from direct sunshine, and that reflected from the sea, or radiated from any hot object. It should be especially guarded from rain and from spray, so that the bulb should never be wetted, also from warm currents of air and from local radiation; completely detached from contact with the ship's side, and fully exposed to the *external* air. In reading it the observer should avoid touching, breathing on, or in any way warming it by the near approach of his person; and in night-observations particular care should be taken not to heat it by approach of the light. The quicker the reading is done the better. At night it should be completely screened from the sky, so as to annihilate all loss of heat by upward radiation; a light frame-case of double wire-gauze will perhaps be found a secure and efficient protection both from injury and obnoxious influences.

The *Self-registering Thermometers* should be placed with the same precautions as the external thermometer, and in similar exposures, and so fastened as to allow one end to be detached and lifted; so that the indices within the tubes may slide down to the ends of the fluid columns, which they will readily do on gentle tapping. They are apt to get out of order by the indices becoming entangled, or by the breaking of the column of fluid. When this happens to the spirit-thermometer, it is easily rectified by jerking the index down to the junction of the bulb and tube; then, by cautiously heating and cooling alternately the bulb, tube, and air-vessel at the top, the disunited parts of the spirit may be distilled from place to place till the whole is collected into one column in union with that in the bulb.

When the steel index of the mercurial thermometer becomes immersed in the mercury, first cool the bulb (by evaporation of ether, if necessary) till the mercury is either fairly drawn below the index, or the column separates, leaving the index with mercury above it. Loosen the index by tapping, by a magnet, or by heating the tube, then apply heat to the bulb, and drive the index with its superincumbent mercury up into the air-vessel. When there, hold the instrument bulb downwards, and, suspending the index by a magnet, effect a union between the globule of mercury and the column below, by continuing to apply the heat till the latter rises into the air-vessel. As the bulb cools the whole mercury should descend in an unbroken column, after which the index may be restored to its place. Much patience and many trials are often required for success. An oil-lamp with a very small clear flame should be used.

Both the self-registering thermometers should be read off at the time of the 9 h. A.M. observation, as it is very improbable that the temperature at that hour should be such as to obliterate either record of the preceding 24 hours. Double maxima and minima, when they occur, if remarkable, should be recorded as supernumerary and separately in a diary, and their accompanying circumstances noted.

The observer should be furnished with several other thermometers, all of sufficient delicacy to allow of estimating tenths of degrees, for observation of the temperature of the sea, or of the earth, (when on shore,) of falling rain, &c., and for a reserve in case of accident. All should be compared with the standard. That in habitual use for the sea temperature should be defended from accident in the act of immersion by a wire guard.

The thermometer for solar radiation should have its bulb blackened with a coat of Indian ink. It should be defended from currents of air by enclosure in a glass tube; and it would add infinitely to the value of a series of observations made with it if this tube were exhausted and hermetically sealed. Its exposure to the sun should be perfectly free and full, and it should be suspended in free air, quite out of reach of any support or object heated by the sun's rays.

Hygrometers are of very various constructions, and

depend on very different principles. That which we recommend in preference to, and as having almost universally superseded in practice every other,* consists of two thermometers, the bulb of the one being dry and of the other wet; being kept so by a covering of muslin, connected by a wet roll of cotton (lamp wick) with a small cup of distilled or rain-water placed close beneath it, so as to absorb and communicate water by capillary attraction. In frosts this arrangement is unavailing, and water must be poured on the muslin envelope, and allowed to freeze into a coat of ice, from which evaporation will still go on and depress the temperature, as if still liquid. They should be placed and observed in such locality as shall afford the best chance for procuring a fair indication of the moisture of the general atmosphere, and by no means in any confined or ill-ventilated situation between decks, where many persons habitually congregate, or which, from any other cause, is usually or periodically damp. The whole instrument should be protected with a cap of wire-work to defend it from injury: this, if it interfere with the readings, should be removed a quarter of an hour before the observation. In reading the thermometers begin with the dry one, and use all the precautions recommended with respect to the "external" thermometer. Enter the simple readings, but at the head of each column place the zero *correction* (with its proper sign) required for its thermometer (a general rule for all thermometric entries), and

* The hair hygrometer is delicate, exceedingly liable to derangement, and, except prepared with extraordinary care, uncertain. Daniell's dew-point hygrometer, excellent in theory, is very costly on account of its great consumption of ether, and scarcely useable in hot climates, owing to the difficulty of preserving that liquid.

leave a blank column for the "hygrometric depression," in calculating which subsequently the zeros must be *applied*. The reduction of the observations to derive the *elastic force of vapour at the dew-point* is effected by the formulæ of Dr. Apjohn:—

$$F = f - \frac{d}{88} \cdot \frac{h}{30} \quad . \ . \ . \ . \ (a); \quad F = f - \frac{d}{96} \cdot \frac{h}{30} \quad . \ . \ . \ . \ (b).$$

(*a*) to be used when the reading of the wet thermometer is above 32, and (*b*) when below. In these d is the hygrometric depression, h the height of the barometer, f the elastic force of vapour for the temperature shown by the *wet* thermometer, to be taken from Table III., Appendix, and F the elastic force of vapour at the dew-point, which (all the other quantities being known) these formulæ enable us to calculate. With F so calculated enter the same table under the column of Force of Vapour, and the corresponding *temperature* is the dew-point, which, however, is not wanted to be known but as a matter of curiosity.

The Rain-Gauge.—This may be of very simple construction. A cubical box of tin or zinc, exactly ten inches by the side, open above, receives at an inch below its edge a square funnel, sloping to a small central hole. On one of the lateral edges of the box, close to the top of the cavity, is soldered a short pipe, in which a cork is *loosely* fitted; the whole should be well painted. The water which enters the reservoir through the funnel hole is poured through the short tube into a cylindrical glass vessel, graduated to cubic inches and fifths of cubic inches. Hence, one inch in depth of rain in the gauge will be measured by

100 inches of the graduated vessel, and a thousandth of an inch may easily be read off. It is very difficult to place the rain-gauge properly on shipboard, and its entries therefore require constant explanatory notes, pointing out causes tending to disturb its influence. In fact, excepting the mast-head (and there upon a gimbal), it seems hardly possible to devise a tolerably *permanent* situation for it. On land, a perfectly open exposure on the ground, or *very* little elevated above it, should be chosen. The quantity of water should be daily measured and registered at 9 A.M., unless the fall of rain be so heavy as to endanger filling the instrument within the 24 hours, when this operation should be performed as often as needed. Snow collected or water frozen in the reservoir should be melted.

The Anemometer.—Lind's would appear to be the only anemometer which can conveniently be used on shipboard. It is adjusted by filling it with water till the liquid in both legs of the syphon corresponds with zero of the scale. It is to be held perpendicularly with the mouth of the kneed tube turned towards the wind, and the amount of depression in the one leg and of elevation in the other are to be noted. The sum of the two is the height of the column of water which the pressure of the wind is able to support: and the force of the wind on a square foot is obtained from this height by Table IV. of the Appendix. In great degrees of cold a *saturated* brine may be used which does not freeze, and whose specific gravity being 1·244, the force given by the table must be multiplied by this factor. In addition to the regular hours of observation this instrument should be observed in storms, white squalls, or other circumstances of interest; the direction of the wind as well as its force should be registered at each observation; and for this it is well to have a small compass with a vane of card, or thin and very moveable sheet *brass*, which may be fastened on the top of the anemometer, and which will indicate the direction in which its opening should be turned. In concluding the direction and the force of the wind from the vane and anemometer readings, a correction depending on the direction and velocity of the ship's motion is required in strictness. But such corrections are not *usually* applied, and it may be doubted whether the observations can be accurately enough made to render it worth while to apply them.

The Actinometer consists of a large hollow cylinder of glass, soldered at one end to a thermometer-tube, terminated at the upper end by a ball drawn out to a point, and broken off, so as to leave the end open. The other end of the cylinder is closed by a silver or silver-plated cap, cemented on it, and furnished with a screw, also of silver, passing through a collar of waxed leather, which is pressed into forcible contact with its thread, by a tightening screw of large diameter enclosing it, and working into the silver cap, and driven home by the aid of a strong steel key or wrench, which accompanies the instrument.

The axis of this screw is pierced to allow the stem of a spirit thermometer to pass out through it, the bulb (a very long one) being within the cylinder, to take the temperature of the enclosed liquid. The graduation is in the stem of the screw, which is prolonged to receive and defend it.

The cylinder is filled with a deep blue liquid (ammonio-sulphate of copper), and the ball at the top being purposely left full of air, and the point closed with melted wax, it becomes, in any given position of the screw, a thermometer of great delicacy, capable of being read off on a divided scale attached. The cylinder is enclosed in a chamber blackened on three sides, and on the fourth, or face, defended from currents of air by a thick glass, removable at pleasure.

The action of the screw is to diminish or increase at pleasure the capacity of the hollow of the cylinder, and thus to drive, if necessary, a portion of the liquid up into the ball, which acts as a reservoir, or, if necessary, to draw back from the reservoir such a quantity as shall just fill it, leaving no bubble of air in the cylinder. The interior thermometer indicates the temperature of the blue liquid approximately for the subsequent reduction of the observation.

To use the instrument, examine first whether there be any air in the cylinder, which is easily seen by holding it level, and tilting it, when the air, if any, will be seen to run along it. If there be any, hold it upright in the left hand, and the air will ascend to the root of the thermometer-tube. Then, by alternate screwing and unscrewing the screw with the right hand, as the case may require, it will always be practicable to drive the air out of the cylinder into the ball, and suck down the liquid, if any, from the ball, to supply its place, till the air is entirely evacuated from the cylinder, and the latter, as well as the whole stem of the thermometer-tube, is full of the liquid in an unbroken column. Then, holding it horizontally, face upwards, slowly and cautiously unscrew the screw till the liquid retreats to the zero of the scale.

The upper bulb is drawn out into a fine tube, which is stopped with wax. When it is needed to empty, cleanse, and refill the instrument, liquid must be first forced up into the ball, so as to compress the air in it. On warming the end, the wax will be forced out, and the screw being then totally unscrewed, and the liquid poured out, the interior of the instrument may be washed with water slightly acidulated, and the tube, ball, &c., cleansed, in the same way, after which the wax must be replaced, and the instrument refilled.

To make an observation with the actinometer, the observer must station himself in the sunshine, or in some sharply terminated shadow, so that without inconvenience, or materially altering his situation, or the exposure of the instrument in other respects, he can hold it at pleasure, either in full sun or total shadow. If placed in the sun, he must provide himself with a screen of pasteboard or tin plate, large enough to shade the whole of the lower part or chamber of the instrument, which should be placed not less than two feet from the instrument, and should be removable in an instant 'of time. The best station is a room with closed doors, before an open window, or under an opening in the roof into which the sun shines freely. Draughts of air should be prevented as much as possible. If the observations be made out of doors, shelter from gusts of wind, and freedom from all penumbral shadows, as of ropes, rigging, branches, &c., should be sought. Generally, the more the observer is at his ease, with his watch and writing-table beside him, the better. He should have a watch or chronometer beating at least twice in a second, and provided with a second hand; also a pencil and paper ruled, according to the form subjoined, for registering the observations. Let him then grasp the instrument in his left hand, or, if he have a proper stand (which is preferable on shore or in a building*), otherwise firmly support it, so as to expose its face perpendicularly to the direct rays of the sun, as exactly as may be.

The liquid, as soon as exposed, will mount rapidly in the stem. It should be allowed to do so for a minute before the observation begins, taking care, however, not to let it mount into the bulb, by a proper use of the screw. At the same time the tube should be carefully cleared (by the same action) of all small broken portions of liquid remaining in it, which should all be drawn down into the bulb. When all is ready for observation, draw the liquid down to zero of its scale, gently and steadily; place it on its stand, with its screen before it, and proceed as follows, first reading off the internal thermometer.

Having previously ascertained how many times (suppose 20) the watch beats in five seconds, let the screen be withdrawn at ten seconds before a complete minute shown by the watch, suppose at $2^h\ 14^m\ 50^s$. From 50^s to 55^s, say 0, 0, 0, ... at each beat of the watch, looking meanwhile that all is right. At 55^s complete, count 0, 1, 2, ... up to 20 beats, or to the whole

* This may consist of two deal boards, eighteen inches long, connected by a hinge, and kept at any required angle by an iron, pointed at each end. The upper should have a little rabate or moulding fitting loosely round the actinometer, to prevent its slipping off.

minute, 2^h 15^m 0^s, keeping the eye not on the watch, but on the end of the rising column of liquid. At the 20th beat read off, and register the reading (12°·0), as in column 3 A, of the annexed form. Then wait, watching the column of air above the liquid, to see that no blebs of liquid are in it, or at the opening of the upper bulb (which will cause the movement of the ascending column to be performed by starts), till the minute is nearly elapsed. At the 50th second begin to watch the liquid rising; at 55^s begin to count 0, 1, 2, up to 20 beats, as before, attentively watching the rise of the liquid; and at the 20th beat, or complete minute (2^h 16^m 0^s), read

1. Date and times of Observation. ———, 1850, Oct. 30.		2. Exposure, sun (⊙) or shade ×.	3. Readings of the Instrument.		4. Change per minute.	5. Internal Therm.	6.	7. Remarks.
Initial.	Terminal.		A. Initial.	B. Terminal.	B—A.			
h. m. s.	m. s.							
2 15 0	16 0	⊙	+12·0	+43·3	+31·3			The times are reduced to *apparent* time, or to the sun's hour angle from the meridian.
16 30	17 30	×	45·2	42·8	− 2·4			
18 0	19 0	⊙	14·8	48·2	+33·4			
19 30	20 30	×	28·0	26·8	− 1·4	75·5		
21 0	22 0	⊙	9·4	43·9	+33·5			
22 30	23 30	×	46·6	45·5	− 1·1			Zero withdrawn.
24 0	25 0	⊙	9·0	43·2	+34·2			Blackened Therm. 106·3.

off, and instantly shade the instrument, or withdraw it *just out* of the sun and penumbra. Then register the reading off (43°·3) in column 3, B, and prepare for the shade observation. All this may be done without hurry in 20 seconds, with time also to withdraw the screw if the end of the column be inconveniently high in the scale, which is often required. At the 20th second prepare to observe; at the 25th begin to count beats, 0, 1, 2, ... 20; and at the 20th beat, *i. e.* at 2^h 16^m 30^s, read off, and enter the reading in column 3, A, as the initial shade reading (45°·2). Then wait, as before, till nearly a minute has elapsed, and at 2^h 17^m 20^s again prepare. At 17^m 25^s begin to count beats; at 17^m 30^s read off, and enter this *terminal* shade reading (42°·8) in column 3, B, and, if needed, withdraw the zero.

Again wait 20^s, in which interval there is time for the entry, &c. At 17^m 50^s remove the screen, or expose the instrument in the sun; at 55^s begin to count beats; and at the complete minute, 18^m 0^s, read off (14°·8), and so on for several alternations, *taking care to begin and end each series with a sun observation,* and to read off the internal thermometer at the end of each set, or, if the observations be continuous, at every fifth sun observation. If the instrument be held in the hand, care should be taken not to change the inclination of its axis to the horizon between the readings, or the compressibility of the liquid by its own weight will produce a very perceptible amount of error.

In the annexed form column 1 contains the times, initial and terminal, of each sun and shade observation. Column 2 expresses by an appropriate mark, ⊙ and ×, the exposure, whether in sun or shade. Column 3 contains the readings, initial and terminal (A and B). Column 4 gives the values of B — A, with its algebraical sign expressing the rise and fall per minute. And here it may be observed, that if by forgetfulness the exact minute be passed, the reading off may be made at the next 10^s, and in that case the entry in column 4 must be not the *whole* amount of B – A, but only $\frac{6}{7}$ths of that amount, so as to reduce it to an interval of 60^s precise. Column 5 contains readings of the internal thermometer; column 6 is left blank for the results when reduced; and in column 7 are entered remarks, such as the state of the sky, wind, &c.; as also (when taken) the sun's altitude, barometer, thermometer, and other readings, &c.

A complete actinometer observation cannot consist of less than three sun and two shade observations intermediate; but five sun and four shade are much better. In a very clear sunny day it is highly desirable to continue the alternate observations for a long time, even from sunrise to sunset, so as to deduce by a graphical projection the law of diurnal increase and diminution of the solar radiation, which will thus readily become apparent, provided the *perfect* clearness of the sky continue,—an indispensable condition in these observations, the slightest cloud or haze over the sun being at once marked by a diminution of resulting radiation. To detect such haze or cirrus, a brown glass applied before the eye is useful, and by the help of such a glass it may here be noticed that solar halos are very frequently to be seen when the glare of light is such as to allow nothing of the sort to be perceived by the unguarded eye.

When a series is long continued in a good sun, the instrument grows very hot, and the rise of the liquid in the sun observation decreases, while the fall in the shade increases; nay, towards sunset it will fall even in the sun. This phenomenon (which is at first startling, and seeming to impeach the fidelity of the instrument) is, in fact, perfectly in order, and produces absolutely no irregularity in the resulting march of the radiation. Only it is necessary in the reduction of such observations to attend carefully to the algebraic signs of the differences in column 4.

Every series of actinometer observations should be accompanied with notices in the column of remarks of the state of the wind and sky generally, the approach of any cloud (as seen in the coloured glass) near to the sun; the barometer and thermometers, *dry* and *wet*, should especially be read off more than once during the series, if a long one, and, if kept up during several hours, hourly. The blackened thermometer for solar radiation should also be read off at the middle of every set, so as to accumulate a mass of comparative observations of the two instruments. The times should be correct to the nearest minute at least, as serving to calculate the sun's altitude; but if this be taken (to the nearest minute or two) with a pocket sextant, or even by a style and shadow, frequently (at intervals of an hour or less) when the sun is rising or setting, it will add much to the immediate interest of the observations. When the sun is near the horizon, its reflection from the sea, or any neighbouring water, must be prevented from striking on the instrument; and similarly of snow in cold regions, or on great elevations in alpine countries.

Every actinometer should be provided with a spare glass, and all the glasses should be marked with a diamond; and it should always be noted at the head of the column of remarks which glass is used, as the coefficient of reduction from the parts of the scale (which are arbitrary) to parts of the *unit of radiation* varies with the glass used.

To reduce provisionally a set of actinometer observations.—If the set consist of only four or five sun observations, with intermediate shades, take the mean of the "changes per minute" in column 4, for all the sun and for all the shade observations separately, attending duly to the signs. Change the sign of the latter mean, and add it to the former. The aggregate will be the uncorrected radiation in parts of the scale. To correct it for the unequal dilatability of the liquid, take the mean of the temperatures shown by the internal thermometer at the beginning and end of the set, and with it enter the table Appendix, Table V., which contains the factor by which the uncorrected radius is to be multiplied. If the series consist of more than a quadruple or sextuple observation, it must be broken into quadruplets, or quintuplets, and each reduced separately as above.

The abstract unit of solar radiation to be adopted in the *ultimate* reduction of the actinometric observations is the *actine*, by which is understood that intensity of solar radiation which at a vertical incidence, and supposing it wholly absorbed, would suffice to melt one millionth part of a metre in thickness from the surface of a sheet of ice horizontally exposed to its action per minute of mean solar time; but it will be well to reserve the reduction of the radiations as expressed in parts of the scale to their values in terms of their unit until some future and final discussion of the observations.

Meanwhile no opportunities should be lost of *comparing*

together the indications of different actinometers under similar and favourable circumstances, so as to establish a correspondence of scales, which in case of accident happening to one of the instruments will preserve its registered observations from loss. The comparison of two actinometers may be executed by one observer using alternately each of the two instruments, beginning and ending with the same; though it would be more conveniently done by two observers observing simultaneously at the same place, and each registering his own instrument. An hour or two thus devoted to comparisons in a calm clear day, and under easy circumstances, will in all cases be extremely well bestowed. In frosty or very cold weather the instrument should be exposed, for some time previous to commencing the observations, to the sun, which, by warming the liquid, increases its dilatability, which at low temperatures is inconveniently small.

Neither should each observer neglect to determine for himself the heat stopped by each of his glasses. This may be done also by alternating quadruplet observations made with the glass on and off, beginning and ending with the glass off, and (as in all cases) beginning and ending each quadruplet with a sun observation. For the purpose now in question a very *calm* day must be chosen, and a great many quadruplets must be taken in succession.

The actinometer is well calculated for measuring the defalcation of heat during any considerable eclipse of the sun. The observations should commence an hour at least before the eclipse begins, and be continued an hour beyond its termination, and the series should be uninterrupted, leaving to others to watch the phases of the eclipse. The atmospheric circumstances should be most carefully noted during the whole series.

Thermometers for Terrestrial Radiation.—The measure of terrestrial radiation is of no less importance to the science of meteorology than that of solar radiation, but no perfect instrument has yet been contrived for its determination. Valuable information, however, may be derived from the daily register of the minimum nocturnal temperature of a register spirit-thermometer, the bulb of which is placed in the focus of a concave metallic mirror, turned towards the clear aspect of the sky, and screened from currents. Such a thermometer may be read off and registered at the regular hours by day as well as by night, but it must be screened from sunshine, and a thermometer beside it also read off at the same times.

Registers.

To keep a meteorological register with due regularity, a skeleton form (No. 1) should be prepared, by ruling broad sheets of paper into columns destined for the reception of the daily and hourly entries in their *uncorrected* state, as read off or otherwise noted. This form may be most advantageously arranged in groups of columns, with general heading (*A*, *B*, *C*, &c.), and particular sub-headings (*a*, *b*, *c*, &c.), so as to class the entries in an order favourable to subsequent comparison and reduction. Thus, the group *A* should carry the general heading *Date*; *B*, *Pressure*; *C*, *Temperature of air*; *D*, *Moisture*; *E*, *Radiation*; *F*, *Temperature of water*; *G*, *Wind*; *H*, *Cloud*; *I*, *Weather*; *K*, *Rain*;

L, *Reference*; and opposite to every page ruled for entries should stand a blank page for remarks.

Under the general heading *A*, the sub-heading *Aa* will indicate the day of the month (marking the Sundays with S, and the days of new, full, and quarters of the moon with their appropriate marks ●, ☽, ○, ☾); and *Ab* will contain the hours of observation in each day, following the civil reckoning of time.

B will contain two sub-headings, *Ba*, *Bb*, corresponding to columns in which are entered respectively the readings of the barometer and its attached thermometer.

C will contain three: viz. *Ca* for entries of the external thermometer; *Cb*, the daily maxima; and *Cc*, the daily minima, placed opposite to the hours at which they are read off on the self-registering thermometer.

D will have two, *Da* and *Db*: viz. the readings of the dry and wet bulbs of the hygrometer.

Under *E* will stand three sub-columns, *Ea* for solar, and *Eb*, *Ec*, for terrestrial radiation. *Ea* will contain the readings of the black-bulb thermometer exposed in the exhausted tube to the sun at such of the regular hours when it can be observed; *Eb* those of the thermometer exposed to clear sky in the metallic reflector; and *Ec* those of a similar thermometer placed close beside it, and in all *other* respects similarly exposed. In these columns may also be entered the observed maxima of these elements, whether obtained by watching the instruments or by self-registering ones; and these observations should be distinguished from the others by enclosing them in parentheses, or underlining them, &c.

F will contain the temperature of the surface-water,

under the first sub-heading, *Fa;* and under the second, *Fb*, that at two fathoms depth: the latter not being taken more than once a day, except when *Fa* indicates some sudden change.

G will contain the *direction* of the wind, per vane and compass, in its first column, *Ga;* and its force, as read off on the anemometer, in *Gb.* If there be an upper and under current of wind, both their directions should be set down above and below a line, like a fraction.

H should have three sub-columns: viz. *Ha* for the amount of cloud in the region from the zenith down to 30° of altitude, and *Hb* for the amount below that altitude, each estimated in eighth parts of the whole respective areas of sky included in the two regions (*which are equal*), according to the best of the observer's judgment. *Hc* will contain the prevalent character of cloud, according to the nomenclature of Howard; denoting by C *cirrus*, by K *cumulus*,* by S *stratus*, and by N *nimbus*, by double letters their combination in transition from one to the other form (as *CS* cirro-stratus), and by letters with interposed commas (thus, *K, S*) the prevalence of one species of cloud in one and another in the other region. Two layers of cloud, one above the other, may be denoted by placing their characteristic letters above and below a line in the manner of a fraction. These forms of cloud are thus characterized:—*Cirrus* expresses a cloud resembling a lock of hair, or a feather, consisting of streaks, wisps, and fibres, vulgarly known as mares' tails. *Cumulus* denotes a cloud in dense convex heaps or rounded

* To avoid the otherwise inevitable confusion of C and c in MS.

forms, definitely terminated above, indicating saturation in the upper clear region of the air, and a rising supply of vapour from below. *Stratus* is an extended continuous level sheet, which must not be confounded with the flat base of the *cumulus*, where it simply reposes on the vapour plane. The cumulo-stratus, or *anvil shaped* cloud, is said to forerun heavy gales of wind. Peculiar aspects of cloud, preceding gales, squalls, or hurricanes, should be specially described in the sheet of remarks, or in a journal. *Nimbus* is a dense cloud, spreading out into a crown of cirrus above, and passing beneath into a shower.

Under the heading *I* will stand a note of the general state of the weather, according to Admiral Beaufort's system of abbreviations, which is as follows:—Numbers from 1 to 12 denote the force of the wind: thus, 0 denotes *calm;* 1, *light air*, just perceptible; 2, *light breeze*, in which a ship, clean full, in smooth water, would go from one to two knots; 3, *gentle breeze* (from two to four knots); 4, *moderate breeze* (from four to six knots); 5, *fresh breeze*, in which a ship *could just carry* on a wind royals, &c.; 6, *stormy breeze* (single-reefed topsails and topgallant-sails); 7, *moderate gale*.(double-reefed, &c.); 8, *fresh gale* (triple-reefed and courses); 9, *stormy gale* (close-reefed, &c.); 10, *whole gale* (close-reefed main-topsail and reefed foresail); 11, *storm* (storm-staysails); 12, *hurricane* (no canvas can stand). These numbers, in the absence of an anemometer, may be entered in column *Gb.* The following abbreviations denote the state of the weather:—

b. *Blue sky*, be the atmosphere clear or heavy.
c. *Clouds.* Detached passing clouds.
d. *Drizzling* rain.
f. *Foggy*.
g. *Gloomy* dark weather.
h. *Hail.*
l. *Lightning.*
m. *Misty* hazy atmosphere.
o. *Overcast.* The whole sky covered with thick clouds.
p. *Passing*, temporary showers.
q. *Squally*.
r. *Rain.* Continued rain.
s. *Snow.*
t. *Thunder.*
u. *Ugly* threatening appearance.
v. *Visibility* of objects; clear atmosphere.
w. *Wet.* (Dew.)

o under any letter denotes a great degree.

K contains only a column, *Ka*, for the quantity of rain, melted snow or hail, collected in the rain-gauge at the regular hour. One entry a day will suffice, except in rains of unusual heaviness or in paroxysmal discharges, which will require special note. There will always be room in this column to note the temperature of the falling rain, if remarkable.

Finally, *L* is a small column at the edge of the page, containing merely *numbers of reference*, from 1 to the number of lines of entry in the page, to connect each entry with the *remarks* on it, or on any phenomenon which may have occurred in the interval since the last entry, which it will be probably necessary to enter on another sheet or interleaved page (carrying at its left-hand edge a similar reference column), or with any more extended notes which may form part of a diary such as every observant traveller or voyager ought to keep, and of which a summary for the month, so far as relates to meteorological subjects, should be appended to each monthly register.

Another skeleton form, No. II., should be prepared

and ruled in corresponding columns, to receive the corrected and reduced results of the raw observations in Form I. This should have the column *A*, as in Form I; *B* will consist of a single column: viz. the barometer, reduced to 32 ; *C*, of the same number as in Form I, containing the *corrected* thermometer readings; under *D* will come two columns, *Da* and *Db*, of which the former will contain the *corrected* difference of the dry and wet bulb readings, and the latter the value of *F*, *the elastic force of vapour* at the dew-point by the formula already given. *E* and *F* will merely contain the *corrected* values of the corresponding entries in Form I; and if there be anything in the remaining columns requiring correction or reduction, it will here, of course, be done; if not, those columns must be either carefully copied or simply referred to. In this form should be entered, *when needed*, the monthly means of the several columns (in calculating which care should be taken to verify the results by repetition); and it is recommended, before adding up the columns, to look down each to see that no obvious error of entry (as of an inch in the barometer, a very common error) may remain to vitiate the mean result. The precaution should also be taken of counting the *entries* in each column, so as to make no mistake in the divisor. The monthly maxima, minima, and ranges of the instruments should also be entered. Except, however, the ship has been nearly stationary during the month, these calculations and their results are of little utility.

Both forms should be headed with the year and month on every page, and should bear the name of the ship and observer.

The observer will find it both interesting and instructive in a high degree to project the reduced observations (as fast as reduced, or monthly) of the barometric pressure, vapour tension (the value of *F* above mentioned), and temperature, in curves, by the aid of a paper of engraved squares, divided into inches and tenths by vertical and horizontal lines.* The comparison of the curves so projected in the case of the pressure and vapour tension is of especial interest, since there is great reason to believe that the diurnal fluctuation of the barometer is mainly, if not entirely, a hygrometric phenomenon arising from the superaddition of a variable hygrometric pressure to the otherwise uniform pressure of the *dry* atmosphere. The course of the barometric curve too will show far better than simple observation the chief maxima and minima which indicate the passage of the crests and troughs of atmospheric waves. And its continuance at a high or low level, or its gradual change, corresponding over long intervals of time with progressive changes of latitude and longitude, will enable the observer to trace out the limits of those deviations from the simple law of statical equilibrium which the researches of Schouw, Humboldt, and others have proved to exist more or less over the whole globe, and which those of Ermann in the Arctic, and King and Ross in the Antarctic regions have shown to rèsult in permanent local depressions to the enormous amount of a whole inch in the mercurial column.

* Such papers may be obtained from Messrs. W. H. Allen and Co., Booksellers to the Honourable East India Company, No. 7, Leadenhall-street, London.

Occasional Observations.

There is much and most valuable matter for meteorological observation and remark which cannot find a place in the regular entries of a register, either from its occasional nature, or from its statement requiring more detail than is consistent with the brevity of such entries. Observations of the Actinometer are of this kind, and require a separate register. Such also are all meteorological phenomena of a transitory nature, as hurricanes, thunderstorms, waterspouts, auroras, &c., of all which special and connected statements should be drawn up (embodying all notes made at the time) as soon after their occurrence as possible, and entered in a diary, carefully noting all circumstances connecting them with the state of the atmosphere preceding and subsequent, and especially every precursory appearance or fact which may have left on the observer's mind the impression of a *prognostic.* Such also are those occasions of which the attentive observer will not fail to take advantage, when particular meteorological sequences of cause and effect stand out in unusual prominence, or when opportunity is offered for the exact or approximate determination of some *datum* of scientific interest. The following hints respecting observations coming under these descriptions will be worth attending to.

Squalls, Storms, and Hurricanes or Cyclones.—It is hardly necessary to impress on the nautical observer the extreme importance of a minute attention to every adjunct of these formidable phenomena. From their first indications they should be attentively watched in all their phases, with a vigilance proportioned to their actual or expected intensity. Nothing in the way of *prognostic* should be left unnoticed. The "ugly threatening appearance" (*u*) should be analysed into its elements—atmospheric, celestial, oceanic, and (if in port) terrestrial signs of all kinds noted—such as the small white advancing cloud expanding into an arch, or the little white spot (*bull's-eye*) suddenly appearing in the zenith—lurid sky—remarkable red colour of clouds and of other objects—bands of light, and distant advancing walls of darkness—portions of cloud driven rapidly and irregularly—appearance of *ascending* lightning—peculiar aspect of stars or planets at night, or of the sun or moon at rising or setting, and *in what that peculiarity consists*—whirlwinds, waterspouts (and the direction in which they turn, whether ↺ or ↻), and peculiar veerings of wind with them, and alternations of calm—singular rises and lulls of wind and moaning or roaring noises, and whether these are *certainly* in the atmosphere—phosphorescent sea—flight of birds, uneasiness of animals—unusual abundance of certain fish. Meanwhile the movements of the barometer and the direction and force of the wind should be watched with unceasing vigilance. Hourly observations should be at once commenced, and the intervals diminished as it becomes more and more certain that a storm is in progress. When fairly established, they cannot be read too frequently, and every sudden rise or fall of the one, every lull or *shift* of the other (as distinguished from *veering*, which is a *gradual* change of direction), should be noted *to the minute of its occurrence*, or rather the watch should be read and noted to the minute, at every entry made.

During the continuance of the storm, and especially if there be reason (from its characters, and its occurrence in the hurricane regions) to consider it as a revolving hurricane, or "cyclone," in the sense insisted on by Mr. Redfield, Colonel Reid, and Mr. Piddington, all the atmospheric appearances and changes should be noted as frequently as possible, particularly at changes of the wind and in the calm centre of the vortex, should this unfortunately reach the ship. Flashes of light appearing in the barometer tube (*not* simply arising from oscillation) —thunder and lightning, particularly at shifts of the wind, and their relation to sudden discharges of hail or rain— temperature of rain and of the sea—form and size of hailstones—whirlwinds or waterspouts occurring *in* the cyclone, their appearance, whirlings, tracks, size, &c.— circles of light overhead in the centre of the gale, to be estimated or measured as to their angular diameter—sun, moon, or stars if seen, and if of peculiar brilliancy or colours—state of the sea as to regularity, rising, falling, breaking, &c., particularly at the centre—veerings or oscillations of wind, and the exact intervals in which they occur —moderating of the wind for an hour or two or more after the gale has appeared to commence, and the state of the instruments and sky at the time. An exact account should be kept of the vessel's coming up or falling off, and the log hove, if possible, to ascertain with the utmost care the ship's drift in lying to. Blasts of hot and cold air—extraordinary light and darkness. Whenever partial clearing of the sky affords an opportunity, pay particular attention to the direction of the upper scud. At the going off of the gale the same attention to all the phenomena as at its coming on. Observations of barometer, &c., to be continued at gradually increasing intervals, and at length hourly till the usual state of things is fully restored—gradual rising of clouds at horizon or zenith and banks forming to be noted, and their altitudes measured. The precise position of the ship before and after the gale to be carefully indicated, and all possible information to be collected of the manner, exact times, &c., in which other ships have been affected by it, and every endeavour used to trace out the path of the centre, the diameter of the vortex, and the direction of its revolution, by subsequent inquiry whenever opportunity may occur.

Hurricanes, revolving storms, or "cyclones," according to the meteorologists above named (now fully established as true representations of fact), differ from mere local and temporary exaggerations of the regular atmospheric currents in this—that they are in the nature of vortices, or circulating movements participated in by masses of air of from 50 to 500 miles in diameter, revolving the more rapidly the nearer the centre, up to a certain distance, or radius, *within which there is a calm.* The place of this centre of rotation meanwhile advances steadily along a definite line upon the globe, with a velocity varying from 2 to 30 or 40 miles per hour, and pursuing a track which in some of the hurricane regions, as in the West Indies, has a singular fixity of geographical situation and geometrical form. But the character which it is of most importance to a seaman to know, and the knowledge of which may often save his ship from disaster, as ignorance of it has repeatedly been the cause of catastrophes which might have been avoided, is this, viz. :—that in the same hemisphere great cyclones *always* revolve *the same way* (so far at least as our present information extends), but that this direction is opposite in opposite hemispheres. In the northern hemisphere their rotation is *retrograde*, *i. e.*, contrary to the motion of the hands of a watch laid face upwards, or in conformity with the motion of the hand in *unscrewing* a screw. In the southern, their rotation is *direct*, conformable to the hands of a watch, or to the motion of the hand screwing *in* a screw into a horizontal board. And this general fact affords the following simple rule by which to know at any given moment *the bearing of the centre of the vortex*, which is the point of extreme danger, by reason of the fury of the wind in its vicinity, its sudden reversal, and the terrible sea which prevails there. When *sure* that you are within the limits of a cyclone, stand erect and look full in the wind's eye, then, if in the northern hemisphere, turn yourself 90° or one quarter of the circle round *to your right* (if in the southern, as much to your left), and you will have the centre of the hurricane facing you. Thus, if in the northern hemisphere the wind at the ship be due north (blow *from* the north), the centre bears due east from the ship's place.

The habitual tracks of hurricanes are but imperfectly known, and all which tends to throw light upon this part of the subject is of the last importance to navigation. The reader may consult the works of Mr. Redfield, Colonel Reid on the 'Law of Storms' (2nd edition), and 'The Sailor's Hornbook for the Law of Storms,' by Mr. Piddington, a work full of interest and information on the subject, and which no navigator should go to sea unprovided with. The direction of the wind *after* the complete passage of the hurricane is a point of interest, as indicating whether the "cyclone" consist in the bodily transfer of a given mass of rotating air, or in the successive transmission of a rotary movement from air to air *in situ*, the air in each point of its track being only transiently agitated.

Winds.

The points most important to remark respecting the wind are,

1st. Its average intensity and general direction during the several portions of the day devoted to observation.

2ndly. The hours of the day or night when it commences to blow from a calm, or subsides into one from a breeze.

3rdly. The hours at which any remarkable changes of its direction take place.

4thly. The course which it takes in veering, and the quarter in which it ultimately settles.

5thly. The usual course of *periodical winds*, or such as remarkably prevail during certain seasons, with the law of their diurnal progress, both as to direction and intensity; at what hours and by what degrees they commence, attain their maximum, and subside; and through what points of the compass they run in so doing.

6thly. The existence of crossing currents at different heights in the atmosphere, as indicated by the courses of the clouds in different strata.

7thly. The times of setting-in of remarkably hot or cold winds, the quarters from which they come, and their courses, as connected with the progressive changes in their temperature.

8thly. The connexion of rainy, cloudy, or fair weather with the quarter from which the wind blows, or has blown for some time previously.

Several of these points of inquiry have especial reference to land winds, and can only be duly studied in port or during residences on shore. In cruises along shore, or on arrival or departure, observe how far the influence of the land extends, and by what gradations the character of the winds changes from terrestrial to oceanic, especially with reference to the difference between the hours when the sun is above and when below the horizon.

Clouds and Fogs.—The dissipation of cloud under the apparent influence of the full moon is a point to which attention has lately been called; the state of the sky on the days and nights of the full moon, and those immediately preceding and following it, should be noted with a view to confirming or refuting this connexion. Hourly observations, commencing before sunset, of the relative proportion of clear and clouded sky would be desirable.

The height of the vapour plane is a datum of importance, especially in tropical regions. At sea it is not easy to determine; but when near a mountainous coast, where the clouds repose at definite levels on the hills, many opportunities may occur to ascertain it with precision. The lower level of unmelted snow may in such localities also be made a subject of inquiry. The average height of the vapour plane at sea under the equator is a meteorological element of much interest. Opportunities of determining it by measuring the apparent altitude of the flat bases of cumular clouds from two ships whose distance is known, or otherwise, should be seized when they offer.

When the sky suddenly clouds over, or when fogs form unexpectedly, the barometer and thermometer should be noted minutely, as such appearances often result from a rapid diminution of atmospheric pressure and consequent absorption of heat by the rarefaction of the air. Fogs prevailing at definite localities indicate a temperature of the sea habitually exceeding that of the air. They often also indicate the vicinity of ice. Whenever fogs are met with at sea, the temperature of the air and of the surface water should be recorded with more than usual care.

Temperature of the Sea.—Shoals cast up water from a lower level to the surface where any current exists, and therefore a sudden change of temperature of the surface may indicate a shoal. In crossing currents coming from warmer or colder latitudes, the surface temperature should be especially attended to, and the maximum of irregularity due to the current watched for. Should opportunities offer of obtaining deep sea temperatures, they should be eagerly seized.

Observations in Port or in Temporary Residences on Shore.—Opportunities should not be lost of ascending lofty eminences, and noting thereon the hygrometric and thermometric conditions corresponding to altitudes measured by the portable barometer, or otherwise known.

The temperature of deep wells should be ascertained, and that of the soil at different depths, which, if made with due care and under favourable circumstances, is an observation of very great interest. Excavations should be made in dry soil, and *under fair exposure to sun and wind*, in which should be buried, at depths of three, six, and nine feet, thermometers well wrapped in woollen cloth, or in pots of pounded charcoal or even of dry sand, enclosed in strong vessels to defend them from damage, and to prevent the possibility of change of temperature in extracting and reading them. The zero points of the thermometers should be most scrupulously ascertained, and their errors *at* the temperatures registered made a subject of special inquiry. The readings should be exact to tenths of degrees. Observations thus made under the equator in various longitudes *with scientific precision*, might furnish data of the utmost value towards determining the constancy or variability of the sun's radiation from year to year. If the thermometers cannot be spared, bottles of water similarly defended may be buried (or of brandy, if in frozen soil), and the temperature of the liquid taken immediately on raising them. In case of prolonged sojourn this should be repeated monthly.

Some localities are remarkable for enormous falls of rain. Thus it is stated on the authority of Captain Roussin, that between the 1st and 24th of February, 1820, there fell *twelve feet* seven inches of rain in the Isle of Cayenne.* In all such localities great attention should be paid to the rain gauge, and pains taken to procure extracts from perfectly authentic registers containing instances of the kind, and information respecting their attendant circumstances. In some geographical localities it is said never to cease raining—in others, that rain never

* Ed. Phil. Journ., viii. 186, from Silliman's American Journal of Science, iv. p. 375.

falls. Local inquiry and consultation of records must here stand in lieu of personal observation, due care being taken to rely on none but unexceptionable evidence

The phenomena of dew are of more interest, and can be better studied on shore than at sea. The amount of dew collected by a given surface of any bibulous radiant, as cotton, &c., in clear nights, in exposures *perfectly* open to the sky, and on the level of the soil, should be registered. If accompanied with observations of the depression of the terrestrial radiant thermometer, and also of the hygrometer, such observations would acquire additional value.

The temperature of the soil under the direct influence of the sun as indicated by a thermometer barely covered with dry earth, is an element of importance to the botanist, and may be recommended as an apt accompaniment to actinometric observations. The thermometer used should have a scale reading at least to 180° Fahr.

Meteorological registers kept by persons of credit at places where the ship may remain or touch, should be enquired for and copied, or the originals procured, and the instruments with which they have been made carefully compared, and the height of their stations above the sea-level ascertained.

Waterspouts, Bull's-eye Squalls, Whirlwinds.—The transition from the mere eddy to the whirlwind, and from the whirlwind to the waterspout (*Trombe*), should be traced if possible. All circumstances from the first trace or prognostic to their final dissipation should be minutely noticed, especially the movements of the sea under their influence, and the direction of their rotation. At what distance is the whirling motion of the air perceptible? What are the indications of the barometer during their approach and recess? Do any and what electrical phenomena accompany them? Does the water really ascend along their axis, and to what height? Is the water they discharge in "bursting" fresh or salt? Note its temperature.

Showers of Dust or Ashes.—When they fall, preserve specimens and examine them microscopically, whether consisting of organized or mineral matter. Note every circumstance, especially the direction of the wind, and whether an upper current differing in direction from the lower, exist. Geographical situation of the ship especially to be exactly ascertained. Enquire for volcanic eruptions within a thousand miles of the place of their occurrence.

Thunderstorms, Lightning, Fireballs, &c.—Note the quarter of the horizon where distant lightning unaccompanied with thunder appears and the extent it embraces. Especially notice any appearance of forked lightning striking *upwards.* In an actual thunderstorm especially attend to the quantity of rain or hail that falls—its intermittences, and its correspondence or the contrary with great bursts of lightning near at hand. Notice the apparent direction taken by the storm, with, or against the wind. Attend to the remarkable reversal in the direction of the wind which often immediately follows the cessation of a thunderstorm. Violent thunder and lightning in the immediate vicinity of the place of observation sometimes, though *very* rarely, take place without rain, or with very little. In such cases notice every particular with the utmost minuteness, and ascertain if possible whether the storm has been *elsewhere* attended with rain. Fireballs are stated to have been occasionally seen running along the surface of the sea, and so reaching, striking and "bursting" on ships—appearances which have been supposed analogous to the electrical phenomenon termed *the glow discharge.* Attend to every circumstance which may favour or oppose this idea—especially the height of the clouds at the time, and whether or no remarkably depressed along the line taken by the fireball.

Should the ship be struck by lightning, if furnished with Sir Snow Harris's conductors (which appear to afford almost complete security against serious damage), examine the magnetism communicated to small steel bars (originally non-magnetic) fixed transversely across the copper conducting plates. Note any luminous appearance seen along the line of conduction. Immediately on the stroke, ascertain, by placing the hand on the conducting plate, whether it is in any degree heated. Notice peculiar noises, and endeavour to trace their origin, also the mode in which the lightning escapes from the ship, and the phenomena attending its escape.* If damage be done, describe minutely the sort of effects produced, and endeavour to trace the direction and character of the forces immediately productive of such as are purely mechanical.

Atmospheric Electricity can hardly be well studied at sea, the masts, sails, and rigging acting as perpetually interfering conductors. Indeed it is said that, except in actual thunderstorms, no indications whatever of atmospheric electricity can be detected in the open ocean.

* For the infinitely varied ways in which lightning may affect a ship when struck we recommend a perusal of Sir S. Harris's short but interesting work, 'Remarkable Instances of the Protection of certain Ships from the Destructive Effects of Lightning, &c.,' London, 1847.

This, however, should not be taken for granted. By going aloft the observer may put himself out of the reach of much of the interfering influence, taking with him an electroscope and a common jointed fishing-rod, having a glass stick well varnished with shellac substituted for its smallest joint to project into the atmosphere. To the end of the glass must be fixed a metallic rod terminating in a point, or carrying a small brass lantern, in which a lamp is burning, and connected with the electroscope by means of a fine copper wire. The electroscope may be either Saussure's pith ball, or Singer's gold leaf electrometer, and when charged the nature of the electricity may be tested by excited glass or sealing wax.

Auroral Phenomena.—All such should be minutely registered, and all their phases, especially the formation, extent, situation, movement and disappearance of arches, or any definite patches or banks of light. An acquaintance with the principal stars of the constellations is necessary to observe such phenomena with effect, and the observer will do well to provide himself for the purpose with planispheres, on which only the more conspicuous stars are indicated, with their allineations. The *exact* time (true at least to the nearest minute) of any such definite body of light being centrally on any known star should be observed, as a means (by the aid of corresponding observations) of determining its real situation and altitude. The slow drifting motion of such masses (in north latitudes generally southward—query if the reverse in south?) should be specially attended to. Pulsations, like waves of light, rushing up from the horizon, should be also particularly remarked, and any appearance of *patches of definite forms rendered visible by such pulsations as it traverses them*, but not otherwise appearing as luminous masses, particularly noticed. When arches or any considerable well-defined cloud-like masses are formed, mark on the chart their situation and extent among the stars at several noted epochs of time, particularizing the brightest portions; observe also the point of convergence of streamers and the formation of the corona, the central point or focus of which should be projected on the chart *with all possible exactness*, and the time of so doing exactly taken, so as to determine by subsequent calculation its altitude and azimuth. Any indication of the *near vicinity* of auroral phenomena, or of their existence at a level below that of ordinary clouds, should be most minutely investigated at the moment, and carefully and circumstantially recorded. The connexion, if any, between auroral masses and cirrous clouds should be traced if opportunity occur. Note also the meteors if remarkable within the auroral region.

Halos, *parhelia*, *mock suns*, and other luminous phenomena of the kind, should be noted, delineated with care if complicated, and their dimensions measured with a sextant, or otherwise, by bringing the limb of the sun or moon (noting which limb) in contact with the two edges of the phenomenon in succession. Their colours also and their order should be described. Light cirro-stratus cloud in the neighbourhood of the sun has been observed to be bordered with three fringes of pink and green colours following the outline of the cloud. This rare and beautiful phenomenon if seen should be most particularly and carefully described. Perhaps in some climates it may be of not unfrequent occurrence. Unusual tints observed in the sky should be noted, and should that extremely rare phenomenon—the sun's disc appearing *of a pale blue colour*, so little luminous as to allow of being gazed at with impunity*—occur, the atmospheric circumstances should be carefully recorded.

The polarization of the light of the sky should be examined *habitually* with a polariscope, and the relation of the points of maximum polarization to the sun, and the observer's zenith, noticed in every variety of climate, and in various states of the sky, and anything apparently abnormal recorded.

Zodiacal light.—In the seasons of its appearance take every opportunity in tropical climates to ascertain with precision the place of its apex among the stars, its breadth and degree of brightness, and whether variable or not.

Meteors.—See § 1. Astronomy—Appendix.

APPENDIX.

TABLE I.—Correction to be added to Barometers for Capillary Action.

Diameter of Tube.	Correction for	
	Unboiled Tubes.	Boiled Tubes.
Inch.	Inch.	Inch.
0·60	0·004	0·002
0·50	0·007	0·003
0·45	0·010	0·005
0·40	0·014	0·007
0·35	0·020	0·010
0·30	0·028	0·014
0·25	0·040	0·020
0·20	0·060	0·029
0·15	0·088	0·044
0·10	0·142	0·070

* It occurred at Bermuda on the 12th and 13th of August, 1831, two days after the great Barbadoes hurricane of that year.

Table II.—Correction to be applied to Barometers with *Brass Scales*, extending from the Cistern to the top of the Mercurial Column, to reduce the observation to 32° Fahrenheit.

Temp.	INCHES.															Temp.
	24	24·5	25	25·5	26	26·5	27	27·5	28	28·5	29	29·5	30	30·5	31	
°	+	+	+	+	+	+	+	+	+	+	+	+	+	+	+	°
0	·061	·063	·064	·065	067	·068	·069	·071	·072	·073	·074	·076	·077	·078	·080	0
1	·059	·061	·062	·063	·064	·065	·067	·068	·069	·071	·072	·073	·074	·076	·077	1
2	·057	·058	·060	·061	062	·063	·064	·066	·067	·068	·069	·070	·072	·073	·074	2
3	·055	·056	·057	·059	·060	·061	·062	·063	·064	·065	067	068	·069	·070	·071	3
4	·053	·054	·055	·056	·057	·058	·059	·061	·062	·063	·064	·065	·066	·067	·068	4
5	·051	·052	053	·054	·055	·056	·057	·058	·059	·060	·061	·062	·063	·065	·066	5
6	·049	·050	·051	·052	·053	·054	·055	·056	·057	·058	·059	·060	·061	·062	·063	6
7	·046	·047	·048	·049	·050	·051	·052	·053	·054	055	·056	·057	·058	·059	·060	7
8	·044	·045	·046	·047	·048	·049	·050	·051	·052	·053	·054	054	·055	·056	·057	8
9	·042	·043	044	·045	·046	·046	·047	048	·049	·050	·051	·052	·053	·054	·054	9
10	·040	·041	·042	·042	·043	·044	·045	·046	·047	·047	·048	·049	·050	·051	·052	10
11	·038	·039	·039	·040	·041	·042	·042	·043	·044	·045	·046	·046	·047	·048	·049	11
12	·036	·036	·037	·038	·039	·039	·040	·041	·042	·042	·043	·044	·045	·045	·046	12
13	·033	·034	·035	·036	·036	·037	·038	·038	·039	·040	·040	·041	·042	·043	·043	13
14	·031	·032	·033	·033	·034	·035	·035	·036	·037	·037	·038	·038	·039	·040	·040	14
15	·029	·030	·039	·031	·032	·032	·033	·033	·034	·035	·035	·036	·036	·037	·038	15
16	·027	·028	·028	·029	·029	·030	·030	·031	·032	·032	·033	·033	·034	·034	·035	16
17	·025	·025	·026	·026	·027	·027	·028	·028	·029	·030	·030	·031	·031	·032	·032	17
18	·023	·023	·024	·024	·025	·025	·025	·026	·026	·027	·027	·028	·028	·029	·029	18
19	·021	·021	·021	·022	·022	·023	·023	·024	·024	·024	·025	·025	·026	·026	·027	19
20	·018	·019	·019	·020	·020	·020	·021	·021	·021	·022	·022	·023	·023	·023	·024	20
21	·016	·017	·017	017	·018	·018	·018	·019	·019	·019	·020	·020	·020	·021	·021	21
22	·014	·014	·015	·015	·015	·016	·016	·016	·016	·017	·017	·017	·018	·018	018	22
23	·012	·012	·012	·013	·013	·013	·013	·014	·014	·014	014	·015	·015	·015	·015	23
24	·010	·010	·010	·010	·011	·011	·011	·011	·011	·012	·012	·012	·012	·012	·013	24
25	·008	·008	·008	·008	·008	·008	·009	·009	·009	·009	·009	·009	·009	·010	·010	25
26	·005	·006	·006	·006	·006	·006	·006	·006	·006	·006	·007	·007	·007	·007	·007	26
27	·003	·003	·003	·003	·004	·004	·004	·004	·004	·004	·004	·004	·004	·004	·004	27
28	·001	·001	·001	·001	·001	·001	·001	·001	·001	·001	·001	·001	·001	·001	·001	28
	—	—	—	—	—	—	—	—	—	—	—	—	—	—	—	
29	·001	·001	·001	·001	·001	001	·001	·001	·001	·001	·001	·001	·001	·001	·001	29
30	·003	·003	·003	·004	·004	·004	·004	·004	·004	·004	·004	·004	·004	·004	·004	30
31	·005	006	·006	·006	·006	·006	·006	·006	·006	·006	·007	·007	·007	·007	·007	31
32	·008	·008	·008	·008	·008	·008	·008	·009	·009	·009	·009	·009	·009	·010	·010	32
33	·010	·010	·010	·010	·011	·011	·011	·011	·011	·012	·012	·012	·012	·012	·012	33
34	·012	·012	·012	·013	·013	·013	·013	·014	·014	·014	·014	·015	·015	·015	·015	34
35	·014	·014	·015	·015	·015	·015	·016	·016	·016	·017	·017	·017	·018	·018	·018	35
36	·016	·017	·017	·017	·017	·018	·018	·019	·019	·019	·020	·020	·020	·021	·021	36
37	·018	·019	·019	·019	·020	·020	·021	·021	·021	·022	·022	·022	·023	·023	·024	37
38	·020	·021	·021	·022	·022	·023	·023	·023	·024	·024	·025	·025	·026	·026	·026	38
39	·023	·023	·024	·024	·024	·025	·025	·026	·026	·027	·027	·028	·028	·029	·029	39
40	·025	·025	·026	·026	·027	·027	·028	·028	·029	·029	·030	·030	·031	·031	·032	40
41	·027	·027	·028	·029	·029	·030	·030	·031	·031	·032	·033	·033	·034	·034	·035	41
42	·029	·030	·030	·031	·031	·032	·033	·033	·034	·034	·035	·036	·036	·037	·037	42
43	·031	·032	·032	·033	·034	·034	·035	·036	·036	·037	·038	·038	·039	·040	·040	43
44	·033	·034	·035	·035	·036	·037	·037	·038	·039	·040	·040	·041	·042	·042	·043	44
45	·035	·036	·037	·038	·038	·039	·040	·041	·041	·042	·043	·044	·044	·045	·046	45
46	·038	·038	·039	·040	041	·042	·042	·043	·044	·045	·045	·046	·047	·048	·049	46
47	·040	·041	·041	·042	·043	·044	·045	·046	·046	·047	·048	·049	·050	·051	·051	47
48	·042	·043	·044	·045	·045	·046	·047	·048	·049	·050	·051	·052	·052	·053	·054	48
49	·044	·045	·046	·047	·048	·049	·050	·050	·051	·052	·053	·054	·055	·056	·057	49
50	·046	·047	·048	·049	·050	·051	·052	·053	·054	·055	·056	057	·058	·059	·060	50

TABLE II.—*Continued.*

Temp.	INCHES.															Temp.
	24	24·5	25	25·5	26	26·5	27	27·5	28	28·5	29	29·5	30	30·5	31	
°	—	—	—	—	—	—	—	—	—	—	—	—	—	—	—	°
51	·048	·049	·050	·051	·052	·053	·054	·055	·056	·057	·058	·059	·060	·061	·062	51
52	·050	·052	·053	·054	·055	·056	·057	·058	·059	·060	·061	·062	·063	·064	·065	52
53	·053	·054	·055	·056	·057	·058	·059	·060	·061	·063	·064	·065	·066	·067	·068	53
54	·055	·056	·057	·058	·059	·060	·062	·063	·064	·065	·066	·067	·068	·070	·071	54
55	·057	·058	·059	·060	·062	·063	·064	·065	·066	·068	·069	·070	·071	·072	·073	55
56	·059	·060	·061	·063	·064	·065	·066	·068	·069	·070	·071	·073	·074	·075	·076	56
57	·061	·062	·064	·065	·066	·068	·069	·070	·071	·073	·074	·075	·076	·078	·079	57
58	·063	·065	·066	·067	·069	·070	·071	·073	·074	·075	·077	·078	·079	·081	·082	58
59	·065	·067	·068	·070	·071	·072	·074	·075	·076	·078	·079	·080	·082	·083	·085	59
60	·068	·069	·070	·072	·073	·075	·076	·077	·079	·080	·082	·083	·085	·086	·087	60
61	·070	·071	·073	·074	·075	·077	·078	·080	·081	·083	·084	·086	·087	·089	·090	61
62	·072	·073	·075	·076	·078	·079	·081	·082	·084	·085	·087	·088	·090	·091	·093	62
63	·074	·076	·077	·079	·080	·082	·083	·085	·086	·088	·089	·091	·093	·094	·096	63
64	·076	·078	·079	·081	·082	·084	·086	·087	·089	·090	·092	·094	·095	·097	·098	64
65	·078	·080	·082	·083	·085	·086	·088	·090	·091	·093	095	·096	·098	·100	·101	65
66	·080	·082	·084	·085	·087	·089	·090	·092	·094	·096	·097	·099	·101	·102	·104	66
67	·083	·084	·086	·088	·089	·091	·093	·095	·096	·098	·100	·102	·103	·105	·107	67
68	·085	·086	·088	·090	·092	·094	·095	·097	·099	·101	·102	·104	·106	·108	·109	68
69	·087	·089	·090	·092	·094	·096	·098	·100	·101	·103	·105	·107	·109	·110	·112	69
70	·089	·091	·093	·095	·096	·098	·100	·102	·104	·106	·108	·109	·111	·113	·115	70
71	·091	·093	·095	·097	·099	·101	·102	·104	·106	·108	·110	·112	·114	·116	·118	71
72	·093	·095	·097	·099	·101	·103	·105	·107	·109	·111	·113	·115	·117	·119	·120	72
73	·095	·097	·099	·101	·103	·105	·107	·109	·111	·113	·115	·117	·119	·121	·123	73
74	·097	·099	·102	·104	·106	·108	·110	·112	·114	·116	·118	·120	·122	·124	·126	74
75	·100	·102	·104	·106	·108	·110	·112	·114	·116	·118	·120	·122	·125	·127	·129	75
76	·102	·104	·106	·108	·110	·112	·114	·117	·119	·121	·123	·125	·127	·129	·131	76
77	·104	·106	·108	·110	·112	·115	·117	·119	121	·123	·126	·128	·130	·132	·134	77
78	·106	·108	·110	·113	·115	·117	·119	·122	·124	·126	·128	·130	·133	·135	·137	78
79	·108	·110	·113	·115	·117	·119	·122	·124	·126	·128	·131	·133	·135	·137	·140	79
80	·110	·113	·115	·117	·119	·122	·124	·126	·129	·131	·133	·136	·138	·140	·143	80
81	·112	·115	·117	·119	·122	·124	·126	·129	·131	·134	·136	·138	·141	·143	·145	81
82	·114	·117	·119	·122	·124	·126	·129	·131	·134	·136	·138	·141	·143	·146	·148	82
83	·117	·119	·121	·124	·126	·129	·131	·134	·136	·139	·141	·143	·146	·148	·151	83
84	·119	·121	·124	·126	·129	·131	·134	·136	·139	·141	·144	·146	·149	·151	·154	84
85	·121	·123	·126	·128	·131	·133	·136	·139	·141	·144	·146	·149	·151	·154	·156	85
86	·123	·126	·128	·131	·133	·136	·138	·141	·144	·146	·149	·151	·154	·156	·159	86
87	·125	·128	·130	·133	·136	·138	·141	·143	·146	·149	·151	·154	·157	·159	·162	87
88	·127	·130	·133	·135	·138	·141	·143	·146	·149	·151	·154	·157	·159	·162	·165	88
89	·129	·132	·135	·137	·140	·143	·146	·148	·151	·154	·156	·159	·162	·165	·167	89
90	·131	·134	·137	·140	·142	·145	·148	·151	·153	·156	·159	·162	·164	·167	·170	90
91	134	·136	·139	·142	·145	·148	·150	·153	·156	·159	·162	·165	·167	·170	·173	91
92	·136	·139	·141	·144	·147	·150	·153	·156	·158	·161	·164	·167	·170	·172	·175	92
93	·138	·141	·144	·147	·149	·152	·155	·158	·161	·164	·167	·170	·172	·175	·178	93
94	·140	·143	·146	·149	·152	·155	·157	·161	·163	·166	·169	·172	·175	·177	·180	94
95	·142	·145	·148	·151	·154	·157	·160	·163	·166	·169	·172	·175	·178	·180	·183	95
96	·144	·147	·150	·153	·156	·159	·162	·165	·168	·171	·174	·178	·181	·183	·186	96
97	·146	·149	·152	·156	·159	·162	·165	·168	·171	·174	·177	·180	·183	·186	·189	97
98	·148	·152	·155	·158	·161	·164	·167	·170	·173	·176	·179	·183	·186	·188	·191	98
99	·151	·154	·157	·160	·163	·166	·169	·173	·176	·179	·182	·185	·188	·191	·194	99
00	·153	·156	·159	·162	·165	·169	·172	·175	·178	·181	·184	·188	·191	·194	·197	100

TABLE III.—Elastic Force of Aqueous Vapour for every Degree of Temperature, from 0° to 103° Fahrenheit.

Temp. Fahr.	Force. Inches of Mercury.	Temp. Fahr.	Force. Inches of Mercury.	Temp. Fahr.	Force. Inches of Mercury.	Temp. Fahr.	Force. Inches of Mercury.
°		°		°		°	
0	0·051	26	0·147	52	0·389	78	0·942
1	0·053	27	0·153	53	0·402	79	0·973
2	0·056	28	0·159	54	0·417	80	1·005
3	0·058	29	0·165	55	0·432	81	1·036
4	0·060	30	0·172	56	0·447	82	1·072
5	0·063	31	0·179	57	0·463	83	1·106
6	0·066	32	0·186	58	0·480	84	1·142
7	0·069	33	0·193	59	0·497	85	1·179
8	0·071	34	0·200	60	0·514	86	1·217
9	0·074	35	0·208	61	0·532	87	1·256
10	0·078	36	0·216	62	0·551	88	1·296
11	0·081	37	0·224	63	0·570	89	1·337
12	0·084	38	0·233	64	0·590	90	1·380
13	0·088	39	0·242	65	0·611	91	1·423
14	0·092	40	0·251	66	0·632	92	1·468
15	0·095	41	0·260	67	0·654	93	1·514
16	0·099	42	0·270	68	0·676	94	1·562
17	0·103	43	0·280	69	0·699	95	1·610
18	0·107	44	0·291	70	0·723	96	1·660
19	0·112	45	0·302	71	0·748	97	1·712
20	0·116	46	0·313	72	0·773	98	1·764
21	0·121	47	0·324	73	0·799	99	1·819
22	0·126	48	0·336	74	0·826	100	1·874
23	0·131	49	0·349	75	0·854	101	1·931
24	0·136	50	0·361	76	0·882	102	1·990
25	0·142	51	0·375	77	0·911	103	2·050

TABLE IV.—Showing the force of the Wind on a square foot for different heights of the Column of Water in Lind's Wind-gauge.

Height of the Column of Water.	Force of the Wind in Avoirdupois Pounds.	Height of the Column of Water.	Force of the Wind in Avoirdupois Pounds.
Inches.		Inches.	
12	62·5	5	26·04
11	57·29	4	20·83
10	52·08	3	15·62
9	46·87	2	10·42
8	44·66	1	5·21
7	36·55	0·5	2·60
6	31·75	0·1	0·52
		0·05	0·26

TABLE V.—Showing the Factors by which the Results of Actinometer Observations have to be Multiplied, to reduce them to a constant dilatability of the enclosed Blue Liquid, viz., that at 60° Fahr.

Internal Ther.	Reducing Factor.	Internal Ther.	Reducing Factor.	Internal Ther.	Reducing Factor.	Internal Ther.	Reducing Factor.
°		°		°		°	
30	2·888	63	0·943	96	0·642	129	0·550
31	2·713	64	0·925	97	0·638	130	0·549
32	2·862	65	0·908	98	0·634	131	0·547
33	2·430	66	0·892	99	0·630	132	0·545
34	2·312	67	0·877	100	0·626	133	0·543
35	2·205	68	0·862	101	0·622	134	0·541
36	2·106	69	0·849	102	0·618	135	0·540
37	2·014	70	0·837	103	0·615	136	0·538
38	1·930	71	0·826	104	0·612	137	0·536
39	1·853	72	0·815	105	0·609	138	0·535
40	1·781	73	0·805	106	0·606	139	0·533
41	1·715	74	0·795	107	0·603	140	0·532
42	1·654	75	0·785	108	0·600	141	0·530
43	1·597	76	0·775	109	0·597	142	0·528
44	1·543	77	0·766	110	0·594	143	0·527
45	1·492	78	0·757	111	0·591	144	0·525
46	1·443	79	0·748	112	0·588	145	0·524
47	1·397	80	0·739	113	0·586	146	0·522
48	1·353	81	0·731	114	0·583	147	0·521
49	1·312	82	0·723	115	0·581	148	0·519
50	1·274	83	0·716	116	0·578	149	0·518
51	1·239	84	0·709	117	0·576	150	0·517
52	1·206	85	0·702	118	0·573	151	0·515
53	1·175	86	0·695	119	0·571	152	0·514
54	1·147	87	0·689	120	0·569	153	0·513
55	1·120	88	0·683	121	0·566	154	0·512
56	1·094	89	0·677	122	0·564	155	0·511
57	1·069	90	0·671	123	0·562	156	0·509
58	1·045	91	0·666	124	0·560	157	0·508
59	1·022	92	0·661	125	0·558	158	0·507
60	1·000	93	0·656	126	0·556	159	0·506
61	0·980	94	0·651	127	0·554	160	0·505
62	0·961	95	0·646	128	0·552		

Section X.

ON ATMOSPHERIC WAVES AND BAROMETRIC CURVES.

By WILLIAM RADCLIFF BIRT, Esq.

In sketching out a system of barometric observation having especial reference to the acquisition of data from which the *barometric character* of certain large areas of the surface of the globe may be determined—inasmuch as such areas are distinguished from each other, on the one hand by consisting of extensive spaces of the oceanic surface unbroken, or scarcely broken, by land; on the other by the proximity of such oceanic surface to large masses of land, and these masses presenting two essentially different features, the one consisting of land particularly characterized as continental, the other as insular, regard has been accordingly had to such distribution of land and water.

As these instructions are intended for officers in Her Majesty's and the mercantile service, observations on land have not been alluded to; but in order that the data accumulated may possess that value which is essential for carrying on the inquiry in reference to atmospheric waves and barometric curves with success, provision is made to mark out more distinctly the barometric effects of the junction of large masses of land and water. It is well known that the oceanic surface, and even the smaller surfaces of inland seas, produce decided inflexions of the isothermal lines. They exercise an important influence on temperature. It has also been shown that the neighbourhood of water has a very considerable influence in increasing the oscillations of the mercurial column in the barometer, and in the great systems of European undulations it is well known that these oscillations increase especially towards the north-west. The converse of this, however, has not yet been subjected to observation; there has been no systematic co-operation of observers for the purpose of determining the barometric affections of large masses of water, such as the central portion of the basin of the northern Atlantic, the portion of oceanic surface between the Cape of Good Hope and Cape Horn, the Indian and Southern oceans, and the vast basin of the Pacific. Nor are we yet acquainted with the character of the oscillations, whether increasing or decreasing, as we recede from the central portions of the oceanic surfaces we have mentioned towards the land which forms their eastern, western, or northern boundaries. This influence of the junction line of land and water, so far as it is yet known, has been kept in view in framing these instructions, and, as it appears so prominently in Europe, it is hoped that additional observations between the four daily readings * to which probably many observers may habitually restrict themselves, making on certain occasions and in particular localities a series of observations at intervals of three hours, will not be considered too frequent when the great importance of the problem to be solved is fully

* See p. 271.

apprehended. It need scarcely be said that the value of these observations at three-hourly intervals will be greatly increased by the number of observers co-operating in them. Upon such an extensive system of co-operation a large space on the earth's surface, possessing peculiarities which distinguish it from others extremely unlike it in their general character, or assimilate it to such as possess with it many features in common, is marked out below for particular observation, occupying more than two-thirds of a zone in the northern hemisphere, having a breadth of 40°, and including every possible variety of terrestrial and aqueous surface, from the burning sands of the great African desert, situated about the centre, to the narrow strip of land connecting the two Americas on the one side, and the chain of islands connecting China and Hindostan with Australia on the other. On each side of the African continent we have spaces of open sea between 30° and 40° west longitude north of the equator, and between 60° and 80° east longitude, in or to the south of the equator, admirably suited for contrasting the barometric affections, as manifested in these spaces of open water, with those occurring in situations where the influence of the terrestrial surface comes into more active operation.

The localities where three-hourly readings are chiefly desirable may be specified under the heads of *Northern Atlantic, Southern Atlantic, Indian* and *Southern Oceans*, and *Pacific Ocean*.

Northern Atlantic. Homeward-bound Voyages.—The discussion of observations made in the United Kingdom and the western border of central Europe, has indicated that off the north-west of Scotland a centre of great barometric disturbance exists. This centre of disturbance appears to be considerably removed from the usual tracks of vessels crossing the Atlantic; nevertheless some light may be thrown on the barometric phænomena resulting from this disturbance by observations during homeward-bound voyages, especially after the vessels have passed the meridian of 50° west longitude. Voyagers to or from Baffin and Hudson bays would do well during the whole of the voyage to read off the barometer every three hours, as their tracks would approach nearest the centre of disturbance in question. Before crossing the 50th meridian, the undulations arising from the distribution of land and water in the neighbourhood of these vast inland seas would receive considerable elucidation from the shorter intervals of observation, and after passing the 50th meridian the extent of undulation, as compared with that observed by the more southerly vessels, would be more distinctly marked by the three-hourly series. Surveying vessels stationed on the north-western coasts of Ireland and Scotland may contribute most important information on this head by a regular, and as far as circumstances will allow, an uninterrupted series either of six-hourly or three-hourly observations. The intervals of observation on board vessels stationed at the Western Isles, the Orkneys, and the Shetland Isles, ought not to be longer than *three* hours, principally on account of the great extent of oscillation observed in those localities. Vessels arriving from all parts of the world as they approach the United Kingdom should observe at shorter intervals than six hours. As a general instruction on this head the series of three-hourly observations may be commenced on board vessels from America and the Pacific by the way of Cape Horn on their passing the 20th meridian, such three-hourly observations to be continued until the arrival of the vessels in port. Ships by the way of the Cape of Good Hope should commence the three-hourly series either on leaving or passing the colony, in order that the phænomena of the tropical depression hereafter to be noticed may be well observed.

Northern Atlantic. Outward-bound Voyages.—Vessels sailing to the United States, Mexico, and the West Indies, should observe at three hours' interval upon passing the 60th meridian. Observations at this interval, on board vessels navigating the Gulf of Mexico and the Caribbean Sea, will be particularly valuable in determining the extent of oscillation as influenced by the masses of land and water in this portion of the torrid zone, as compared with the oscillation noticed off the western coast of Africa, hereafter to be referred to.

Southern Atlantic. Outward and homeward bound.—Without doubt the most interesting phænomenon, and one that lies at the root of the great atmospheric movements, especially those proceeding northwards in the northern hemisphere and southwards in the southern, is the equatorial depression first noticed by Von Humboldt and confirmed by many observers since. We shall find the general expression of this most important meteorological fact in the Report of the Committee of Physics and Meteorology, appointed by the Royal Society in 1840, as follows:

"The barometer, at the level of the sea, does not indicate a mean atmospheric pressure of equal amount in all parts of the earth; but on the contrary the equatorial pressure is uniformly less in its mean amount than at and beyond the tropics." Vessels that are outward bound should, upon passing 40° north latitude, commence the series of three-hourly observations, with an especial reference to the equatorial depression. These three-hourly observations should be continued until the latitude of 40° south has been passed: the whole series will then include the minimum of the depression and the two maxima or apices forming its boundaries. (See Daniell's 'Meteorological Essays,' 3rd edition.) In passages across the equator, should the ships be delayed by calms, opportunities should be embraced for observing this depression with greater precision by means of *hourly* readings; and these readings will not only be valuable as respects the depression here spoken of, but will go far to indicate the character of any disturbance that may arise, and point out, as nearly as such observations will allow, the precise time when such disturbance produced its effects in the neighbourhood of the ships. In point of fact they will clearly illustrate the diversion of the tendency to rise, spoken of in the Report before alluded to, as resulting in ascending columns and sheets, between which wind flaws, capricious in their direction and intensity, and often amounting to sharp squalls, mark out the course of their feeders and the indraft of cooler air from a distance to supply their void. Hourly observations, with especial reference to this and the following head of enquiry, should also be made off the western coast of Africa during the homeward-bound voyage.

Immediately connected with this part of the outward-bound voyage, hourly observations, as often as circumstances will permit, while the ships are sailing from the Madeiras to the equator, will be extremely valuable in elucidating the origin of the great system of south-westerly atmospheric waves that traverse Europe, and in furnishing data for comparison with the amount of oscillation and other barometric phænomena in the Gulf of Mexico and the Caribbean Sea, a portion of the torrid zone essentially different in its configuration and in the relations of its area to land and water, as contra-distinguished to the northern portion of the African continent; and these hourly observations are the more desirable as the vessels may approach the land. They may be discontinued on passing the equator, and the three-hourly series resumed.

There are two points in the southern hemisphere between 80° west longitude, and 30° east longitude, that claim particular attention in a barometric point of view, viz., Cape Horn and the Cape of Good Hope; the latter is within the area marked out for the three-hourly observations, and too much attention cannot be paid to the indications of the barometer as vessels are approaching or leaving the Cape. The northern part of the South Atlantic Ocean has been termed the *true Pacific Ocean of the world;* and at St. Helena a gale was scarcely ever known; it is also said to be entirely free from actual storms (Col. Reid's 'Law of Storms,' 1st edition, p. 415). It may therefore be expected that the barometer will present in this locality but a small oscillation, and ships in sailing from St. Helena to the Cape will do well to ascertain, by means of the three-hourly observations, the increase of oscillation as they approach the Cape. The same thing will hold good with regard to Cape Horn: it appears from previous observation that a permanent barometric depression exists in this locality, most probably in some way connected with the immense depression noticed by Captain Sir James Clark Ross, towards the Antarctic Circle. The general character of the atmosphere off Cape Horn is also extremely different from its character at St. Helena. It would therefore be well for vessels sailing into the Pacific by Cape Horn, to continue the three-hourly observations until the 90th meridian is passed.

Before quitting the Atlantic Ocean it may be well to notice the marine stations mentioned in my Third Report on Atmospheric Waves,* as being particularly suitable for testing the views advanced in that report and for tracing a wave of the south-westerly system from the most western point of Africa to the extreme north of Europe. A series of hourly observations off the western coast of Africa has already been suggested. Vessels staying at Cape Verd Islands should not omit to make observations at three hours' interval *during the whole of their stay*, and when circumstances will allow, hourly readings. At the Canaries, Madeiras, and the Azores, similar observations should be made. Vessels touching at Cape Cantin, Tangier, Gibraltar, Cadiz, Lisbon, Oporto, Corunna, and Brest, should also make these observations while they are in the localities of these ports. At the Scilly Isles we

* Reports of the British Association for the Advancement of Science, 1846, p. 139.

have six-hourly observations, made under the superintendence of the Honourable the Corporation of the Trinity House. Ships in nearing these islands and making the observations already pointed out, will greatly assist in determining the increase of oscillation proceeding westward from the nodal point of the two great European systems. We have already mentioned the service surveying vessels employed on the coasts of Ireland and Scotland may render, and the remaining portion of the area marked out in the report may be occupied by vessels navigating the North Sea and the coast of Norway, as far as Hammerfest.

In connection with these observations, having especial reference to the European system of south-westerly atmospheric waves, the Mediterranean presents a surface of considerable interest both as regards these particular waves, and the influence its waters exert in modifying the two great systems of central Europe. The late Professor Daniell has shown from the Mannheim observations, that small undulations, having their origin on the northern borders of the Mediterranean, have propagated themselves northward, and in this manner, but in a smaller degree, the waters of the Mediterranean have contributed to increase the oscillation as well as the larger surface of the northern Atlantic. In most of the localities of this great inland sea six-hourly observations may suffice for this immediate purpose ; but in sailing from Lisbon through the Straits of Gibraltar, in the neighbourhood of Sicily and Italy, and in the Grecian Archipelago, we should recommend the three-hourly series, as marking more distinctly the effects resulting from the proximity of land ;

this remark has especial reference to the passage through the Straits of Gibraltar, where, if possible, hourly observations should be made.

The Indian and Southern Oceans. Outward and homeward bound.—On sailing from the Cape of Good Hope to the East Indies, China, or Australia, observations at intervals of three hours should be made until the 40th meridian east is passed (homeward-bound vessels should commence the three-hourly readings on arriving at this meridian). Upon leaving the 40th meridian the six-hourly observations may be resumed on board vessels bound for the Indies and China until they arrive at the equator, when the readings should again be made at intervals of three hours, and continued until the arrival of the vessels in port. With regard to vessels bound for Australia and New Zealand, the six-hourly readings may be continued from the 40th to the 100th meridian, and upon the vessels passing the latter, the three-hourly readings should be commenced and continued until the vessels arrive in port. Vessels navigating the Archipelago, between China and New Zealand, should make observations every three hours, in order that the undulations arising from the configuration of the terrestrial and oceanic surfaces may be more distinctly marked and more advantageously compared with the Gulf of Mexico, the Caribbean Sea, and the northern portion of the African continent.

The Pacific Ocean.—As this ocean presents so vast an aqueous surface, generally speaking observations at intervals of six hours will be amply sufficient to ascertain its leading barometric phænomena. Vessels, however, on approaching the continents of North and South America, or sailing across the equator, should resort to the three-hourly readings, in order to ascertain more distinctly the effect of the neighbourhood of land on the oscillations of the barometer, as generally observed, over so immense a surface of water in the one case, and the phænomena of the equatorial depression in the other: the same remarks relative to the latter subject, which we offered under the head of South Atlantic, will equally apply in the present instance. The configuration of the western shores of North America renders it difficult to determine the precise boundary where the three-hourly series should commence; the 90th meridian is recommended for the boundary as regards South America, and from this a judgment may be formed as to where the three-hourly observations should commence in reference to North America.

In the previous sketch of the localities for the more important observations, it will be seen that within the tropics there are three which demand the greatest regard.

I. The Archipelago between the two Americas, more particularly comprised within the 40th and 120th meridians west longitude, and the equator and the 40th degree of north latitude. As a general principle we should say that vessels within this area should observe the barometer every three hours. Its eastern portion includes the lower branches of the storm paths, and on this account is peculiarly interesting, especially in a barometric point of view.

II. *The Northern portion of the African Continent, including the Sahara or Great Desert.*—This vast radiating surface must exert considerable influence on the waters on each side northern Africa. Vessels sailing within the area comprised between 40° west and 70° east and the equator and the 40th parallel, should also make observations at intervals of three hours.

III. *The great Eastern Archipelago.*—This presents a somewhat similar character to the western; like that, it is the region of terrific hurricanes, and it becomes a most interesting object to determine its barometric phenomena; the three-hourly system of observation may therefore be resorted to within an area comprised between the 70th and 140th meridians, and the equator and the 40th degree of north latitude.

The southern hemisphere also presents three important localities, the prolongations of the three tropical areas. It is unnecessary to enlarge upon these, as ample instructions have been already given. We may, however, remark, with regard to Australia, that three-hourly observations should be made within the area comprised between the 100th and 190th meridians east, and the equator and the 50th parallel south, and hourly ones in the immediate neighbourhood of all its coasts.

Storms, Hurricanes, and Typhoons.

The solution of the question—How far and in what manner are storms connected with atmospheric waves?—must be extremely interesting to every one engaged in either the naval or merchant service. It is foreign to the purpose of these instructions to enter into any examination of the views that exist on this head. Our great object here will be to endeavour to mark out such a line of observation as appears most capable of throwing light, not only on the most important desiderata as connected with storms, but also their connection or non-connection with atmospheric waves. We shall accordingly arrange this portion of the instructions under the following heads:—*Desiderata; Localities; Margins; Preceding and Succeeding Accumulations of Pressure.*

Desiderata.—The most important desiderata appertaining to the subject of storms, are certainly their origin and termination. Of these initial and terminal points in the course of great storms we absolutely know nothing, unless *the white appearance of a round form* observed by Mr. Seymour on board the Judith and Esther, in lat. 17° 19′ north and long. 52° 10′ west (see Col. Reid's 'Law of Storms,' 1st edit. p. 65), may be regarded as the commencement of the Antigua hurricane of August 2, 1837. This vessel was the most eastern of those from which observations had been obtained; and it is the absence of contemporaneous observations to the eastward of the 50th meridian that leaves the question as to the origin of the West Indian revolving storms unsolved. Not one of Mr. Redfield's storm routes extends eastward of the 50th meridian; this at once marks out, so far as storms are concerned, the entire space included between the 20th and 50th meridians, the equator, and the 60th parallel, as a most suitable area for observations, under particular circumstances hereafter to be noticed, with especial reference either to the commencement or termination of storms, or the prolongation of Mr. Redfield's storm paths.

Localities.—The three principal localities of storms are as follows:—I. The western portion of the basin of the North Atlantic; II. The China Sea and Bay of Bengal; and III. The Indian Ocean, more particularly in the neighbourhood of Mauritius. The first two have already been marked out as areas for the three-hourly observations; to the latter, the remark as to extra observations under the head of Desiderata will apply.

Margins.—Mr. Redfield has shown that on some occasions storms have been preceded by an unusual pressure of the atmosphere; the barometer has stood remarkably *high*, and it has hence been inferred that there has existed *around* the gale an accumulation of air forming a margin; barometers placed under this margin indicating a much greater pressure than the mean of the respective localities. With regard to the West Indian and American hurricanes—any considerable increase of pressure, especially within the space marked out to the eastward of the 50th meridian, will demand immediate attention. Upon the barometer ranging *very high* within this space, three-hourly observations should be immediately resorted to; and if possible, *hourly* readings taken, and this is the more important the nearer the vessel may be to the 50th meridian. Each observation of the barometer should be accompanied by an observation of the wind—its direction should be most carefully noted, and the force estimated according to the scale in p. 300, or by the anemometer. It would be as well *at the time* to project the barometric readings in a curve even of a rough character, that the extent of fall after the mercury had passed its maximum might be readily discernible by the eye. A paper ruled in squares, the vertical lines representing the commencement of hours, and the horizontal tenths of an inch, would be quite sufficient for this purpose. The *force* of the wind should be noted at, or as near to the time of the passage of the maximum as possible. During the fall of the mercury particular attention should be paid to the manner in which the wind changes, should any change be observed; and should the wind continue blowing steadily in *one* direction, but gradually *increasing* in force, then such increments of force should be most carefully noted. During the fall of the barometer, should the changes of the wind and its increasing force indicate the neighbourhood of a revolving storm, (independent of the obvious reasons for avoiding the focus of the storm) it would contribute as much to increase our knowledge of these dangerous vortices to keep as near as possible to their margins as to approach their centres. The recess from the centre towards the margin of the storm, will probably be rendered apparent by the *rising* of the mercury, and so far as the observations may be considered valuable for elucidating the connection of atmospheric waves with rotatory storms (other motives being balanced), it might be desirable to keep the ship near the margin—provided she is not carried beyond the influence of the winds which characterize the latter half of the storm—until the barometer has nearly attained its usual elevation. By this means some notion might be formed of the general direction of the line of barometric pressure preceding or succeeding a storm.

Should a gale be observed commencing without its

having been preceded by an unusual elevation of the mercurial column, and consequently no additional observation have been made; when the force of the wind is noted in the usual observations at or above 5, then the three-hourly series should be resorted to, and the same care taken in noting the direction, changes, and force of the wind as pointed out in the preceding paragraph.

The foregoing remarks relate especially to the central and western portions of the North Atlantic; they will however equally apply to the remaining localities of storms. Under any circumstances, and in any locality, a *high* barometer not less than a low one should demand particular attention, and if possible, *hourly* readings taken some time before and after the passage of the maximum: this will be referred to more particularly under the next head.

Preceding and Succeeding Accumulations of Pressure.—Mr. Redfield has shown in his Memoir of the Cuba Hurricane of October, 1844, that two associated storms were immediately preceded by a barometric wave, or accumulation of pressure, the barometer rising above the usual or annual mean. We have just referred to the importance of *hourly* observations on occasions of the readings being *high* as capable of illustrating the marginal phenomena of storms, and in connection with these accumulations of pressure in advance of storms we would reiterate the suggestion. These strips of accumulated pressure are doubtless crests of atmospheric waves rolling forwards. In some cases a ship in its progress may cut them transversely in a direction at right angles to their *length*, in others very obliquely; but in all cases, whatever section may be given by the curve representing the observations, too much attention cannot be bestowed on the barometer, the wet and dry bulb thermometer, the direction and force of the wind, the state of the sky, and the appearance of the ocean during the ship's passage *through* such an accumulation of pressure. When the barometer attains its mean altitude, and is rapidly rising above it in any locality, then *hourly* observations of the instruments and phenomena above noticed should be commenced and continued until after the mercury had attained its highest point and had sunk again to its mean state. In such observations particular attention should be paid to the direction and force of the wind preceding the barometric maximum— and the same phenomena succeeding it, and particular notice should be taken of the time when, and amount of any change either in the direction or force of the wind. It is by such observations as these, carried on with great care and made at every accessible portion of the oceanic surface, that we may be able to ascertain the continuity of these atmospheric waves, to determine somewhat respecting their length, to show the character of their connection with the rotatory storm, and to deduce the direction and rate of their progress.

SEASONS FOR EXTRA OBSERVATIONS.

In reference to certain desiderata that have presented themselves in the course of my researches on this subject (see Report of the British Association for the Advancement of Science, 1846, p. 163), the *phases* of the larger barometric undulations, and the *types* of the various seasons of the year, demand particular attention and call for extra observations at certain seasons: of these, three only have yet been ascertained—the type for the middle of November—the annual depression on or about the 28th of November—and the annual elevation on or about the 25th of December. The enunciation of the first is as under: "That during fourteen days in November, more or less equally disposed about the middle of the month, the oscillations of the barometer exhibit a remarkably symmetrical character, that is to say, the fall succeeding the transit of the maximum or the highest reading is to a great extent similar to the preceding rise. This rise and fall is not continuous or unbroken; in some cases it consists of *five*, in others of *three* distinct elevations. The complete rise and fall has been termed the great symmetrical barometric wave of November. At its setting in the barometer is generally low, sometimes below twenty-nine inches. This depression is generally succeeded by *two* well-marked undulations, varying from one to two days in duration. The central undulation, which also forms the apex of the great wave, is of larger extent, occupying from three to five days; when this has passed, two smaller undulations corresponding to those at the commencement of the wave make their appearance, and at the close of the last the wave terminates." With but slight exceptions, the observations of eight successive years have confirmed the general correctness of this type. On two occasions the central apex has not been the highest, and these deviations, with others of a minor character, form the exceptions alluded to. This type only has reference to London and the south-eastern parts of England; proceeding westward, north-westward, and northward, the symmetrical character of this type is considerably departed from; each locality possessing its own type of the barometric movement during November. The desiderata in immediate connection with the November movements, as observed in the southern and south-eastern parts of England, that present themselves, are—the determination of the types for November, especially its middle portion, as exhibited on the oceanic surface within an area comprised between the 30th and 60th parallels, and the 1st and 40th meridians west. Vessels sailing within this area may contribute greatly to the determination of these types by making observations at intervals of three hours from the 1st of November to the 7th or 8th of December. The entire period of the great symmetrical wave of November will most probably be embraced by such a series of observations as well as the annual depression of the 28th. For the elevation of the 25th of December the three-hourly observations should be commenced on the 21st, and continued until the 3rd or 4th of the succeeding January.

With respect to the great wave of November, our knowledge of it would be much increased by such a series of observations as mentioned above, being made on board surveying and other vessels employed off Scotland and Ireland; vessels navigating the North Sea; vessels stationed off the coasts of France, Spain, Portugal, and the northern parts of Africa, and at all our stations in the Mediterranean. In this way the area of examination would be greatly enlarged, and the *differences* of the curves more fully elucidated; and this extended area of observation is the more desirable, as there is some reason to believe that the line of greatest symmetry *revolves* around a fixed point, most probably the nodal point of the great European systems.

It is highly probable that movements of a somewhat similar character, although presenting very different curves, exist in the southern hemisphere. The November wave is more or less associated with storms. It has been generally preceded by a high barometer and succeeded by a low one, and this low state of the barometer has been accompanied by stormy weather. We are therefore prepared to seek for similar phenomena in the southern hemisphere, in those localities which present similar states of weather, and at seasons when such weather predominates. We have already marked out the two capes in the Southern hemisphere for three-hourly observations: they must doubtless possess very peculiar barometric characters, stretching as they do into the vast area of the Southern Ocean. It is highly probable that the oscillations, especially at some seasons, are very considerable, and vessels visiting them at such seasons would do well to record with especial care the indications of the instruments already alluded to. At present we know but little of the barometric movements in the Southern hemisphere, and every addition to our knowledge in this respect will open the way to more important conclusions.

Section XI.

ZOOLOGY.

By RICHARD OWEN, F.R.S., Hunterian Professor to the Royal College of Surgeons of England.

Instructions for Collecting and Preserving Animals.

As water is the element in which the greater number of the classes of animals exist, and as the sea is the scene of such existence and the field of research which will be most commonly presented to those for whom the following instructions for collecting and preserving animals have been drawn up, they will commence with the marine species and the lowest forms of animal life.

Algæ, Sponges, Corallines, and Corals.

The line of demarcation between the vegetable and animal kingdoms is so obscurely marked in the lowly organized marine species, and the modes of collecting and preserving these are so similar, that the kindred groups above-named are associated together as the subjects of the following remarks.

Algæ, commonly called sea-weeds, may be divided, for the convenience of the collector, into three kinds, according to their colour:—

1. Olive-coloured (*Fuci*); generally of large size and leathery texture, rarely gelatinous; usually laminate or leafy, rarely filamentous or thready.

2. Red-coloured (*Florideæ*); firm, fleshy, or gelatinous; usually filamentous, sometimes membranaceous.

3. Green (*Chlorosperms*); membranaceous or filamentous; rarely horny.

Sponges are bodies usually adherent in irregular or amorphous masses, rarely in the form of hollow reticulate cones; composed of a soft, jelly-like tissue, supported by siliceous or calcareous spiculæ, or by horny filaments. They are divided accordingly into horny, or "keratose," "siliceous," and "calcareous" sponges. Their soft organic substance is commonly diffluent, and drops from the firmer basis when removed from the water, or it is easily washed away. It exhibits no sign of sensibility; no contraction or retraction when touched or otherwise stimulated. The evidence of life is afforded, as in the corallines and algæ, by the flow of currents of water through canals, entering by pores, and in the sponges escaping by larger orifices; and an appearance of animal life is given to both algæ and sponges by the locomotion of the sporules or gemmules.

Corallines are plants coated with a calcareous covering, either red or green when fresh, becoming white and brittle on exposure to the air.

Corals, though called "zoophytes," are true animals; the currents which permeate them enter by "mouths," always provided with a crown of feelers or seizers, called tentacles, and communicating with digestive sacs or "stomachs," into which the pores of the nutrient canals open. The tentaculated mouths are called "polypes." Their fleshy tissue, as well as that which connects them together into an organic whole when the coral is compound or has more than one mouth, is "sensitive," or retracts and shrinks when touched. For the purposes of the collector corals may be divided into the "fleshy" (*Polypi carnosi*), in which the flesh has no firm supporting part; the "horny" or "flexible," usually having this supporting substance as an external tube; and the "calcareous," in which the supporting substance is usually covered by the animal matter or flesh, forming an internal skeleton, usually of one piece, rarely jointed.

The above-defined classes of organized beings, which all present the "habit" or outward form, more or less, of plants, are found from the extreme of high-water mark to the depth of from 50 to 100 fathoms. Living algæ rarely descend below 50 fathoms, but corals of the genera *Lepralia*, *Retepora*, and *Hornera* have been dredged up from 270 fathoms, and fragments of dead coral from 400 fathoms.* Specimens within the reach of the tide are to be collected at low water, especially of spring tides: the most interesting species occur at the verge of low-water mark. Those that dwell at greater depths must be sought by dredging, or by dragging after a boat an iron cross furnished with numerous strong hooks. One or more strong glass bottles with wide mouths, or a hand-basket lined with japanned tin, should be provided for the purpose of bringing on board the smaller and more delicate species in sea-water, and they should be kept in it, the "*Florideæ*" more especially, until they can be arranged for drying, or other modes of permanent preservation can be attended to.

In collecting algæ, corallines, or the branched, horny,

* Capt. Sir James C. Ross, 'Antarctic Voyage,' Appendix, No. IV.

or calcareous corals, care should be taken to bring the *entire specimen* with its *base or root*. With respect to the coarser algæ, it is merely requisite, for the purpose of transmission, to spread the specimens immediately on being brought fresh from the sea, without previous washing, in an airy situation to dry, but not to expose them to too powerful a sun: if turned over a few times they will dry very rapidly. When thoroughly dried they may be packed loosely in paper bags or boxes, and will require only to be re-moistened and properly pressed, in order to make cabinet specimens. For the purpose of transmission it is better *not* to wash the specimens in fresh-water previous to drying, as the salt they contain tends both to preserve them and to keep them pliable, and more ready to imbibe water on re-immersion. With respect to the delicate algæ:—"The collector should have two or three flat dishes, one of which is to be filled with salt water and two with fresh; in the first of these the specimens are to be rinsed and pruned, to get rid of any dirt or parasites, or other extraneous matter; they are then to be floated in one of the dishes of fresh water for a few minutes, care being taken not to leave them too long in this medium, and then one by one removed to the third dish, and a piece of white paper, of the size suited to that of each specimen, is to be introduced underneath it. The paper is to be carefully brought to the surface of the water, the specimen remaining displayed upon it, with the help of a pair of forceps or a porcupine's quill, or any fine-pointed instrument; and it is then to be gently drawn out of the water, keeping the specimen displayed. These wet papers, with their specimens, are then placed between sheets of soft soaking-paper, and put under pressure, and in most cases the specimen adheres in drying to the paper on which it is laid out. Care must be taken to prevent the blotting-paper sticking to the specimens and destroying them. Frequent changes of drying-paper (once in six hours), and cotton rags laid over the specimens, are the best preservatives. The collector should have at hand four or five dozen pieces of unglazed thin calico (such as sells for *2d.* or *3d.* per yard), each piece about eighteen inches long and twelve inches wide, one of which, with two or three sheets of paper, should be laid over every sheet of specimens as it is put in the press. These cloths are only required in the first two or three changes of drying-papers; for, once the specimen has begun to dry, it will adhere to the paper on which it has been floated in preference to the blotting-paper laid over it."*

For dried specimens of corallines, corals, and sponges, it is advisable to soak the specimen for a time in fresh water before drying. They may then be packed among the rough-dried sea-weeds in boxes; but the more delicate specimens should be placed in separate chip-boxes with cotton.

With regard to corals, etc., it must be remembered that dried specimens are but the skeletons of those animals, and that only the "horny" and "calcareous" species can be so preserved. The "fleshy" kinds, commonly known as "polypes," "sea-anemones," or "animal-flowers," must be preserved entire in alcohol or saline

* Dr. Harvey, in Mr. Ball's 'Report on the Dublin University Museum,' p. 3.

solution, and of the latter the following (No. I. of Goadby's recipes) has been found successful :—

SOLUTION No. I.

Bay salt	4 oz.
Alum	2 oz.
Corrosive sublimate	2 grains.
Rain-water	1 quart.

In order to preserve the specimens expanded they should be removed and placed alive in a dish of sea-water; and when they have protruded and expanded their tentacles, the solution should be slowly and quietly added to the sea-water, when the animal may be killed and fixed in its expanded state. So prepared, the specimens should be transferred to a bottle of fresh solution.

In like manner the minute polypes of the flexible or horny corals may be preserved protruded from their cells and expanded. If a small piece of corrosive sublimate is put into the vessel of sea-water containing such living polypes, it will kill or paralyse them when protruded, as it slowly dissolves; but they must be removed as soon as they have lost their power of retraction, otherwise their tissue is rendered fragile or is decomposed. The polypes or animal part of the calcareous kinds, called "madrepores," "millepores," "fungiæ," "red coral," "gorgoniæ," &c., require for their preservation, in connection with their supporting basis, the following solution (No. II.) :—

SOLUTION No. II.

Bay salt	$\frac{1}{2}$ lb.
Arsenious acid, or white oxide of arsenic . .	20 grains.
Corrosive sublimate	2 grains.
Boiling rain-water	1 quart.

All the polypes concerned in the formation of coral-reefs, atolls, or coral-islands, may be preserved in the above solution, provided they be killed by its gradual application as above described, and be afterwards transferred into fresh solution. With regard to the structure and formation and mode of observation of coral islands and reefs, the work by Charles Darwin, Esq., *On the Structure and Distribution of Coral Reefs* (8vo., 1842), should be consulted.* Never fail to ascertain, if possible, to what depth below the surface of the sea the corals descend, and on what basis they rest; and for particular instructions with reference to coral reefs, see Mr. Darwin's remarks under the head of 'Geology.'

INFUSORIAL ANIMALCULES (*Polygastria, Polythalamia, Phytolitharia*).

Some idea of the value and importance of attending to the collection of these microscopical organized beings may be had by reference to Ehrenberg's Observations forming Appendix No. V. of Captain Sir James C. Ross's 'Antarctic Voyage,' vol. i. p. 339; a better idea by the perusal of Ehrenberg's numerous communications to scientific journals, some of which have been translated in Taylor's 'Annals of Science;' and the best idea by the study of Ehrenberg's great work, 'Entwickelung, Lebensdauer und Struktur der Magenthiere und Räderthiere,' &c., fol., 1832. The important relations of these minutest forms of animal life to great questions in geology, to the alteration of coast-lines, and to the phenomena

* See also, on this subject, Lieut. Nelson's paper 'On the Geology of the Bermudas;' Geological Transactions, 2nd Series, vol. v. pp. 103–123.

of oceanic luminosity, make it indispensable to include them in directions for collecting facts in natural history.

Whenever the surface of the sea presents a difference of colour and density, in the form of pellicles, streaks, or shining oil-like spots, lift up portions by dipping in thin plates of mica or stout paper, and raising them horizontally: dry these and preserve them in a book, noting the latitude and longitude, the time of day, and the temperature of the sea. The animalcules remain attached to the pieces of paper or mica employed in their capture, and may be determined by subsequent microscopical observation.

Where the sea seems pure and colourless a bucketful may be raised and strained through fine linen; by repeating this act a portion will commonly remain on the filter, which is then generally rich in invisible animalcules, and should be preserved in small glass bottles or tubes, with a bubble of air between the cork or stopper and the water. Any visible gelatinous acalephæ should be removed and placed in spirit of wine, or the solution No. I. Specimens of sea-water thus saturated with animalcules should be prepared at each degree of latitude and longitude traversed on the voyage, by which means the geographical distribution of these minute organisms may be ascertained, when the species so collected are determined, after the voyage, by microscopic observation.

Small bottles or tubes of the water of each mineral spring or hot-spring should be preserved for the same purpose. In a deposit from melted pancake ice from the Barrier, in 78° 10 S. lat., 162° W. long., brought home in Ross's Antarctic voyage, Ehrenberg detected of siliceous-shelled Polygastria fifty-one species, including four new genera; siliceous Phytolitharia twenty-four species; and of calcareous-shelled Polythalamia four species. Small packets of the sand of each coast that may be visited, and of the sand or mud brought up with the anchor or the sounding-line, should be preserved; the localities, or latitude and longitude, being precisely noted in each case.

ACALEPHÆ (*Sea-blubber or Medusæ, Portuguese Men-of-war, Jelly-fish, and other floating marine gelatinous animals*).

The brilliant but evanescent hues of many of this class of animals can only be preserved by coloured drawings executed at the time of capture. The solution No. I. will suffice for the preservation of the animals themselves, provided it be changed after they have remained in it about twenty-four hours, for most of the gelatinous animals, especially the medusæ, contain a great quantity of fluid, which, mixing with the preserving liquid, dilutes it, and renders it unfit for long-continued preservation. The best preserved specimens of these delicate animals are those that have been placed immediately after capture in the solution No. I. diluted with an additional pint of rain-water, and which have been afterwards transferred to fresh solution of the proper strength. Glass-stoppered bottles with wide mouths are the best adapted for the larger Acalephæ.

ECHINODERMS (*Star-fish* [*Asterias*], *Sea-urchins* [*Echinidæ*], *Trepang or Sea-cucumbers* [*Holothuriæ*]).

For the preservation of the entire animal with the soft parts of a star-fish (*Asterias*), or a sea-urchin (*Echinus*), the arsenical solution (No. II.) is preferable: the softer trepangs (*Holothuriæ*) may be preserved in either solution. It should be gradually added to the vessel of water in which the living specimen is at rest, in order to kill it, with the soft appendages protruded or elongated. This is particularly requisite in the case of the *Holothuriæ*, which, if plunged suddenly in solution, are apt to squeeze out and rupture their viscera. With regard, however, to long and slender star-fishes (*Ophiuræ*), sometimes called "brittle stars," from their habit of breaking themselves into pieces when captured, these should be instantly plunged into a large basin of cold fresh water, when they die in a state of expansion, and too quickly for the acts of contraction by which the rays are broken off. After lying for an hour or so in the fresh water they may be transferred to the solution: if preserved dry they should be dipped for a moment in boiling water, then dried in the sun or in a current of air, and packed in paper. When the specimens have soaked in solution one or two days, according to the temperature, they should be removed into fresh solution. The Echini should be sewed up each in a separate bag of muslin, and not be crowded so as to press upon each other in the same bottle. The starfish and sea-urchins that are preserved dry should be emptied of their viscera or soft contents by the mouth or larger (lower) aperture, and should then be soaked in fresh water, changed two or three times, for so many hours, or until the saline particles of their native element have been extracted, before they are dried. The Echini should be wrapped up in cotton and sewed up, each in its separate bag, in order to preserve the spines, which may become detached in the course of a voyage, and are apt to become so if the precaution of soaking away the saline particles be not previously taken. All Echini and star-fish should be examined for small shells (*Stylifer* of Broderip, for example), which nestle in and among the rays and at the roots of the spines, and for other parasites.

Recent Pentacrini (Lily-stars), especially their bases, will be valuable acquisitions. They may be dredged up of large size in tropical seas, as those of Guadaloupe, for example.

ENTOZOA (*intestinal worms and other internal parasites*).

These are to be preserved either in solution No. I. or in colourless proof-spirit. This class of animals has been too much neglected by collectors. Every animal that is opened and dissected, especially fishes, may present rare or undescribed species of Entozoa. The eyes of fishes are often the seat of such: the noses of sharks are frequently infested by them. They may be found not only in the alimentary canal, but in the tissues of most of the organs. When the parasite is adherent, the part to which it adheres should be removed with it, care being taken to secure the whole mouth or proboscis of the parasite. When it is encysted in an organ, the cyst is to be removed entire with the surrounding tissue of the organ. Portions of muscle or other tissue which appear speckled with minute white spots should be preserved, as these may be occasioned by the cysts of *Trichinæ*, or allied microscopic Entozoa. The number attached to the specimen should correspond with that in the list having reference to the animal and part or organ infested by the parasite.

EPIZOA (*Lerneæ or Fish-lice, and other external parasites*); ANNELIDES (*Leeches, Worms, Nereids, or Sea-centipedes, Tube-worms, &c.*).

The exterior surface, the mouth, and the gills of all fishes should be examined for parasitic animals, some of which exhibit the most extraordinary forms and combinations of structure, as, *e. g.*, the *Diplozoon* of NORDMANN, a genus of Entozoa, from the gills of the bream. When the parasites adhere firmly to the part they should be cut out with the adhering organ entire, which sometimes penetrates to a great depth in the flesh. The exterior surface of porpoises, grampuses, and the larger species of the whale tribe should be scrutinized for adherent parasitic animals. Rare kinds of leeches may be found on fishes, as, for example, the *Branchellion* of the Torpedo. A species of leech with external tufted gills, *Hirudo branchiata*, has been detected on a marine tortoise or turtle in the Pacific, the anatomical examination of which is especially recommended by Cuvier. Leeches and all the various kinds of sea-worms comprehended under the class name "Annelides," and including the Nereids, or sea-centipedes, usually found amongst sea-weed or under stones, sometimes attaining the length of twelve feet;*

* See the specimen, from Bermuda, of *Leodice gigantea*, No. 253 *a*, Museum, College of Surgeons, London.

and the tube-worms usually crowned with brilliant coloured tentacles, may be preserved in the solution No. I. or in colourless spirit. Those, however, as the *Serpulidæ*, that form calcareous tubes, should be preserved in the solution No. II. In all cases it is desirable that the specimens should be allowed to die gradually in the water they inhabit, when they commonly display their natural external form and appendages in a relaxed state; they should then be immediately put into the solution or spirit to prevent putrefaction, which otherwise takes place rapidly.

CIRRIPEDIA (*Barnacles and Acorn-shells or Crown-shells*).

The Barnacles or pedunculated Cirripeds, with soft stalks, should be preserved in the solution No. II. or in spirit; they are commonly attached to floating timber, and the smaller species to seaweed, shells, &c. The sessile kinds (acorn-shells, &c.), which encrust the coast-rocks all over the world, and are found parasitic on turtles, whales, &c., should likewise be preserved in spirit or solution No. II., as the included animal is necessary in some genera for the recognition of the species. The colours of the pedunculated kinds should be noted whilst fresh. If the sessile kinds are preserved dry the included animal ought never to be taken out. In removing all the kinds from their points of attachment care must be taken that in some specimens, at least, the base, which is either membranous or calcareous, be preserved. It is particularly desirable that some young as well as large specimens should be collected. In the tropical seas certain corals and shells contain embedded in them singular forms of cirripeds, which, presenting externally little more than a simple aperture, are easily overlooked; such kinds had better be preserved in the coral. Others live embedded in sponges; two genera live on whales' skin (*Coronula* and *Tubicinella*), the development of which needs to be studied by specimens of the ova and young; another less known genus (*Chelonobia*) lives partly embedded in the skin of turtles; a third attaches itself to the manatee or sea-cow; and some small and interesting species of barnacle are parasitic on sea-snakes. Lobsters, crabs, bivalve and other shells, as well as floating pieces of wood, or even net-corks, become the habitat of animals of the class *Cirripedia*. It should always be noted to what animals these parasitic cirripedes are attached, as well as any circumstances that may determine the period during which they have remained attached.

Crustacea (*Shrimps, Sea-mantises, Cray-fish, Lobsters, Crabs, and King-crabs*).

All the animals of this class are most profitably preserved in spirit or solution. If they be defended by a soft, flexible, or horny covering, the solution No. I. answers well; if by hard, calcareous plates, the solution No. II. is preferable. They vary in size from microscopic minuteness to upwards of a yard in length. The larger and middle sized specimens should be kept by themselves, or sewed up in a bag if placed with others in the same jar or bottle. Rare and beautiful kinds, with transparent glass-like shells, may be captured by the towing-net in tropical seas. The minuter kinds have been commonly neglected, especially those of fresh water: any such species observed darting about in the fresh water of foreign countries should be preserved in tubes, in spirit or solution No. I. The larger kinds of marine crustacea should be suffered to die in fresh water before immersion in the preserving liquor. The different kinds of king-crab (*Limulus*) usually found on sandy or muddy coasts are particularly worthy of preservation in spirits or solution with the ova or young.

In preparing crustacea for drying care is to be taken to preserve all their external parts as perfect and as expressive of the natural progressive action as possible. Crabs and lobsters should be cleaned out as soon as practicable, *i. e.*, the soft internal parts and the flesh should be removed, and they should be soaked in fresh water previous to drying. The claws when large require to be separated at each joint for the purpose, and then refixed, or a small piece may be neatly removed and afterwards replaced. When dried, the specimens should be wrapped in very soft paper and then packed in cotton, so as not to allow of their being displaced in the case nor to touch one another. It is desirable, with regard to brilliantly-coloured crabs, to wash them over, after they are dried, with a thin coat of the following varnish:—

VARNISH FOR CRABS, EGGS, &c. No. I.

Common gum	4 oz.
Gum tragacanth	¼ oz.

Dissolve these in three pints of water, add to the solution 20 grains of corrosive sublimate, and 20 drops of oil of thyme, dissolved in 4 oz. of spirit of wine; mix it well, and let it stand for a few days to separate: the clearer part is to be used as varnish; the thicker part forms an excellent cement.

A very important subject of investigation is the development of the crustacea from the earliest period at which they can be observed to the assumption of the mature or parent form. The eggs, usually of some bright colour, attached beneath the tail of the female crab, lobster, or shrimp, should be examined for this purpose: the embryo if in course of development may be readily seen by opening the egg under a moderately magnifying power (see the note on Microscopes). Drawings of the different forms or stages of the embryo should be made, if possible, and the eggs and embryos preserved in spirit or solution in small glass tubes.

Insecta.

Some specimens of all kinds of insects should be preserved for anatomical examination in spirit or the solution No. I. Many of the softer kinds of insects and spiders can only be profitably so preserved. Care must be taken that the softer kinds of insects are not put into the same bottle with the harder kinds. Gauze nets must be used for catching the *Lepidoptera* (butterflies and moths) on the wing, and a fine muslin net, like a landing-net, for the water insects. Many species may be taken by spreading a cloak, or placing an open umbrella reversed under trees or bushes, and shaking or beating the latter. Caterpillars should be carefully placed in a perforated box with the leaves of the plants on which they are found feeding: they will often undergo their metamorphosis in this captivity, and no *lepidoptera* are more perfect than those thus *bred*, as it is termed, if carefully watched. The perfect insect should be accompanied, if possible, by its *larva* (caterpillar) and *pupa* (chrysalis or cocoon), together with a specimen of the plant on which it is found feeding. The latter should be kept in an herbarium set apart for the purpose, and should have a number corresponding with that of the insect. *Larvæ* and *pupæ* may be preserved in spirit or solution, as well as a specimen of every perfect insect that can be spared, with a view to anatomical investigation. It must be remembered that the *larvæ* will very soon lose their colours when so treated, and, in order to retain these, a specimen or two of the larger ones and of their pupæ may be opened, the viscera removed, and the inside, after it has been brushed with arseniate soap, stuffed with cotton. Boxes lined with cork are the best conveyances for dried butterflies, moths, and indeed for insects in general: or they may be pinned in the crown of the hat until they can be transferred to a place of safety. The more delicate insects, such as butterflies, moths, sphinxes, the different species of mantis, the locusts, dragonflies, &c., after being killed by pressure on the thorax, should be pinned down, while in a relaxed state, with the wings and legs kept close to the body, to save space and prevent collision. The pin should be greased or oiled to prevent rust, and if pointed at both ends the specimen more readily admits of being turned. The pin should be made fast so as to allow of the motion of the box in all directions, and the fastening must be adjusted to the weight of the insect. The harder winged insects may be killed by immersion in hot water, and after having been dried on blotting-paper, may be laid carefully in boxes upon cotton, so as not to interfere with or injure each other. A ready mode of preserving beetles (*Coleoptera*), when found in abundance on any foreign coast, is to put them, when dried, in a box, on the bottom of which a layer of fine dry sand has been strewed. When the layer is overspread with beetles they must be covered with another layer of sand, and the packer must proceed with layers of beetles and sand alternately, till the box, which should be water-tight, is quite full, when it should be screwed down and pitched at the seams. Mr. Darwin preserved all his dry specimens of insects, excepting the lepidoptera, between layers of rag in pill-boxes, placing at the bottom a bit of camphor, and they arrived in an excellent state.

MOLLUSCA (*Cuttles, Squids, Snails (land and sea), Slugs (land and sea), Shell-fish, Cowries, Limpets, and Bivalves, as Mussels, Oysters, &c.*).

"A superficial towing-net, another so constructed as to be kept a fathom or two below the surface, and the deep-sea trawl, are the principal agents for capturing these animals. But when the tide is at the lowest, the collector should wade among the rocks and pools near the shore, and search under overhanging ledges of rock as far as his arms can reach. An iron rake, with long close-set teeth, will be a useful implement on such occasions. He should turn over all loose stones and growing sea-weeds, taking care to protect his hands with gloves, and his feet with shoes and stockings, against the sharp spines of *echini*, the back fins of weevers (sting-fishes), and the stings of *medusæ* (sea-nettles). In detaching chitons and *patellæ* (limpets), which are all to be sought for on rocky coasts, the surgeon's spatula * will prove a valuable assistant.

* A case knife, *in experienced hands*, is even a better instrument; but great care must be taken not to wound the ligamentous border of the shell of the chitons, and not to injure the edges of the limpets.

Those who have paid particular attention to preserving chitons have found it necessary to suffer them to die under pressure between two boards. *Haliotides* (sea-ears) may be removed from the rocks to which they adhere by throwing a little warm water over them, and then giving them a sharp push with the foot sideways, when mere violence would be of no avail without injuring the shell. Rolled madrepores and loose fragments of rock should be turned over. *Cypræce* (cowries) and other *testacea* are frequently harboured under them. Numbers of *mollusca*, *conchifera*, and *radiata* are generally to be found about coral reefs."—*Broderip.*

Among the floating mollusca likely to be met with in the tropical latitudes is the *spirula*, a small cephalopod with a chambered shell. An entire specimen of this rare mollusk is a great desideratum; and if it should be captured alive, its movements should be watched in a vessel of sea water, with reference more especially to the power of rising and sinking at will, and the position of the shell during those actions. The chambered part of the shell should be opened under water, in order to determine if it contain a gas; the nature of this gas should likewise, if possible, be ascertained. As a part of the shell of the *spirula* projects externally at the posterior part of the animal, this part should be laid open in the living *spirula*, in order to ascertain how far such mutilation would affect its power of rising or sinking in the water.

In the event of a living pearly nautilus (*Nautilus Pompilius*) being captured, the same observations and experiments should be made on that species, in which they would be attended with more precision and facility, as the species is much larger than the *spirula*, and its shell external.

The towing-net should be kept overboard at all practicable periods, and drawn up and examined at stated intervals, as some of the rarest marine animals have been taken by thus sweeping the surface of the sea.

A sketch or drawing of molluscous and radiate animals, of which the form and colour are liable to be materially altered by death, or when put in spirit, will aid materially in rendering the description of the species useful and intelligible.

Some of each species should be preserved in spirit or the solution No. II. If they have died with their soft parts protruded, they should be suspended so as to prevent distortion from pressure. If the shell be of the spiral form, the whorls should be perforated with a fine awl so as to allow the spirit or solution to enter; otherwise, as the main body of the animal fills up the whole mouth of the shell, the deeper seated and softer parts would become putrid before the preserving liquor could get to them.

Where the animal has been detached from its shell, the soft parts and the shell should be marked with corresponding numbers. When the animal is furnished with an *operculum* (the little door which closes the mouth of many turbinated shells), it should be carefully preserved; and if detached from the animal should be so numbered as to prevent the possibility of its being attributed to the wrong species. Shells should never be cleaned, but should be preserved as they come from the sea, taking care only to fill the mouths of those which are turbinated with tow or cotton to prevent fracture. It may be sometimes requisite to put a live shell into hot water and boil it a minute or two, in order to dislodge the animal, which may then be removed with a crooked pin.

The land-shells are found in various situations; as in humid spots covered by herbage, rank grass, &c.; beneath the bark or within the hollows of old trees, crevices of rocks, walls, bones, &c.; about the draînage of houses, or in the dry season by digging near the roots of trees. Early in the morning, especially in rainy weather, is the best time for taking them. The freshwater kinds may be sought for in quiet inlets, on the sides of lakes, rivers, and brooks: the greater number of univalves occur at or near the surface, under the leaves of aquatic plants and among decayed vegetables; while the bivalves and certain univalves keep at the bottom, and are often more or less imbedded in the sand or mud, from which they may be raked into a landing-net.

With regard to the marine bivalves, rocks, submarine clay-banks, piles, stones, and indurated sand, should be carefully inspected for *Pholades*, *Lithodomi*, and other boring species. If the collector should find any of these perforators in the ruins of an ancient temple, or in the remains of any ancient works of art, or any adhering shells (*serpulæ* for instance) attached to the surface of such works, the specimens become doubly interesting, especially in a geological point of view. In such cases, the situation should be accurately noted, as well as the distance of the perforations from the surface of the sea, either above or below.

By digging with a wide-pronged fork in sand-banks, at

low water, many bivalves, such as *Solens*, *Cardia*, *Tellinæ*, &c., will be procured alive ; and, if the inhabitants of the coast be accustomed to diving, their services should be secured for deeper water. Care must be taken not to separate the ligament which binds the hinge. When the animal is dead the shell will gape, and the soft parts may then be removed without injury. Attempts to open bivalves, while the animals are alive, generally terminate in great injury to the shells.

For deep-sea shells the dredge is indispensable. Dredging requires experience to judge of the length of rope to be used ; if there be too much on a sandy bottom, the dredge will bury itself; if too little, it will not scrape properly ; on rocky bottoms the rope must be kept as short as possible ; in deep water the dredge can only be made to act effectually by placing a weight on the line, which, as a rule, may be about one-third of the weight of the dredge, and placed on the line at about two-thirds of the depth of the water ; the object is to sink the rope, and counteract the tendency it has to float the dredge. The contents of the dredge are best examined by means of sieves, of which three should be used, one over the other, first a riddle, next a wheat sieve, and third an oat sieve ; these may be fastened together, the contents of the dredge being emptied into the riddle, and water being poured upon them, the mud, &c. will be washed off, and the contents separated, so as to be very easily examined ; by this plan a hundred fold more will be discovered, than can be found by searching in mud or sand in the usual manner. Besides shells, numbers of crabs, star-fishes, sea urchins, worms, corals, zoophytes, algæ, &c. are procured by the dredge.

VERTEBRATA.

Fishes.

All specimens not too large to be preserved entire should be immediately plunged into spirit or solution. In the case of cartilaginous or soft-spined fishes the solution should be No. I. Fishes with hard spines should be preserved in the solution No. II. It will be found to be convenient to have a common receptacle for the fresh-caught specimens, and to transfer them, after soaking a day or two, into the vessel, with fresh spirit or solution, in which they are to be sent home. As the colours are more or less evanescent, it is desirable that they should be accurately noted before death.

With regard to large specimens of the shark or ray kind, John Hunter recommended that " the abdomen should be first opened, then the head taken off by dividing the fish below the heart across the upper part of the liver, by which means the mouths of the oviducts, if it be a female, the heart, and head are all preserved together.

" The tail, if a thick one, as that of a shark, may be taken off a little below the anus, and the trunk alone preserved for examination. If the trunk be too large, it should be cut through above the pelvis, and the parts contained in the hinder portion, as the claspers of the male, should be preserved in spirit.

" If a female, separate the two oviducts through their whole length, where they run along the abdomen, on each side of the spine; but keep them attached to the cloaca and surrounding parts, and preserve the whole.

" If with young, or eggs, take the whole out in the same way, without opening the oviducts.

" The peculiarities of the fœtus in these animals should be attended to.

" If not of the ray or shark kind, take out such parts from the abdomen as are uncommon or singular.

" If fish of the roe kind (*i. e.* osseous and cyclostomous fishes), then cut transversely through the fish near the lower part of the roe, some way above the anus. This saves part of the roe, with the connection between it and the anus, the principal parts concerned in generation.

" The tail may be cut off some inches below the anus.

" The stomach and intestines may be saved, if anything particular is observed in them. They should be examined for the presence of entozoa, which, if adherent to the coats of the intestine, should be preserved with the part to which they are attached.

" Eyes of fishes are proper objects of preservation.

" Separate and preserve the heads of such fishes as have anything singular about the teeth or gills, and are too large to be preserved entire."

Preserve the jaws and teeth, together with the backbone, or some of the vertebræ, of every shark or large ray which is not otherwise preserved, being careful to keep the teeth and vertebræ of each individual attached together. Such specimens would be of great service in the determination of fossil teeth and vertebræ. A section of the jaws and teeth with part of the vertebral column should be preserved in spirits or the solution No. II.

Amongst the more interesting fishes of the Southern Ocean is the Port Jackson shark (*Cestracion Phillippi*).

Moderate sized specimens of this species should be preserved entire; and the head, vertebræ, with the dorsal spines, viscera, and especially the impregnated oviduct, should be placed in spirit or solution. The Southern Chimæra (*Callorhynchus antarcticus*) merits also the especial attention of the naturalist.

Certain rivers of Africa, *e. g.* the Gambia, and of South America, contain a peculiar eel-like fish, the *Lepidosiren*, with filaments for fins, which burrows and becomes torpid in the mud during the dry season. The male and female of this fish, and the ova and young in different grades of development, preserved in spirits or the solution No. I., are much wanted, in order to complete its anatomical and physiological history.

With regard to most fishes preserved in spirit or solution it is desirable to inject some of the preserving liquor into the alimentary canal, and, if the fish be large, to make a small opening into the belly. The more delicate specimens should be sewed or wrapped in linen, in order to preserve the scales.

For dry specimens the larger kinds may be skinned, and the skin should be washed on the inner side with the arsenical soap, and then loosely filled with cotton, wool, or tow. With regard to the smaller or moderate-sized specimens, the Curator of the Dublin University Museum states :—" An excellent mode of preserving fishes, easily accomplished, may be thus described: Lay the fish on a table, with the side up which you wish to preserve, then with scissors cut it, so as to separate the fins, skin of one side, mouth, and tail, from the body and viscera; spread the skin so obtained on a linen cloth, fold it over it, and subject it to some small pressure; remove the cloth, and take away any portions of flesh which may appear easily removeable; then fold it in a dry cloth and subject it again to pressure—a board and a few weights or stones will do if no other press be at hand; repeat the operation at intervals, until the skin becomes quite dry, then wash it well at both sides with the varnish No. I. When dry, sew it on strong paper, and you will have as it were a coloured drawing of your fish. The great advantages of this plan are the ease with which it is done, and the small space specimens occupy when finished; a large collection does not require more room than so many dried plants."

Reptiles (*Crocodiles, Tortoises, and Turtles, Lizards, Snakes, Toads, Frogs, Salamanders, and Newts*).

All these animals are best preserved, particularly the smaller kinds, in spirit or solution No. II. Both preserving liquors require to be changed once at least, if not twice; a piece of linen being wrapped round each specimen preserves the scales; this is requisite at least for the smaller lizards and snakes. In skinning lizards the operator must be very careful not to break the tail. The larger snakes may require to be skinned, when care should be taken to preserve the head attached to the skin, and the skins with the heads attached should be put into spirits. In flaying serpents great care must be taken not to damage the scales; and the operator should be cautious, for his own sake, when employed about the head of the poisonous species: a scratch from a fang of a rattlesnake or of a *cobra di capello* soon after death may be fatal. The heads of both poisonous and innocuous species should be preserved for the examination of their teeth.

Tortoises and turtles may be prepared in a dry state, the breastplate being separated by a knife or saw from the back, and, when the viscera and fleshy parts have been removed, restored to its position. The skin of the head and neck must be turned inside out as far as the head, and the vertebræ and flesh of the neck should be detached from the head, which, after being freed from the flesh, the brain, and the tongue, may be preserved with the skin of the neck. In skinning the legs and the tail, the skin must be turned inside out, and the flesh having been removed from the bones, they are to be returned to their places by re-drawing the skin over them, first winding a little cotton or tow round the bones to prevent the skin adhering to them when it dries.

When turtles, tortoises, crocodiles, or alligators, are too large to be preserved whole in liquor, some parts, as the head, the whole viscera stripped down from the neck to the vent, and the cloaca, should be put into spirit or solution. The bones of such specimens are especially desirable: they may be separated and scraped clean: all those of the same individual should be packed in a bag or box with bran, paper-cuttings, hay or dried seaweed. The bones of the smaller species need not be separated. After detaching as much of the flesh as is practicable, the entire skeleton may be suffered to dry in a naturally connected state, and then may be laid in a box on cotton, tow, or other soft material, and covered with the same.

The eggs, at different stages of development, of crocodiles, turtles, and tortoises, and also of the larger snakes, should be preserved in spirit or solution, as also the young animals. As the colours of most reptiles are much altered by spirit, a coloured sketch should be made, when practicable, of them either during life or immediately after death.

The batrachia or amphibia should be obtained in the different stages of their metamorphoses. The different species of the burrowing snake-like genus called *Cæcilia* are especially desirable in the young state. The gravid oviducts of these and of the viviparous kinds of salamander should be preserved in spirits or the solution No. I., together with the young of the perennibranchiate amphibia of the United States, called menopoma, amphiuma, menobranchus, siren.

Birds.

All the rarer kinds, especially the smaller species, should be preserved in spirits or the solution No. II. for anatomical examination. Of such as are too large to be preserved entire, the gullet, stomach, or gizzard, liver, intestines, ovary, oviduct, or the male organs, should all be taken out as low as the anus, and with the cloaca should be preserved in spirit or the solution No. I. The tongue and trachea with the lower larynx should be preserved wet by themselves; and if more than two specimens of a rare bird are captured, the head of one should be preserved in strong spirit, a small portion of the cranium being removed to allow the spirit to get to the brain.

The most common as well as convenient mode of preserving birds for zoological purposes is by removing and preparing the dry skin, with the head and feet attached, and a few words on the mode of performing this operation may be found of use. First put some cotton or bits of blotting-paper into the mouth of the bird to absorb the blood that may be there, and then tie the bill close by passing a thread with a needle through the nostril, and round the lower mandible; then, after parting the breast-feathers, the incision for skinning should be made from the lower point of the sternum, or breast-bone, to the tail, care being taken not to cut into the body. Whilst removing the skin, thrust cotton-wool between it and the body, at the parts not being operated upon, to keep the feathers clean, and prevent them from coming in contact with the moist parts. Having detached the skin of those parts on each side, the legs are next to be pushed through and cut off at the joint that protrudes; and then follows the more difficult process of separating the vertebræ near the tail. Having detached, however, the legs, and leaving the flesh upon them for the present, the operator must continue to separate the skin from the hind part of the body as well as he can, and then very carefully cut through the vertebral column near the tail, without injuring the skin above it; that of the back is then detached with much ease, and a little practice is now necessary to keep back the feathers of the breast while the skin is drawn over the shoulders; the wings should then be separated at the shoulder joint, and the skin pulled over the neck, and very gently and carefully over the head, taking especial caution not to enlarge the auditory orifices or those of the eyes. With the majority of birds the skin may be drawn back over and from the head without much difficulty; but there are some, as woodpeckers and ducks, in which the head is larger than the neck, and consequently could not be drawn through that part without stretching the skin: it is advisable to make an incision in the skin at one side of the head, and thus uncover the skull to remove the fleshy parts, not forgetting the tongue, eyes, and brain. In small birds a quill cut in a slanting manner will be found useful to scoop out the brain; a little wool may afterwards be wound round it to remove any moisture that may remain in the hollow parts of the skull. Whilst skinning the head, upon reaching the eye it will be necessary to cut the tough membrane that surrounds that part. The brain and flesh being thoroughly removed, and the skin anointed with the arsenical soap,* the limbs are easily drawn back, a little cotton or tow being previously wrapped round the thigh-bones, and

* *Receipt for Arsenical Soap.*

Camphor	5 ounces.
Arsenic in powder . . .	2 pounds.
White soap	2 pounds.
Salts of tartar	12 ounces.
Lime in powder	4 ounces.

Cut the soap in thin small slices, as thin as possible, put them in a pot over a gentle fire, with very little water, taking care to stir it often with a wooden spoon: when it is well melted, put in the salts of tartar and powdered chalk. Take it off the fire, add the arsenic, and triturate the whole gently. Lastly, put in the camphor, which must first be reduced to powder in a mortar by the help of a little spirits of wine; mix the whole well together. This paste ought then to have the consistence of flour paste. Put it into china or glazed earthen pots, taking care to put a ticket on each.

When it is to be used, put the necessary quantity into a preserve-pot, dilute it with a little cold water until it has the consistence of cream; cover this pot with a lid of pasteboard, in the middle of which bore a hole for the handle of the brush.

The three first ingredients in the above receipt may be used, if the whole cannot be readily obtained.

care being taken that no feathers adhere to the interior of the skin of them and are drawn in with it, and then (after putting some cotton into the cavities of the orbits) the head must be pulled forth by means of the bill, an operation requiring much caution, so as not to tear the very tender skin of the sides of the neck; this is frequently a rather difficult matter with beginners, and may be much facilitated by partially crushing the skull, which is easily put back into shape when the skin is again over it; bits of cotton should also be freely used to prevent the feathers being anywhere soiled by adhering to the skinned body, as they are extremely apt to do, despite all care, unless some such precaution be resorted to. Having now returned the skull within the skin, a little art is necessary in arranging the feathers of the head properly, which is best done with a large needle; the eyelids should be neatly placed, and not stretched too large, the feathers covering the ears disposed as they originally were, and the orifice of the ears contracted to its proper form; the feathers before and over the eye should also be set naturally; and lastly, the skin of the crown and occiput should be loosened or lifted from the skull, and not be pulled too tightly backward. The arsenical soap is to be sparingly applied to the inside of the skin, and the legs and beak brushed with a solution of corrosive sublimate. As regards the rest, it is as well to tie together, but not too closely, the bones of the two wings, to put a little cotton around these and the bones of the legs, and, in stuffing the bird, to avoid stretching the skin by putting in too much cotton, especially to avoid puffing out the neck, in which it is enough to prevent the skin of its two sides adhering together, and lastly, to mind that the bird is restored to its original length and proportions, and that the feathers are laid down as smooth as possible. In large birds, more especially, it will be found useful to put a reed or thin bit of stick up the neck, around which the stuffing of the neck may be wound, for this will prevent the tender skin of the neck bursting, when dry, upon the specimen not being handled with sufficient care: and in large birds, it is also necessary to make an incision above the elbow-joint of the wings extending along their under surface, and to remove from thence the muscles of that part. In general, it will be found more easy to skin birds, after one or two trials, to the complete satisfaction of the operator, than to put them nicely into shape afterwards, in the form they are to take on drying: and upon being dried thoroughly, they are to be rolled up in paper and tied round with a string.

Birds should be skinned as soon as they are cold; they cannot be kept so long as quadrupeds, and as soon as decomposition begins the feathers are affected, and, if the operation of skinning be deferred till it take place, they will drop off. The os coccygis, or rump-bone, should be left with the skin, otherwise the tail-feathers will be liable to fall out.

The nest, eggs, and young should be procured if possible.

To preserve the eggs of birds with their nests, each nest should be put into a round box just large enough to contain it. After having made a small perforation at each end of the eggs, and expelled their contents, some cotton should be laid upon them to keep them from being moved about, and the whole covered with the lid.

Large eggs, as those of the ostrich and cassowary, at different periods of incubation, should be preserved in spirit.

To each bird attach a note—1. The colour of the eyes, bill, and legs, before they fade. 2. The season of the year when killed, and in what locality. 3. If known, state whether male or female.

The skeletons of birds may be prepared in a short time for sending home by removing the viscera, cutting away all the soft parts, breaking down the brain with a probe or stick and washing it out by the "foramen magnum," or hole for the exit of the spinal marrow, and drying the skeleton with its parts naturally connected, except the head, which may be packed in the thorax; and the whole, when dry, packed in bran or sawdust. Admit the bones of only one individual into each bag or box, taking care to label it with the same number as that attached to the skin. The viscera and any other soft part which appears curious should be preserved in spirit or the solution No. I.

Mammals (*Hairy Quadrupeds, Seals, Porpoises, Grampuses, Whales*).

The smaller kinds, as bats, shrews, mice, may be preserved entire, in spirit or the solution No. II., an opening being made in the skin of the belly to give the preserving liquor access to the viscera, and care being taken not to crowd too many specimens in the same vessel. In all cases, since the preserving liquor becomes diluted and deteriorated by the blood and other fluids of the recent specimen, such specimen should be removed after a few days, according to the temperature, into fresh spirit or solution.

The larger mammals must be skinned, taking care that the head and feet remain attached to the skin according to the directions subsequently given. Such skins, if transmitted either in spirits or the arsenical solution No. II., usually arrive in excellent condition, and may be mounted as well as if recently taken off the animal, which is never the case with such as have been dried. If the circumstances under which the animal is taken will admit of preserving the skeleton, that ought to be done; for its importance is of great moment in a physiological point of view, not only as relating to the organization of the animal, but as a measure of comparison with other living species, and with those which are extinct and only found in a fossil state. If want of space or other circumstances forbid the preservation of the entire skeleton, the skull is the most valuable part to select, and it should be preserved whenever the opportunity occurs.

The mode of preparing the skull of a mammal for the museum is to place the head in a jar of water until the soft parts become detached by maceration and putrefaction; being then washed clean, care being taken to prevent the loss of the small ear-bones, tongue-bone, or loose teeth, it should be placed in fresh water, and the water frequently changed, until the skull becomes free from offensive smell: it should then be exposed to the sun and air, and will in a few days become beautifully white. But this process is not requisite for the mere preservation and transmission of skulls: if the brain be broken down and extracted by means of a small flattened stick through the "foramen magnum," and the soft parts cut away, it may be simply dried, with the lower jaw and hyoid bone attached, and packed in bran, sawdust, or dried sea-weed.

When the entire head of a duplicate mammal is preserved in strong spirit, for the examination of the brain and organs of sense, a small portion of the cranium should be removed, and the membranes of the brain carefully cut to give the alcohol access to that organ.

The œsophagus and stomach should be preserved in spirit or the solution No. I., with a portion of the duodenum; and the cæcum, if any, with a small portion of the ileum and colon. If the animal be not too large, it will be preferable to cut off from the mesentery the jejunum and ileum, which (after their length and circumference and the nature of their contents have been ascertained and noted) may be thrown away, and then to strip down from the spine the contents of the abdomen, beginning at the diaphragm, so as to have the liver, stomach, spleen, pancreas, colon, &c., all with their attachments, taken out together as low as the rectum, where it lies in the pelvis, and, after being cleansed and the contents examined, put into spirits or solution No. I.

The heart and lungs may be preserved together, or, if too large, the heart alone with the large blood-vessels.

The contents of the pelvis, viz., the bladder and rectum, with the internal parts of generation, both male and female; also the external parts, not separated from the internal, with a large portion of the surrounding skin, should be left attached in their natural state, and preserved in spirit or solution.

If the female parts are in a state of impregnation, the whole are to be taken out as before described, without opening the uterus unless for the purpose of admitting the spirit for the preservation of its contents, where of large size.

The young of very large animals, as whales, seals, the walrus, elephants, &c., and all fœtuses or abortions, should be preserved entire: but if a young cetaceous animal be too large, the tail may be cut off below the anus, and the body put into spirit; and if this should be too big for one cask, the head may be taken off and preserved in another.

Of a full-grown whale, or other large animal, the following parts should be preserved:—

The eyes, with the surrounding external skin, their muscles and fat, in an entire mass. The organs of hearing. The brain. Sections of the spinal chord. The supra-renal glands. The ganglions of the sympathetic nerve. The beginning of the aorta and pulmonary artery, for the valves.

The mammæ of the female, with part of the surrounding skin; also the ovaria and uterus. The fœtus, when found in the belly, to be taken out with the whole of the uterus, vagina, ovaria, &c.

The penis of the male taken off as far back as to include the anus with it.

In skinning quadrupeds the skull and leg bones should always be retained. The first incision should be made from the breast along the middle of the abdomen; the skin is then easily separated from the body by the finger, occasionally helped with the knife. Upon reaching the legs, they should be cut through, the fore legs by the shoulder bone, and the hind legs by the base of the thigh bone. The whole of the leg bones are to be left in their places, until the operation with the other part of the body is completed. In skinning the neck and head the skin must be turned inside out, great care being taken, in separating the skin from the head, that the ears and eyelids be not cut. The skin being drawn off the head as far as the ears, the head should be separated from the neck, and then freed from every particle of flesh, such as the tongue, &c.; and the brain taken out by making an opening at the back of the skull. The next thing is to skin the legs and the tail. In these parts, as in the neck, the skin must be turned inside out; all the flesh then being removed from the bones, they are to be returned to their places by re-drawing the skin over them, first winding a little cotton or tow round the bones to prevent the skin adhering to them when it dries.

In animals of moderate or large size, it will be necessary to skin the face upwards, commencing from the lips, in order that all the flesh may be removed from the bones of the face.

In most colonies native assistants may be soon taught this process, and nothing more is necessary beyond washing and then wiping the skin tolerably dry, if it is to be put into spirit or solution: but if intended to be sent home dry, then the interior surface, with the bones, must be anointed with arsenical soap, and likewise the nostrils, ears, and lips, internally; and the hair or fur ought to be wetted with a weak solution of corrosive sublimate. The skin should then be stuffed with tow or cotton, but *not tightly* so as to stretch it.

In warm climates of course it is necessary to skin the animal immediately after death, and it is very desirable that the skin be kept in the shade. Large quadruped skins should be immersed in a strong solution of alum, in which they may remain three or four days, and when taken out of the alum-water they should be washed on the inner side with arsenical soap,* especially about the skull and bones of the feet; a painting-brush may be used for this purpose, and the soap should be mixed with water until it has the consistence of cream: a very small quantity of soap is sufficient; it should not be used too freely. When it is inconvenient to use alum-water, the powdered alum may be used in a dry state, and should be well rubbed over the whole of the inner side of the skin.

Packing.—Great care should be taken in packing skins that they be thoroughly dry. They should be packed in wooden boxes, and some pieces of camphor must be placed with them in order to prevent the attacks of moths. Tobacco is often used, but does not always answer the intended purpose. When soldered up in tin boxes specimens often become mouldy, and are sometimes perfectly destroyed by the damp.

Labelling specimens.—The labels or numbers should never be placed on the paper or wrapper in which a specimen is enclosed; in this case they often become accidentally transferred, especially in the examination which the specimens undergo at the custom-house, &c. Small parchment labels, with the locality of the specimens, should be securely tied to the legs or some other convenient part;

* See p. 372.

a number corresponding with the collector's note-book should also be attached; this number may be stamped on a small piece of sheet-lead or trebly thick tin foil; when specimens are preserved in spirit the latter must be used, since the former will corrode and injure the specimens. A set of steel dies from 0 to 9, with a small punch, should be got, when the numbers may at any time be stamped in a line, with a hole punched in front of each, and then cut off with a pair of scissors as wanted.

Notes.—The collector should note down the colour of the eyes or *irides*, and the form of the pupil, and the colours of those parts, the naked parts, *e. g.* which are likely to be altered in drying: also the form of the head and muzzle, the habitual position of the ears and tail. The exact locality in which the several specimens were procured is of great importance in the determination of the laws of geographical distribution of mammalia: and not only the country, but the nature of the country, its elevation and geological character, as nearly as can be ascertained. Also the degree of commonness of the animal and any of its known habits, and the native name.

Neither shape nor colour can be preserved in the dried skins of whales, porpoises, &c., nor can they be ascertained from skins alone, without the aid of drawings taken from the specimens in a fresh state. Skins of the cetaceans (whale and porpoise tribe), and of seals, are, nevertheless, great desiderata for public museums, and with the addition of sketches and notes of the recent animal, are especially recommended to the attention of the naturalist voyager. The skulls or skeletons of all the species of the southern cetaceans and seals should be preserved, the sex being noted.

As the greater portion of the smaller mammals are of nocturnal habits they can seldom be procured without the aid of traps, which must be baited some with flesh, or a dead bird, some with cheese, bread, fruits, &c.; small pits, widest at the bottom, and baited, often serve to entrap small quadrupeds.

Necessary materials for determining Species.—In almost all cases the zoologist is desirous of examining more than one specimen—in fact, of having before him at least a specimen of the male, female, and young animal, and also one or two skulls, before he can give a satisfactory description of a new species, that is, such a description that the animal may be with tolerable certainty identified through its means. When one specimen only can be procured, the skull should not be injured; a little extra time is well spent in removing the brain through the occipital opening, the back part of the skull being of *great importance.* When the species are small, and several specimens can be procured, one at least should always be preserved in spirit or solution.

The Human Race.

The chief points to which the attention of the philosophic and zoological voyager should be directed towards the advancement of this most important branch of Natural History are included in the following queries:—

What is the average or general stature and weight of the individuals, and the extreme cases?

Is there any prevailing disproportion in the size of the head? of the upper or lower extremities?

What is the prevailing complexion, and the colour of the eyes?

What is the colour of the hair, and its character, as fine or coarse, straight, curled, or woolly?

Is the head round or elongated in either direction? Is the face broad, oval, lozenge-shaped, or of any other marked form? (A profile and also a front view should be given.)

Does infanticide occur, and to what causes is it to be referred?

What is the practice as to dressing and cradling children? Are there any circumstances connected with it tending to modify the form of particular parts, *e. g.*, the head or the feet?

Are the children easily reared?

At what age does puberty take place?

Are births of more than one child common? What is the proportion of sexes at birth and among adults?

To what age do the females continue to bear children? And for what period are they in the habit of suckling them?

What is the menstrual period, and what the time of utero-gestation?

What are the ceremonies and practices connected with marriage?

Is polygamy practised, and to what extent?

Is divorce tolerated, or frequent?

How are widows treated?

What is the prevailing food of the people? Describe their modes of cooking.

What number of meals do they make, and what is their capacity for abstinence, and for temporary or sustained exertion?

Describe the kind and materials of dress; and any practice of tattooing or otherwise modifying the person for the sake of ornament or distinction.

Do the people appear to be long or short lived? State the ascertained cases of extreme old age.

What is the general treatment of the sick, and the superstition, if any, connected with it?

What are the prevailing forms of disease?

Do Entozoa prevail, and of what kind?

How are the dead disposed of?

What is the received idea respecting a future state?

What are the kinds of habitations in use amongst the people?

Have they any monuments; and of what kind and for what object?

What are the domestic animals, if any? Whence derived, and whether degenerated or modified?

Note down any illustrative particulars of the government, policy, religion, superstitions, or sciences of the people; their mode of noting or dividing time; their mode of carrying on war, and favourite weapons.

In these researches collect and preserve the skeletons, both human and of other animals that may be buried or preserved with man: with any works of art or implements.

Besides the *skeletons*, or, at least, the *skulls* of aborigines of foreign countries, plaster casts of the head, the hands, and the feet should be taken; and wherever the opportunity may occur, the *brain* should be preserved in strong alcohol.

Important aids to the advancement of zoology may be rendered by the transport of living animals, and more especially their transmission to the Menagerie of the Zoological Society of London, for which purpose the following remarks have been contributed by a former Vice-President of the Society, William John Broderip, Esq., F.R.S. :—

"In the endeavour to bring a captured animal home alive, it will be well to remember that the younger quadrupeds and birds are,—provided they are of an age to be separated from the mother with safety,—the greater will be the chance of success in bringing them home in a thriving state. There is hardly any young vertebrated animal which judicious kindness will not render familiar. The captive should be kept clean, and should be fed sparingly; that is, it should have only sufficient to sustain it in health;—all trash should be kept out of its reach—and it should not be subjected to the capricious kindness or ill-treatment of strangers.

"Herbivorous quadrupeds, and hard-billed or seed-eating birds, are obviously most easily accommodated during a voyage; but carnivorous animals and insectivorous birds may be transported without much difficulty by paying attention to their food and habits.

"It would be far from impracticable for ingenuity to

devise a mode of introducing even humming-birds alive into this country. A strict attention to temperature, and the aid of an artificial florist, might effect this. If it be found that the birds will not feed out of little troughs, quills, or tubes of coloured paper,* the flowers which are observed to be their favourites might be imitated, and liquefied honey, or even sugar and water, might be placed in a little reservoir in the site of the nectarium. To take these brilliant creatures alive is not difficult, if the following method be adopted. Some plant (the aloe for instance), the flowers of which are particularly attractive to the humming-birds, being selected, all the bunches of blossom, save one or two, should be broken off in the evening after the birds have retired. These bunches should be enclosed in light bamboo trap-cages, with large open falling doors kept up by strings, to be held by a person in concealment. A little before the usual time of the appearance of the humming-birds, the bird-catcher must be in his hiding-place with the door-strings in his hand, and when he finds his prize busily employed about the enclosed flower, he must drop the door and secure his prey. Mr. Bullock tried this plan with great success; and, while on this subject, it may not be irrelevant (as connected with their diet) to state, that he saw these birds frequently take insects out of the spiders' webs, where they lay entangled, and swallow them; and that

* Captain Lyon, in his 'Journal of a Residence, &c., in the Republic of Mexico,' p. 212, states that he kept a humming-bird for nearly a month on sugar and water, slightly impregnated with saffron. It greedily sucked this mixture from a small quill; and the Captain adds, that he is sure that, with constant attention, these little creatures might be kept for a long time.

Mr. George Loddiges has observed the remains of insects in the crops of some of those species which he has opened.

"Reptiles are so tolerant of hunger, and are gifted with such tenacity of life, that they bear a voyage extremely well. Turtles (*Chelones*) and alligators are brought over without difficulty, and tortoises and terrapenes (*Testudines*) may be imported almost without trouble. It is not uncommon for those who touch at the Gallapagos, where great land-tortoises abound, to put them into dry casks, one over the other, without any provision; and, after many weeks, they are found not only alive, but in excellent condition for the table, where they are said to exceed turtle in delicacy of flavour.

"Guanas, chamæleons, together with others of the lizard tribe, and all serpents, bear abstinence from food for a long time, and are brought from their native countries with little trouble.

"Insects may be taken in the caterpillar stage when about to enter the chrysalis state, and, in this manner, may attain their imago or perfect development, either on the voyage, or after their arrival, by attention to their habits, and to the temperature of their natural locality.

"The terrestrial or pulmoniferous *Mollusca* (land-shells) may be brought over alive with ease. When they show a disposition to hybernate, by sticking firmly to the side of the box or vessel wherein they may be, and at the same time throwing out the thick parchment-like secretion, which serves many of the species instead of a true operculum, they should not be disturbed, but must be kept dry, and, if possible, excluded from the air. Many species have been thus accidentally imported. *Bulinus undatus* was brought sticking to timber from the West-India islands into Liverpool, and is now naturalized in the woods near that town. The author possessed living specimens of *Bulinus rosaceus*, which had been brought to England by Captain King and Lieutenant Graves, R.N., from Chiloe, and were in full vigour, though the animals had been packed up in cotton, with the collection of shells, one for eighteen months, and another for two years. He now possesses one in good health, brought home by Mr. Cuming, which had been packed up for a longer time. *Testacellus* and other species had been imported previously.

"By strict attention to changing the sea-water, which very soon becomes unfit for respiration when put into a vessel, marine *conchifera*, *mollusca*, and *crustacea* might be brought home alive, and an opportunity given of studying their organization much more satisfactorily than can be done by a mere post-mortem examination.

"The land-crab of the West Indies has been brought over with success. A pair of them were exhibited, in full vigour, for a few weeks at the end of summer, in one of the enclosures open to the air, at the Zoological Gardens."

Specimens of Fossils should always be accompanied with part of the rock in which they occur, whether stone, clay, or sand, &c.

All fossils, without exception, may be brought home, in large number and quantity.

When fossils occur in sand or gravel, note their condition as to freshness or decomposition; whether they resemble those of the adjacent coast or seas; the mode of their occurrence; whether regularly interstratified with clay or stone; in patches, or continuously, on the sides of hills, or of cliffs, and at what altitudes above the sea.

State whether the fossils found in any given situation are all marine, terrestrial, or fluviatile; mixed, or in distinct groups; and, if mixed, in what proportions.

Observe whether there be any intermixture of fresh-water and marine shells in bodies of water near the shore, or in lakes at a distance from it.

Inquire whether bones of mammalia occur among them.

Note the brackishness of the water;—whether it communicates or not with the sea. State any differences of the animals from those of purely fresh, and of salt water.

Observe carefully the *position* of fossils in the beds which afford them. If corals, whether vertical or inclined? If shells, are they disposed in layers parallel to the strata?

Notice whether testacea are carried up to cliffs by birds; the quantity of shells thus accumulated, and their state of preservation.

Notice the relative numbers of shells of the same species on the shores.

Seek for and preserve all traces of *fossil bones.*

Be careful to ascertain that they are imbedded in the alluvium, not loose or intermixed with the *recent* detritus.

If any bones should be *dredged* up, note the place of their occurrence, latitude and longitude, its distance from any great rivers, whether within currents.

Bones in caves.—Examine the materials forming the bottom of caves for bones.

Bone-breccia.—Search for this in crevices.

Observe all indications of coal, and collect specimens;

note any traces of vegetable impressions in the rocks, and preserve them carefully.

Seek with the microscope for infusorial animals, both in a fossil and recent state.

On the Use of the Microscope on board Ship.

The following remarks embody the experience of Mr. Charles Darwin, F.R.S., on this subject, the importance of which increases as the science of zoology advances.

The facility in examining the smaller invertebrate animals, either alive or dead, depends much more on the form of the microscope used than would be at first expected. The chief requisite of a simple microscope for this purpose is strength, firmness, and especially a large stage; the instruments generally sold in this country are much too small and weak. The stage ought to be firmly soldered to the upright column and have no movement; besides the strength thus gained, the stage is always at exactly the same height, which aids practice in the delicate movements of the hand. The stage should be able to receive saucers, three inches in internal diameter. A disc of blackened wood, with a piece of cork inlaid in the centre, made to drop into the same rim which receives the saucers, is useful for opaque and dry objects: there should also be a disc of metal of the same size, with a hole and rim in the centre to receive plates of glass, both flat and concave, in diameter one inch and a half, for dissecting minute objects; a plate of glass of three inches diameter lets in too much light and is otherwise inconvenient. Close under the stage there should be a blackened diaphragm, to slip easily in and out, in order to shut off the light completely; in this diaphragm there may be a small orifice with a slide, to let in a pencil of light for small objects. The whole microscope should be screwed into a solid block of oak, and not into the lid of the box as is usual.

The mirror should be capable of movement in every direction, and of sliding up and down the column; on one side there must be a large concave mirror, and on the other a *small* flat one; these mirrors ought to be fitted water tight in caps, made to screw off and on; and two or three spare mirrors ought undoubtedly to be taken on a long voyage, as salt water spilt on the mirror easily deadens the quicksilver. A small cap is very convenient to cover the mirror when not in use, and often saves it from being wet. The vertical shaft by which the lenses are moved up and down should be triangular (as these work much better than those of a cylindrical form), and there should be on both sides *large* milled heads; with such, there is no occasion for fine movements of adjustment, which always tend to weaken the instrument. The horizontal shaft should be capable of revolving, and should be moved to and fro by two milled heads (for the right and left hands), but the left milled head must be quite small, to allow of the cheek and eye approaching close to the lenses of high power. The horizontal shaft must come down to the stage.

The most useful lenses are doublets of 1 inch and 6-10ths of an inch (measured from the lower glass of the doublet) in focal distance; a simple lens of 4 or 5-10ths of an inch is a very valuable power; and, lastly, Codrington lenses (of the kind sold by Adie of Edinburgh), of 1-10th, 1-15th, and 1-20th focal distances, have been found most useful by two of the most eminent naturalists in England. With a little practice it is not difficult to dissect under the 1-10th lens, and some succeed under the 1-20th. A person not having a compound microscope might procure a 1-30th of an inch Codrington lens. All the lenses (except the largest doublet) should be made to drop, *not screw*, into the same ring; the large doublet may slip off and on the opposite end of the horizontal shaft. The best saucers have a flat glass bottom, with thin upright metal sides (silvered within); there should be at least four of them, being in depth (inside measure) 3-10ths, 5-10ths, 7-10ths, and a whole inch. Circular discs of fine-textured cork, of the size of the saucers (with one or two circular springs of steel-wire to keep the cork at the bottom of the water), serve for fixing objects to be dissected by direct instead of transmitted light. For this end short fine pins and lace-needles should be procured; wherever it is possible, the animal ought to be fixed to the cork under water. Of the smaller plates of glass of an inch and a half in diameter, some should be flat and some slightly concave; the latter are very useful—saucers of this small diameter are inconvenient.

The simplest and most useful instruments for minute dissection are the triangular glove-needles, which with a little cotton wool and sealing wax can be easily fixed into pieces of large-bored thermometer tubes; a stock of tubes and needles should be taken on a voyage. With these needles (by keeping the object only just immersed in a drop of water, which can be regulated by the suction of blotting-paper), wonderfully minute objects can be dissected; needles bent at their tips are convenient for some purposes. Arm supports are useful in minute dissections; two blocks of wood with inclined surfaces, coming up a little below the level of the stage, and resting partly on the stand of the microscope, can be made by a common carpenter. As it is often rather dark in the cabins of ships, a large bull's-eye glass on a stand (such as are sold with most compound microscopes) would be most useful to condense the light from a lamp on an opaque object, or to increase it when transmitted. Besides the needles, fine pointed forceps, pointed scissors, and eye scalpels are requisite. The French use an instrument called a microtome, and consider it most useful; others prefer finely pointed scissors, with one leg long and thick, to be held like a pen, and the other quite short, to be pressed by the fore finger, and kept open by a spring. A live-box to act as a compressor, or still better a proper compressor closed by a screw, and both made to drop into the rim of the stage, are valuable aids for making out the structure of transparent animals or organs. The observer should be provided with three slips of glass, or still better with three circular plates, made to drop into the stage of his microscope, and graduated into tenths, hundredths, and thousandths of an inch, to serve as micrometers, on which to place and measure any object he is examining. Some watch-glasses are very useful as temporary receptacles for small sea-animals. Minute parts after dissection can be preserved for years in very *weak* spirits of wine, by covering them, when placed on slips of glass, by small portions of very thin glass (both sold for this purpose), and cementing the edges with gold-size.*

When time and opportunity concur for the anatomical examination of an animal, the following notes or heads of observation will guide the dissector to the facts which it is most desirable to determine and note down.

No. Date 18

Notes of Dissections performed at

Animal's Name

Sex Age Weight

Length of body, from extremity of jaws to root of tail

——— of head of tail

Situation of testes

——— of preputial orifice

——— of vaginal orifice

——— of anus

——— and number of mammæ

Abdominal muscles

——— ring

Stomach	simple	length *Observations.*	greatest circumference
	complex	number of sacs *Obs.*	relative size

Omentum

Mesentery

Intestines	length	greatest circumference
	—— of small	—— of small
	—— of cæcum	—— of cæcum
	—— of large	—— of large
	Observations	

Anus glands

* A microscope such as here described, and most of the apparatus, can be seen at Messrs. Smith and Beck's, opticians, of Colman Street, London.

Cloaca

Liver
- situation
- number of lobes
- weight
- *Observations*

Gall-bladder, size   situation
——— - —— structure

Bile, enters intestine

Pancreas
- form
- situation
- its secretion, enters intestine

Spleen
- situation
- form
- weight

Lungs
- situation
- length   breadth, right   left
- weight
- number of lobes, right   left
- structure, air cells, &c.

Branchiæ

Heart
- situation
- weight
- length   breadth
- shape and structure

Venæ cavæ
Aorta, primary branches
Trachea, number of rings   structure
Larynx
Pharynx
Epiglottis
Thyroid Glands
Salivary glands
Tongue, length   papillæ
Nostrils
Eye-lids
Eye
Pupil, form
Lachrymal gland
Ear
Brain, weight   form, &c.
Spinal cord, length
Supra-renal glands

Kidneys { situation
form length breadth
weight of both
papillæ, number and form }

Ureters terminate

Urinary bladder { situation
size
shape }

Testes { size
structure }

Vasa deferentia terminate

Vesiculæ seminales { size
structure
terminate }

Prostate { size
structure
terminate }

Cowper's glands { size
structure
terminate }

Penis length muscle

Urethra

Ovaries { situation
size
shape
Observations }

Uterus { length of cornua
—— of Fallopian tubes
—— of body
position }

Vagina

Oviduct { length
form
termination }

Peculiarities of muscles
———— air-sacs
———— glandular organs

Morbid appearances
Calculi
Entozoa
Epizoa

*General Directions to be observed during a Voyage.**

The towing-nets should be kept overboard whenever it is practicable, and the dredge should be used perseveringly in soundings.

The anchor should be inspected as soon as it arrives at the surface, especially if the holding ground be mud. The finest shells have been lifted on the flukes of anchors. The cable should also undergo an examination.

Let the arming of the lead be narrowly observed, and let the men have orders to preserve anything that may be sticking to the arming, the lead itself, or the lead line.

Floating masses of sea-weed, especially *sargasso*, should be carefully searched; and if one of those tangled natural rafts, which are often carried adrift from great rivers, should be seen, it should be examined minutely, and the animals, plants, and seeds which it may be transporting to colonize some newly-formed island, should be preserved, if possible, or, at all events, accurately noted.

Whenever a new marine species, or one whose habits are unknown, is obtained, it should be placed in sea-water, and, if practicable, a drawing should be made of it while yet alive, with a note stating whether it is gregarious or solitary—phosphorescent or not—and giving the locality, the temperature, the state of the weather, the depth of water, and the time where and when it was captured. The sea water in which living marine animals are confined should be often changed; for it speedily becomes unfit for life.

* From 'Hints for Collecting,' &c., by Wm. John Broderip, Esq., F.R.S.

If a turtle (*Chelone*) be taken, the shell should be examined for parasitic barnacles (*Chelonobia*) and other adhesions. The specimens ought not to be scraped off; but the plate of shell to which they are affixed should be taken out, and the whole should be preserved together. Whales should be searched for *Coronulæ*, *Tubicinellæ*, &c.: they should be left, as they are found, in the skin and blubber of the animal, and the piece with its contents should be plunged in spirits.

The stomachs and intestines of those fishes and birds which are killed during the voyage should be inspected before they are thrown away; not only for the purpose of noting their food, but for the chance of finding undigested shells, &c., and in search of *Entozoa*. The feathers of birds should be examined with a view to ascertain whether any parasitic insects, any ova of fish or *testacea*, or any seeds of plants, adhere to their plumage. Their crops will often be found stored with fruits and seeds, which they disseminate in their flight.

Particular attention should be paid to the appearance of birds or insects, as well as to the direction whence they seem to come, with a view to the elucidation of their migration.

By placing in the sea clean planks of wood, the rate of growth of *Teredo navalis*, and of the *Cirripedia*, together with the ravages made by the former, in a given time, may be ascertained. *Serpulæ* will probably be found on the board also, and, perhaps, other shells. This experiment should be repeated whenever an opportunity occurs, and in different localities and climates. Some of the planks should be painted, others covered with pitch, others studded closely with copper and other nails, and some should be in their natural state.

When on shore in search of terrestrial *mollusca* (land-shells), the collector must not be content with a close examination of the trunks, leaves, and stems of trees, and other plants, but must turn up all decayed vegetable substances, especially in moist places, and there dig into the earth, more particularly about the roots of trees, and under overshadowing bushes and shrubs. Stones must be lifted,—herbaceous plants must be pulled up and their roots inspected,—and, if the boat's crew be at hand, fallen trunks of trees should be turned over with hand-spikes. All ova must be preserved; and the height above the level of the sea at which the specimens were taken, and the plants on which any of them were feeding, must be noted. In the latter case the plants should be preserved in an herbarium, and numbered as directed under the head of insects.

No boggy places, especially where streamlets ooze out, should be passed without examining the rushes and other plants there growing, for fresh-water *testacea*. At the proper season their ova may be found adhering to living and dead stems of plants, leaves, &c.

No bird, insect, shell, nor any other zoological specimen should be neglected because it does not strike the eye as beautiful, or because it is small and appears to be insignificant. Such objects are often the most interesting.

When a box or barrel of specimens is once securely packed, it should never be opened till it arrives at the place of its destination. If it is wished to have a few duplicates at hand, for the purpose of exchange with other collectors who may be met during the voyage, some specimens should be set aside for that purpose. All observations should be noted down while the impression is warm; and, if possible, with the subjects actually before the observer.

When an animal is seen afloat, and is remarkable for its magnitude or other peculiarity, and is not captured, its nearest approach to the ship, its mode, course, and rate of progression, and the parts actually visible, should be noted at the time with the utmost accuracy. If the observer feel conscious that he has not the zoological knowledge requisite for determining the species from the phenomena, he should abstain from giving the animal any special name. A shot fired, if it do not hit, may so alarm the creature as to cause some sudden movement which may reveal more of its true nature.

Section XII.

BOTANY.

By SIR WILLIAM HOOKER, K.H., D.C.L., F.R.S., &c. &c.,
and Director of the Royal Gardens of Kew.

Botany is a science which requires to be studied at home as well as in the field. For this reason it is highly desirable that persons visiting foreign countries should not only obtain information on the spot respecting the plants and their uses and properties, but that they should transmit to this country ample collections of *well-dried specimens*, with the rarer *fruits* and *seeds*, and all sorts of interesting vegetable *products*. By the latter expression we mean not only *gums and resins, drugs and dye-stuffs*, but whatever may be remarkable of vegetable origin in food and clothing, for building (the various kinds of woods), utensils, &c. We therefore first offer a few plain instructions for collecting and transporting plants in foreign lands.

Living Plants for Cultivation.

Plants for cultivation in our European gardens may be introduced either as *seeds*, *bulbs*, *tubers*, *cuttings*, or *rooted plants*.

Seeds, *bulbs*, and *tubers* are easily collected, and as easily transmitted to Europe from very distant countries.

The first, *seeds*, require to be gathered quite ripe ; to be wrapped, a quantity of each, in paper (common brown paper is as good as any), done up in a parcel, and kept, if possible, while on board-ship in an airy part of the cabin. *Bulbs* and *tubers* should be taken up when the foliage has withered, and, if well dried, they may be packed in the same way as seeds.

Cuttings.—Generally speaking it is vain to attempt sending *cuttings of plants* to a distance: they soon perish. But this is not the case with the greater number of succulent plants, those with thick and firm fleshy stems and leaves. Such are many of the Cactus tribe in South America ; the various succulents of South Africa, as *Aloes*, *Euphorbias*, *Stapelias*, *Mesembryanthemums* or *Fig-Marygolds*, the *Houseleek* kind, &c. Many of the *Bromelia*, or Pine-apple tribe, and the *Agaves*, or *American Aloes*, will survive a long time as cuttings. The cuttings should be taken off, if possible, where there is a contraction or articulation of the stem, or at the setting on of a branch : the wound ought to be dried by exposure to the sun ; and all such cuttings may be packed in a box, with paper wrapped about them, or any dry elastic substance to keep them steady.

Rooted Plants.—Some few of these, namely, such as are of a succulent nature, small *Cactuses*, *Aloes*, *Bromelias*, *Tillandsias*, and *Zamias*, &c., and (which are now highly valued in European stoves) the various *Epiphytes* or *Air-plants*, those numerous *orchideous plants* and others of the *Arum tribe*, which clothe the trunks and branches of trees in tropical countries :—all these will bear a long voyage if removed with their roots and stowed in a box, like the cuttings above described, the larger kinds surrounded with dry straw. But plants when taken up with their roots (and young ones should be preferred) can only be securely transported if placed in earth, in Ward's plant-cases, now generally known and most deservedly esteemed. These cases are glazed at the top or roof, so as to be in fact portable greenhouses. The plants should be established in the cases a few days before sending them off, secured by splines, so as to confine the roots in the soil in the event of the box being overturned, and moderately watered: the lid is then fastened with putty and screws, and the case being placed on the deck of a vessel so as to be exposed to the light, which is an indispensable requisite, will require no watering nor any attention (unless the glass happens to be broken) during the entire voyage.

On Preserving Plants for the Herbarium.

This is by no means the difficult process which many have imagined. The object is to prepare specimens in such a manner that their moisture may be quickly absorbed, the colours, so far as possible, preserved, and such a degree of pressure imparted, that they may not curl in drying.

For these purposes provide a quantity of paper of moderate folio size and rather absorbent quality—brown or stout grey paper answers the purpose exceedingly well. The best of all, and it is not expensive, is Bentall's botanical paper, 16 inches by 10, which costs (folded) 15*s*. a ream, or, if required larger, may be had, 20 inches by 12, at 21*s*. per ream. It is sold by Newman, Great Devonshire Street, Bishopsgate Street, London. Two boards are requisite, of the same size as the paper, one for the top, the other for the bottom of the mass of papers. Some pieces of pasteboard (or millboard) placed between the specimens, if these are numerous or particularly thick and woody, are very useful. For pressure nothing is better than a heavy weight on the topmost board, or, while travelling, three leathern straps and buckles, two to bind the boards transversely, and one longitudinally. Thus provided, gather your specimens, if the plant be small, root and stem; if large, take off portions of the branches, a foot or rather more in length, always selecting those which are slender and in flower, or in a more or less advanced state of fruit. Long, slender plants, as *grasses*, *sedges*, and many *ferns*, may be doubled once or twice. Place them, as quickly after being gathered as you can, side by side, but never one upon the other, on the same sheet of paper, taking care that one part be not materially thicker than the other, and lay over the specimens one, two, three, or more sheets of paper, according to the thickness of your paper and of your plants; and so on, layer above layer of paper and specimens, subjecting them to pressure. In a day or two, according to the more or less succulent nature of the plants and the heat and dryness of the soil and climate, remove them into fresh papers, twice or oftener, till the moisture be absorbed, and dry the spare papers in the sun or by a fire for future use.

If the specimens cannot be laid down as soon as gathered, they should be deposited in a tin box, which indeed is essential to the botanist when travelling:

there they may remain uninjured for a day and night, supposing the box to be well filled and securely closed to prevent evaporation. Some very succulent plants, and others with fine but rigid leaves—the heath and pine tribe, for instance—require to be plunged for an instant into boiling water ere they are pressed. In this case the superabundant moisture must be absorbed by a cloth or by blotting-paper.

When sufficiently dry the specimens should be put into dry papers, one sheet or folio between each layer of plants, except they be unusually woody (which is the case with oaks and pines), and then more paper must be employed, care being used to distribute the specimens pretty equally over the sheets, and thus a great many may be safely arranged in a small compass. They are now ready for transport, either packed in boxes or covered with oil-cloth.

Mosses and cryptogamous plants may be generally dried in the common way: those which grow in tufts should be separated by the hand to form neat specimens. Sea-weeds require a slight washing in fresh water, and common blotting-paper is the best for removing the moisture from this tribe of plants.

It is almost needless to add that *all plants*, whether living or dried, ought to be transmitted to Europe with the least possible delay: the latter, especially in hot or moist climates, are often soon destroyed by the depredations of insects.

The above short instructions refer solely to the collecting and despatching *living plants* and *dried specimens;* in other words, the means of furnishing our gardens and the herbarium. Another important branch of the science comes to be mentioned, hitherto much neglected, but towards which travellers will do well to contribute—we mean the museum of vegetable products, or it may be called the "Museum of Economic Botany." Its design is to bring together in one spot and to exhibit those interesting vegetable products from all parts of the world which cannot be shown in the living plants of a garden or the preserved ones of an herbarium. The public may now see growing in our Botanic Gardens the rare *Lace-tree* of Jamaica, the yet rarer *Ivory Palm-nut* of the Magdalena, and the *Cow-tree* from the Caraccas. The interest of these is greatly enhanced when, in the same establishment, the curious and beautiful lace of the first, the fruit and ivory-like seeds of the second, and the cream-like substance of the third, used as nourishment by the Indians, can be inspected.

Among the objects, therefore, which are to be collected for the museum are—

1. *Fruits and seeds*, especially those which are of large size and possess any peculiarity of form and structure entitling them to notice, such as *Pine-cones*, the various fruits of *Palms*, &c. &c. Many of them are naturally dry, and require little care (except to be freed from moisture) previous to packing. Those which are about to burst open into valves, or to separate by their scales (*Pine-cones* and *Araucarias*), should be bound round with a little packthread. The soft and fleshy kinds can only be preserved in wide-mouthed bottles or jars, or casks (according to size), in alcohol, as rum, arrack, or in diluted pyroligneous acid.

2. *Flowers* which are very large or particularly fleshy, and therefore unsuited to the *Hortus Siccus.* These should be preserved in alcohol or pyroligneous acid. Among those which would be much prized are, for instance, the flowers of the *Victoria* or *Gigantic Water-lily*, from the still waters of tropical South America, portions of the flowering branches of *Palms*, &c., and the larger kinds of *Orchidaceæ.*

3. *Entire plants*, or parts of them. Many have a very fleshy nature, and must be preserved whole in alcohol, or portions of the stem and branches, according to their size, with flower and fruit; such are the rare kinds of *Stapelias, Orchidaceæ, Misseltoe, Rafflesia, Mesembryanthemum, Cactus, Aphyteia, Balanophora, soft Parasites*, and others of a similar sort.

4. *Trunks of trees*, portions and sections, particularly when they exhibit any remarkable structure, as Palms, and many other monocotyledonous plants, *Tree-ferns, Zamias, Cycas*, and parasitical stems, when the latter display their union with the tree whereon they grow.

5. *Woods.* Specimens of the kinds used in commerce, for veneering, cabinet-work, or other useful purposes; or such as recommend themselves by their beauty, hardness, or any other valuable quality. Specimens of wood should be truncheons, five or six inches long, and of such diameter as the plant allows. Generally speaking it is advisable that a small branch dried and pressed, with flowers (and fruit, if convenient), should accompany the specimen of wood, in proof of the *precise* tree or plant from which the latter is derived.

6. *Gums and resins*, eminently those employed in the arts or in domestic economy.

7. *Dye-stuffs* of various kinds.

8. *Medicinal substances.* These are of vast importance, and merit the attention of travellers in every country. With respect to many, it is not yet known, except to the natives who collect and prepare them, what are the particular plants that yield them. It is hoped the present application may be the means of dispelling this ignorance among scientific Europeans; and that travellers will endeavour to procure the substances and well-dried flowering specimens of the plants from which they are obtained.

9. *General products of vegetables.* It is extremely difficult, perhaps impossible, to enumerate all of these which a museum ought to contain, but the enlightened traveller will form a tolerably correct judgment. Such as are useful to mankind cannot fail to be interesting. It were of course idle to exhibit every well-known object of this description, *tea*, *sugar*, *coffee*, *cocoa*, *chocolate*, *paper*, *clothing*, &c.; but there are states even of these familiar substances which would prove both useful and instructive. The *cane yielding sugar*, for instance, is advantageously exhibited. *Paper*, again, is made from an infinite variety of vegetable substances; and the different sorts are well worth collecting, from that afforded by the *papyrus of the ancients* (which gives the name) to what is manufactured of the inner bark of an East Indian *Daphne* (or *Spurge-laurel*), and another from the pith of an unknown plant in China (the so-called *rice-paper*), or the leaves of a *Palm* in India, or *straw* in North America. Of all such, the several stages of preparation should be collected, not only as objects of curiosity, but because they exemplify the progress of art and science.

A question will naturally suggest itself to the traveller not previously versed in the vegetable productions of different parts of the globe, " In what regions can I most effectually serve the cause of botany?" The answer is ready: In almost every portion of our world the inquiring mind will find objects for study; though assuredly the less the country has been tracked by Europeans and men of science, the more fertile it may be expected to prove in novelty. But even where the coast has been visited and tolerably accurately investigated, the interior, especially if mountainous (and the loftier the mountains the more varied the vegetation), will afford an ample field for research. Even with regard to many frequented spots it has been truly observed that few persons visit them " with their eyes open." Thus, while some travellers boldly assert that the vicinity of Aden is utterly destitute of vegetation, others have detected plants of very peculiar structure, and admirably adapted to such a locality; and one estimable naturalist, Pakenham Edgeworth, Esq., has actually published a *Florula* entitled 'Half an Hour's Botanizing Excursion at Aden,' giving an account of forty species gathered during that brief time, eleven of which are considered new to science.

One has only to glance at a map of the world and it will be instantly seen that much of it is unknown alike to the botanist and the geographer. The extensive interior of South America, particularly towards the sources of the great rivers, the deserts and mountain-chains of Africa, the table-lands of Thibet, with the northern declivities of the Himalayan Mountains, the Chinese dominions, and many of the numerous islands of the Malayan Archipelago, are still a *terra incognita* to the naturalist. But it is not to such little traversed realms alone that we need look for new and interesting vegetable productions, particularly of the more useful kinds. The remainder of this article shall be devoted to a mention of some of the many plants, or peculiarities relating to the *substances* * derived from them, which are yet unknown, or very imperfectly known to us, albeit they come from frequented lands. We shall arrange them, according to their countries, as follows :—

Asia (*including Australia*).

[N.B. Being sent by way of the Red Sea, it may be remarked that some of the products, enumerated under this head, are derived from Abyssinia, Arabia, or the East coast of Africa.]

Ammoniacum.—Determine the true origin of this gum-resin by specimens of the plant yielding it in Persia, forty-two miles south of Ispahan. Another kind is equally worthy of inquiry from Morocco in Africa, with the gum-resin and exact locality.

Sagapenum.—A gum-resin : its source ? It is said to come from Persia, and to be derived from a *Ferula.* Specimens of the plant with the gum-resin which it affords are desirable.

Galbanum.—Whence obtained ? It is brought to us from Singapore and Persia.

Gamboge.—Specimens in flower, and fruit of the plant affording the gamboge of Siam, and the mode of extracting this and other kinds of gamboge, such as that of Ceylon, &c.

* For much of this Catalogue of Desiderata and Inquirenda I am indebted to notes from Dr. Royle, Dr. Pereira, and Dr. R. D. Thomson.

Ammi Gum, or *Piney Varnish*, said to be produced by *Vateria Indica.*

Copal.—The origin of this gum-resin in India?

Bdellium.—The source of the Persian and African *false myrrhs* of this name, the localities producing them, the native names, and specimens both of the products and the plants.

Olibanum.—The above remarks apply to Olibanum.

Elemi.—The source of the five varieties of Elemi, viz.: 1. Holland Elemi. 2. Brazilian Elemi. 3. East Indian Elemi, in bamboos. 4. Macula Elemi; and 5. Mexican Elemi. Samples from the various countries, with the plants and native appellations, should be transmitted for verification. Is any *Elemi* procured from Ceylon?

Tragacanth.—The tragacanths of Mount Ida and Mount Libanus have never been correctly traced to the plants which yield them, nor has Tournefort's relation of the formation of this substance in the bark been confirmed. It is still more important to ascertain if the tragacanth of Erzeroom is brought into British commerce, and whether it is yielded by the *Astragalus strobiliferus.*

Senna.—The source of the East Indian or Mocha Senna. Is it really the foliage of *Cassia lanceolata?*

Catechu.—To observe the processes by which the various kinds of *Catechu, Cutch, Terra Japonica,* and *Gambir* are obtained; and if from trees, whether from others besides *Acacia Catechu, Areca Catechu*, and *Uncaria Gambir.* We want to identify the trees with the respective extracts.

Cubebs.—Does *Piper Cubeba* or *Piper caninum* in Java yield cubebs? If both, which gives the best?

Cassia.—Botanical specimens of the plants *seen* to yield *Cassia bark* in Kwagse, China, Malabar, Egypt (and Brazil).

Cassia-buds of the grocers' shops.—To procure specimens of the bark in Cochin-China and Japan, and flowering specimens for the Herbarium.

Rhubarb.—The true source of the medicinal rhubarb, and especially of the Batavian rhubarb. Strange to say, we are still in the dark respecting the real origin of this most valuable drug! In this and all such cases the drug should be procured by one who is an eye-witness to its being gathered, and specimens of the foliage and fruit should accompany it, and be carefully dried for the Herbarium on the spot.

Arrow-root.—The sources of the East Indian arrow-root. It is made largely at Travancore.

Salep.—The different plants which yield salep in Asia Minor, Persia, and especially the best kinds.

Aloes.—The true sources of the Soccotrine, Clear, Bombay, Hepatic, East Indian, and Mocha Aloes.

Minia Batta, or *Stone Oil,* from Borneo.—Whence is this solid oil or fat obtained? Is it abundant or rare?

Gutta Percha.—That of Singapore is ascertained to be the product of a new plant, *Isonandra Gutta* of Hooker, in the 'London Journal of Botany,' vol. vi. p. 331, 463, tab. 17. The appearance of the inspissated gum which is imported from Borneo under that name indicates a different source. Other Malay islands are said to afford *Gutta Percha*, but probably from yet other plants. This should be inquired into: the chemical characters of the juice in a fresh state should be ascertained, and compared with those of caoutchouc.

Green Tea.—Is indigo or any other vegetable dye used to colour the green tea in the northern provinces of China? Specimens of the plant and dye so employed are desiderata. Is turmeric or any yellow vegetable dye used in conjunction with it, or with Prussian blue?

Japan Wax.—The true source of this wax.

Assafœtida.—From what species of *Ferula* is this extracted, and how? Does the same species yield the *Tear Assafœtida* and the *lump?* Specimens of the one or several assafœtida plants should be procured, with the gum-resin produced by each species.

Patchaouli, or *Puchá Pát.*—A well-known perfume, of comparatively recent introduction to Europe. It is referred to a plant now described by botanists under the name of *Pogostemon Patchaouli;* but we are ignorant of the mode of its preparation and the exact locality where it is produced.

Sago of Japan.—Is it from a *Cycas,* and what species? Also specimens of sago in different stages of manufacture, with the trees yielding them, from the various parts of the Indian Archipelago; so that we may identify the particular kinds of sago yielded by the several sago-palms. Is the Ceylon sago the granulated pith of the *talipot-palm* (*Corypha umbraculifera*)?

Korarima.—A large kind of cardamom, or aromatic fruit (an *Amomum?*), found in the markets of Shoa, but probably the produce of a country farther to the west.

Scammony.—Particulars of the manufacture—or, to speak more correctly—the adulteration of Scammony, carried on at Smyrna. What is the purgative resin or gum-resin (if any) which is added, with the view of increasing the bulk and the medicinal activity of the mixture?

Camphor Oils.—There are two sorts: one, brought from the East, does not deposit crystals by keeping; the other does. Is the former the produce of *Dryobalanops Camphora?* If so, what is the source of the latter? Is it obtained from the foliage of *Laurus Camphora* (*Camphora officinarum*, Nees), or is it an artificially manufactured article? The camphor deposited is said to be similar to common camphor.

Kino.—All particulars of details about the manufacture of East Indian Kino (*Pterocarpus Marsupium*) are desirable.

Turmeric.—The several commercial sorts of turmeric differ so much in external character as to throw doubt on the identity of their origin. Are they not the produce of several species of *Curcuma?* Well-dried specimens, accompanying the root, should be transmitted from different parts of India.

The Grass-Oils.—The grasses used in India for affording the oils imported as grass-oils, lemon-grass oil, and essence of verbena, or verbena-oil, to be ascertained, and samples sent home, with details of the manufacture.

Storax of commerce is supposed to be obtained from the *Liquidambar orientale* of Cyprus, and comes by the Red Sea from the Persian Gulf; but the subject requires investigation, for others believe the plant grows in Cobross, an island of the upper end of the Red Sea. Dr. Pereira has ascertained that the *liquid storax* comes to us by way of Trieste; the storax of the Indian Archipelago is yielded by the *Liquidambar Altingia* of Blume.

Adelaide Resin.—What is the source of the red resin from this colony of South Australia?

(N.B. The various gums yielded by many trees and barks affording tannin in Australia require careful investigation.)

Is the true *cinnamon* of Ceylon the production of one species, or are other kinds employed? What occasions the red colour of the oil of cinnamon from Ceylon?

Tacamahaca of Ceylon.—Specimens obtained from *Calophyllum Inophyllum* are desirable, in order to aid pharmacologists in accurately determining the Tacamanaca of European commerce.

Ceylon (*Long or Wild*) *Cardamom.*—What is the plant so called in Ceylon, and named by Mr. Moon *Alpinia Granum Paradisæ?* Can it be identical with the true *Grains of Paradise* of the western coast of Africa?

Rice-paper of China.—This has been incorrectly referred to the Shola (*Æschynomene asperata*); but we are still quite ignorant of the origin of this familiar and exquisitely delicate substance. It is the pith of some plant, but what?

Africa (*including Arabia and Abyssinia*).

Cape Aloes.—What is the particular species of *Aloe* affording the drug of this name? What is the kind used at Bethelsdorp, near Algoa Bay?

Madagascar Cardamom.—Is it *Amomum angustifolium?* Specimens of the plant and fruit should be sent home.

Scitamineous Fruits of Western Africa.—A full collection of these (comprising the various kinds of cardamom), the plants, with roots and fruit, should be transmitted home, with the native names appended to them.

Myrrh.—Is the myrrh of commerce produced by one vegetable species? or several? If by several, specimens of each kind and of the plant affording it are desirable, accompanied by the native appellations. It is particularly important to know whether the myrrh of commerce be the growth of Arabia or of Abyssinia and the adjacent parts of Africa.

Euphorbium Gum.—What is the species of *Euphorbia* affording the substance thus called in commerce, and which comes from Mogadore? The stems found in the commercial Euphorbium are not those of the plant figured in Jackson's 'Morocco,' nor yet are they those of *Euphorbia officinarum* or *E. Canariensis.*

Shea Butter.—Living plants and specimens in flower and fruit are required.

Galam Butter.—Is this identical with the Shea butter of Park?

Camwood.—The source of the dye-wood so called, from the Gold Coast, with specimens of the tree, are a desideratum.

Bucku of the South-African Hottentots.—To determine the different kinds collected by the natives.

Senna.—What plant yields the African Senna? Richardson says it is brought from Ghat, in the Sahara.

African Oak, or *African Teak.*—This wood, though largely imported by our royal dock-yards from the western coast of Africa, is totally unknown botanically.

Ichaboe Resin.—The Ichaboe ships did, on more than

one occasion, bring from the adjacent shores of Africa a gum-resin, constituted of the dead stems of a *Geranium*, allied to, if not identical with, the South African *Geranium spinosum* of Linnæus (*Monsonia Burmanni*, D. C.). An account of this substance appeared in 'Eden's Voyage in search for Nitre and the true nature of Guano. London, 1846.' None is now to be procured in England, and it is believed that the nature and property of this singular gum-resin were not examined. Perfect specimens of the gum-resin and the plant are desired.

N.B. Much information remains to be obtained respecting the useful woods, gums, dye-stuffs, &c., of western Africa.

America.

Sarsaparilla.—To ascertain the plants yielding the several sorts of commerce, especially those called Jamaica Sarsaparilla (said to be the produce of the Mosquito Shore, and to be brought to England *viâ* Jamaica), Lima Sarsaparilla, which comes to us from Costa Rica (can this be extracted from the plant which yields the Jamaica Sarsaparilla ?), Honduras Sarsaparilla, and the Brazilian Sarsaparilla.

Balsam of Copaiva is imported from several parts of Brazil: it varies somewhat in properties, and is considered to be the produce of several species of the genus *Copaifera*. It is desirable to obtain the balsam of each species, with a specimen in flower and leaf, and, if possible, in fruit, of the tree affording it, and the name of the district where the tree grows, and its *native* appellation there.

Balsam of Tolu and *Balsam of Peru.*—It may be said of these balsams, as of the *Storax* and *Liquidambar* of the East, that much confusion exists with regard to the substances so called, the same name being applied in different places to different substances, and different names to the same substance. The trees respectively yielding, the countries producing, the mode of obtaining, the native names and the local ones of these articles should be carefully sought and ascertained: samples of the products must accompany the specimens of the trees.

Yellow-Bark, Royal Yellow-Bark or *Calisaya*, the produce of La Paz, requires to be identified with the tree which yields it. Specimens of the tree in fruit, in flower, and leaf, of the stems with the bark on and the bark removed, should be sent along with the description.

Ipecacuan, False.—From Brazil (the native country also of the true Ipecacuan, *Cephaelis Ipecacuanha*) a "false ipecacuan" is derived. What plant produces the latter? What is its native name, and may it not be equally valuable with the true?

Cascarilla Bark.—It is important to trace the cascarilla bark of English commerce (from Jamaica, or the Bahamas?) and that of Continental commerce (probably of Vera Cruz) to their true source. Are they derived from a species of *Croton?* and what particular kind?

Coca of Peru.—Ascertain the facts as to the consumption and effects of this substance (the foliage of *Erythroxylon Coca:* see a long account in the *Companion to the Botanical Magazine*, vol. i. p. 161; and vol. ii. p. 25, tab. 21). A quantity would be interesting for chemical examination, not less than some pounds.

Quassia.—Why is *Surinam* Quassia superseded by that from *Jamaica?* A comparative examination of the two barks chemically is required.

Angostura Bark.—To trace the different sorts to their true sources in Orinoco.

Tous les Mois.—A nutritive fecula, prepared at St. Kitts. From what species of *Canna* is it derived?

Contrayerva.—Is the Brazilian article yielded solely by *Dorstenia Contrayerva,* or by other species too?

Copaiva.—To determine what species of plants afford the best and worst kinds respectively in Brazil.

Cinchona.—To trace accurately the true kinds to the trees *seen* to produce the barks in Peru.

False Cinchonas.—To obtain authentic samples of *Quinquina Colorada* of Brazil, St. Lucia bark of Mexico, and Piavi bark (of Brazil).

Cevadilla.—To determine whether the *Helonias officinalis* or *Veratrum Sabadilla,* both Mexican plants, is the best adapted for the preparation of Veratrine.

Cohoon Oil.—The source is unknown of this solid oil, which resembles that from the cocoa-nut: it is said to be extracted from an *Attalea* (a kind of Palm) in Mexico.

Barbados Aloe.—What species affords the drug thus called? Is it not a plant imported from Africa? No true Aloe appears to be an original native of the New World, any more than a true *Cactus* is aboriginal of the Old World.

Many gums and drugs are said, in 'Edwards's Voyage up the Amazon River,' to be exported from Pará in Brazil. Whither are they sent? What are they? And what are the plants which yield them?

By this list of Inquirenda it will be seen that they principally refer to the productions of warm and dry climates, where indeed the most valuable vegetable substances are found. Temperate regions, however, afford ample scope for the researches of an intelligent naturalist. In every country, too, certain localities offer peculiar facilities for tracing the articles of commerce to their sources. Such are the capitals and principal sea-ports: Rio Janeiro, for example, for the products of Brazil; Senegambia for the interior of North-Western Africa; and Smyrna for Asia Minor. The rich products of Eastern India find their way to Europe by various channels, especially Calcutta and Ceylon; those of North-Western India, some by the Levant or Trieste, but much the greater quantity by Aden, Mocha, and, above all, Madras. The latter places are also frequently the outlets for the productions of Persia, Arabia, Abyssinia, and the east coast of Africa. Again, at Kurrachee, and probably better at Hyderabad, Dr. Royle states that a detailed list of the articles brought down that river, and which are highly important to be ascertained, may probably be had. Travellers, who visit the localities where any interesting vegetable substances abound, will confer a great service upon science by procuring not only these articles, but the identical plants from which they see them collected. And it is desirable to do this, even in the case of common and well-known productions, for many of them need to be verified, and to have the opinion of preceding observers confirmed. It is only by this means that we shall become acquainted with the *African Oak* or *Teak*, and with the various kinds of *Cinchonas* or Medicinal Barks of Peru. Of the plants affording the latter, we are as ignorant as when this invaluable medicine was first introduced into Europe by the Jesuits in 1632. It is a frequent and just subject of regret, that naturalists who make collections in remote and little frequented countries, are apt to neglect the procuring of authentic specimens of the more useful plants for the Arts, Medicine, and Domestic Economy.

It is difficult to recommend *botanical books*, useful to an universal traveller, and which shall form a portable library. Many of the best collections of plants have been made without books; and, unfortunately, there is no *complete* work published. De Candolle's '*Prodromus Systematis Naturalis Regni Vegetabilis*' is the best, and it is advancing towards a termination; and Walpers' '*Repertorium Botanices Systematicæ*' is a valuable supplement to it. Don's '*Dictionary of Gardening and Botany*' is useful, so far as it goes: it is in 4 vols., 4to., published at 14*l.* 8*s.*, but now offered by many booksellers at 1*l.* 8*s.*! Dr. Lindley's various introductory works on Botany ought to be in the hand of every student, and, above all, his '*Vegetable Kingdom.*' '*Loudon's Encyclopædia of Plants,*' with its numerous wood-cuts, is an excellent travelling companion. And for any particular country it is desirable to ascertain the books that may happen to describe the vegetation of it. They are too numerous to allow of a list being here given.

APPENDIX.

Extract of a Letter to Capt. W. A. B. Hamilton, R.N., Secretary to the Admiralty, on the collection of information respecting Foreign Timber useful for Naval and other purposes, from A. F. B. Creuze, Esq. (Principal Surveyor at Lloyd's), dated Lloyd's, 2, White Lion Court, 26th January, 1848.

A good account of the timber-produce of the world, or at least of our own possessions, is greatly wanted. Timber for masting and ship-building purposes is annually becoming more scarce. In proof of this assertion, compare the ease with which large topmasts could be got at the time of my early service in Her Majesty's Yards, say 25 to 35 years ago, with the present difficulty; or, compare the present supply of trees, which will convert for main pieces of rudder, stern-posts, stems, floors, second futtocks, &c. of large ships, with that of those years. I might multiply instances, but these will suffice in illustration.

The "Memorandum" should, I submit, call the attention of naval officers, especially those of surveying ships, to the timber of every country which they visit. They should be instructed that whenever they meet with timber which they consider available for the naval service, or generally for ship-building purposes, they are to procure information upon it; to the nature of which information the following heads may serve as some guide:—

The average contents per tree, obtained by measuring, say ten trees which have apparently attained the perfection of their growth, that is, are of the average size, without evident decay in the trunk or main limbs, and of which the foliage exhibits in every part equal freshness. The method of measuring may be obtained from a book called 'Hoppus's Measurer.'

The average distance from the ground to the lowest branch.

The diameter of the trunk below and clear of the insertion of this branch, and the diameter above and clear of the swell of the roots.

A rough sketch of the growth of the main stem, and of ten or twelve feet in length of the principal limbs. The example selected for the sketch should be the tree which will give the best idea of the *general* growth of the timber.

A specimen of the flower and of the foliage. The first may be preserved by placing it, stem downwards, in a jar partially filled with dry sand (not from salt water, unless, previously to being dried, it has been thoroughly cleansed from salt), and then gently filling up the jar with fine dried sand till the specimen is completely covered; the foliage may be placed between the leaves of a blotting-book.

The fruit, or seed-vessel, which may be preserved according to its nature.

A section from the trunk of a tree, say six inches in thickness (like a cheese), with the bark and all perfect.

If possible the specific gravity or weight of a very exactly cut cubic foot of the *perfect* wood of the tree taken from the butt end, and of another cubic foot taken from the top end of the trunk, when first felled; these weights marked upon the specimens, and the specimens brought home to be again weighed and measured when seasoned.

The nature of the soil in which the trees apparently flourish most, and whether the weeds, so far as can be ascertained, are such as grow in moist or dry localities.

General observations as to the appearance of the decay in any trees of the species which may have fallen, and may be lying about decaying; also, if the country be inhabited, the local uses to which the timber is put, and the state of any of it which can be ascertained to have been in long use for any purpose which is in its nature trying to the durability of timber, as alternations of exposure to wet and dry.

Observations as to the probable supply, and of the facilities or otherwise afforded by the nature of the country for conveying the logs to water-carriage.

A knowledge of the particulars required under these heads would enable a correct judgment to be formed of the nature of the timber, and of the purposes for which it might be available, and therefore, whether it would be advantageous to import any for Her Majesty's service.

I am, my dear Sir,

Yours very sincerely,

AUGUSTIN F. B. CREUZE.

SECTION XIII.

ETHNOLOGY.

By the late J. C. PRICHARD, Esq., M.D.

THERE are few subjects that can engage the attention of intelligent travellers, more worthy of interest, or on which any additions to our previous stock of information will be more generally appreciated, than ethnology. Under that term is comprised all that relates to human beings, whether regarded as individuals or as members of families or communities. The former head includes the physical history of man; that is, an account of the peculiarities of his bodily form and constitution, as they are displayed in different tribes, and under different circumstances of climate, local situation, clothing, nutrition, and under the various conditions which are supposed to occasion diversities of organic development. The same expression may also, in a wide sense, comprehend all observations tending to illustrate psychology, or the history of the intellectual and moral faculties, the sentiments, feelings, acquired habits, and natural propensities. To the second division of this general subject, viz., to the history of man as a social being, must be referred all observations as to the progress of men in arts and civilization in different countries, their laws and customs, institutions—civil and religious, their acquirements and traditions, literature, poetry, music, agriculture, trade and commerce, navigation; and, which of all things affords the most important aids in all researches as to the origin and affinities of different tribes or races, their languages and dialects.

On almost every topic now enumerated our acquaintance with remote nations is at present much more extensive than it was a quarter of a century ago; but on all it is still very defective. We shall touch upon the different subjects of this investigation in a very brief manner, with a view to point out what remains to be done in each particular, and to offer some suggestions as to the best method of proceeding.

I. *Of the Physical Characters of Nations.*

The physical description of any tribe or race must commence with an account of the more striking and obvious characteristics of complexion, features, figure, and stature.

In reference to the complexion or colour, it is not enough to know generally whether it is black or white, or brown. The exact shade of colour should be described as it prevails in the majority of persons in any tribe, and all the variations should be noted which occur in individuals. If a great difference of colour should be observed in the people of the same community, care should be taken, by repeated inquiries, to ascertain, if possible, whether such diversities are merely accidental varieties, or are connected with any distinction of tribe or caste. In many countries tribes exist who, while they preserve their stock distinct, by avoiding intermarriages, continue to differ from each other in colour and other particulars, though in other instances great varieties are observed within the limits of the same race, which appear as if they were capricious and accidental deviations, analogous to those varieties which appear in cattle and other domestic animals. A careful inquiry as to the history of individuals and families will sometimes determine how far the phenomena alluded to may be referable to either of these observations.

The shape of the features and the form and expression of the countenance should be described. For this purpose words afford but very imperfect means of communicating correct ideas. It will be advisable in all instances to obtain, if possible, correct portraits of persons of both sexes, and these should be coloured so as to represent the complexion as well as the form of the countenance. If no artist should be present who is capable of taking a likeness, the form of the features may at least be described by a profile or shaded outline.

The colour of the eyes should be noted, as well as the direction of the eyebrows; whether oblique, as in the Chinese and some Tartar races, and standing upwards towards the temples, or straight and parallel to the axes of the orbits, as in most European heads.

The hair, whether woolly and crisp, or curled and wavy, or straight and flowing, should be described, and specimens obtained of it. Notice should be taken of any varieties of the hair which occur in any particular tribe, there being great varieties in the nature of the hair in some races, while in others it is nearly uniform. Its colour should also be remarked.

An account should be taken of the average stature and weight in both sexes. This can only be obtained by the actual measurement and weighing of a considerable number of individuals, and the number and extent of the measurements should be mentioned. The proportional stature of the different sexes differing in different races, an account should be taken of this fact. Extreme cases should be noticed.

The proportion between the length of the limbs and the sternum, and the height of the body and the breadth of the pelvis, should be ascertained, and the length of the fore-arm in proportion to the stature of the body. This is known to be much greater in some races than in others.

Particular attention should be paid to the shape and relative size of the head, since this forms one of the principal characters distinguishing the several tribes of the human family from each other. The most authentic testimony in regard to this particular, and one which will be very acceptable to scientific men in this country, will be afforded by bringing home a collection of skulls, if they can be procured. In that case it would be necessary to select those skulls for specimens which afford the best idea of the prevailing form of the head in the particular tribe; and if several forms are observed in any race of people, which is the case in some islands of the Pacific Ocean, specimens should be sought which serve to identify every leading variety. If skulls cannot be procured, the best substitute will be casts of heads. Failing these, it will be requisite to take measurements. Such measurements should state the proportion between the longitudinal and transverse diameters of the skull, which will show whether the skulls of the tribe belong to the elongated form or to a rounder one. The facial angle may also be taken, formed by two lines, one of which falls from the forehead slanting over the edge of the upper jaw-bone, and the other passing from the meatus auditorius to the basis of the nose. The breadth of the face should also be taken by measuring the space between the zygomatic arches. In well-formed heads of the European type, the lateral surfaces of the zygomatic arches are parallel to the temples or the lateral surface of the frontal bone; so that the breadth of the forehead above the eyes is equal to the breadth of the face from cheek-bone to cheek-bone, measured by a line passing across the bridge of the nose. But in the Turanian type, common to the Chinese, Mongolians, and other nations of High Asia, the forehead is so much more narrow than the face as to give the upper part of the head almost a pyramidal form. An account should be taken of these characteristics, which most obviously distinguish the High Asiatic from the European type, and likewise of the extent of the upper and lower jaws, an excess of which is the chief peculiarity in the head of the Negro, and of other races approaching the Negro type. The oval, pyramidal, and prognathous types, as above described, constitute the three leading varieties in the form of the human head, but, together with the description of these characters, notice should be taken of every peculiarity that can be detected on a careful inspection of the cranium, or of the heads of living persons, when skulls cannot be obtained.

Observations on the form and structure of the body

should be followed by inquiries which belong to the department of physiology, which includes all that relates to the functions of life. Under this head we must mention inquiries respecting the senses or sensorial faculties. It is well known that there are differences between the different tribes of men in regard to the perfection of these faculties, and that some of the nomadic nations of High Asia, for example, have a remarkably acute sight and hearing, while other nations are equally noted for the perfection of taste and smell. Observations on these particulars belong to the physical character of each tribe.

Attempts should likewise be made to obtain information as to the relative degrees of muscular strength in various races. An instrument invented for this purpose has been termed the dynamometer. If it should not be at hand, the same purpose may be answered by experiments showing what weights a given number of men can raise by their individual efforts.

Other physiological characteristics should be investigated when opportunity can be found of obtaining information that may be satisfactory respecting them: such are the average length of life in any tribe; the ages of puberty and of the cessation of child-bearing, and all other facts connected with the animal economy, such as the number of children in families. Various questions have been raised by physiologists as to the phenomena connected with the functions of the female, whether they are subject to similar laws in the different races of human beings; and although, generally speaking, the result of such inquiries has been to show that no important difference exists, it is still right to pursue the inquiry in regard to newly discovered tribes, whenever opportunity is afforded by the accidental residence of medical persons in any place, or other contingent causes may promise to afford accurate results.

Pathological observations are nearly connected with physiology. It behoves the traveller to collect whatever information he can acquire as to the diseases prevalent in any tribe of people, or among the inhabitants of any country which it is his fortune to explore.

II. *Characteristics of the State of Society, &c.*

Questions which have regard to men in their social state, or as members of tribes or communities, take a much wider scope than the personal history of individuals. The ordinary habits of life and the modes of obtaining subsistence are the first topics that present themselves when we proceed to this branch of the subject. The rudest or most simple stage of human society is not without its appropriate arts. Some of these indicate as much enterprise and ingenuity, and as great activity of the intellectual faculties, as the practices of more civilized men. People who subsist on the spontaneous fruits of the earth, without pasture of cattle or cultivation of the soil, must exercise great ability in merely obtaining the means of subsistence. This is called the hunting state. It is not always a primitive condition of men. The history of the South African nations proves that tribes of people may sink into it from a higher state. The Bushmen once resembled the pastoral Hottentots: and even the African bushmen, as well as the Australian savages and the most destitute of the Esquimaux and other American tribes, display as much ingenuity in following their respective pursuits as nations of much more refined and artificial habits of life. The arts and customs of nations in this state form an interesting chapter in the history of mankind, and in the ethnography of particular branches of the human family.

Races inhabiting high steppes and open plains, such as Great Tartary and the plateau of Southern Africa, are generally nomadic herdsmen. Their habits of life are very different from those of the hunting tribes, and many of them differ from the latter in physical organization. The pastoral nations, wandering through open plains and enjoying a life of leisure and contemplation, have cultivated astronomy and a simple kind of poetry. Their history presents features of great interest to those who have opportunities of observing them.

Some of the rude and hunting nations have practised agriculture to a limited extent, but this pursuit is precluded by the locomotive habits of the nomadic nations. The indolence of savages generally throws this labour, as is well known, on the females of the tribe. In this state of things hunting continues to be the main occupation, and the habits of the tribe are not greatly changed by the introduction of a scanty tillage. But when the cultivation of the soil becomes the chief means of subsistence, the people must cease to be hunters or wandering herdsmen; they become fixed on particular spots, and separated into small communities. Hence agricultural tribes differ from each other in language, and likewise in physical characters, more than the nomadic races. It is highly desirable to inquire in every country into the facts connected with this transition, and to observe how far the introduction of agricultural habits has been connected with agrestic slavery. The change from the free and wandering life of pastoral nomades to the toilsome drudgery of the agriculturist is so great a change, that it has probably never taken place except under circumstances of peculiar kind. The earliest agriculture of most countries appears to have been connected with slavery. In many places there were "adstricti glebæ," who performed the laborious part. In many instances these were a conquered people reduced to the condition of serfs. Such were the Sudras of India, conquered by the twice-born classes. The Helots of the Spartans and perhaps the *γεωργοι* of the Egyptians were the descendants of captives. In every country where the soil is cultivated, as it often is, by a particular tribe, it will be advisable to make accurate inquiries into the history of such races. In these the traveller will often find the descendants of aboriginal inhabitants, the genuine people of the land, and among them he will discover the ancient and primitive language of the country, while the lords or feudal masters of the soil, the dominant people, will be found to be merely late immigrants from some foreign country.

The methods of agriculture anywhere practised should be noted as well as the kinds of grain which are found to be in use. The whole of the esculent plants used by any tribe of people should be described. Few races of men, however rude and insulated from the rest of mankind, have been found without some exotic vegetables. It has been observed that there is scarcely a hamlet in the most inclement parts of Lapland where some garden-plant may not be discovered which has been imported from places of more genial climate. The esculent plants in the possession of any remote and secluded people may often afford a clue as to the origin and family relations of the tribe.

Light also has been thrown on this subject by the kinds of instruments used in agricultural works. Notice should be taken of the forms of the plough and of the different instruments used in tillage, and of the peculiar methods of cultivation anywhere found to be in use.

The mechanical arts practised by various nations are to be carefully observed, such as their preparation of clothing, their architecture, or the manner in which they construct their dwellings and their household furniture.

A subject worthy of particular inquiry is their metallurgy, and the degrees of skill displayed in the arts of mining and making metallic implements. Many rude nations are known to have had some knowledge of the precious metals, of gold and silver, and even to have smelted copper long before they learned to know the use of iron. Various ornaments of silver and gold are found in the tombs of many northern nations, who were far too rude to invent the manufacture of steel, and who never dug the iron ore which abounded in their own mountains. The western nations of Europe are supposed to have made hatchets and celts and swords of copper, long before they made a similar use of iron. In most countries we trace the remains of a barbarous age, when cutting implements of various kinds were made of flint or stone, and when even ornaments were manufactured from bone, or amber, or ivory, before the use of metals was discovered. The names given to metals should be noted, since these names will often afford a clue as to the countries from which they were imported.*

The art of war, as practised by various nations, affords a wide field of observation. The weapons used, whether bows and arrows, spears, or clubs, or swords, are often common to scattered tribes of the same kindred, and will serve to identify nations, or at least to suggest inquiries as to the probability of their relationship. The ancient Gauls were known by their gæsa or javelins, the Germans by their saga or military cassocks, and the Australians by their woomerangs or throwing-sticks, and the poisoned arrows of the Bushmen are noted among the South African nations.

The sort of clothing used by simple nations, as well as that of the more cultivated, should be described—whether made of the skins of animals, as among the most savage nations, and especially those of arctic countries, or of cloth prepared by weaving, or otherwise preparing the fibrous parts of plants, as cotton, flax, or other vegetable productions. Attention should be paid to the modes of cultivating such plants as contribute the material of clothing.†

* Thus it has been observed, that the Greek name for tin, 'κασσιτερον,' resembles the Indian (Sanscrit) name Kast'hered of the same metal, and it has been inferred that tin was first brought to Europe from India before the British mines were explored. The tin-mines of Tama-Malaga, or Malucca, were celebrated at a very early period.

† A curious mistake was made by the ancients in regard to silk. They imagined that it was prepared from beautiful flowers.

'For clothes the barbarous tribes of Seres use,
Nor oxen hides, nor wool of fatted ewes;

In every newly discovered country it will be an interesting subject of inquiry, what domesticated animals are in the possession of the natives—where they obtained such as they are found to possess—whether they are known as wild animals of the same region, or were brought from some foreign land.

Inquiry should be made as to the art of navigation practised by different races. Some nations appear to have a greater aptitude for maritime pursuits than others. The Polynesians in some places are almost amphibious, while the American natives and the Australians rarely venture upon the sea. Several South American nations are, however, expert navigators in their inland lakes and vast rivers.

The crude notions entertained by uncivilized nations on subjects within the scope of physical science are matters worthy of inquiry. Science they can be hardly said to possess, though this was scarcely true with the ancient Mexicans. All nations observe the changes of the moon, and measure the lapse of time with a greater or less degree of accuracy by the movements of some of the heavenly bodies. Inquiry should be made whether the motions of the planets are observed, and whether their bodies are distinguished from fixed stars, whether attempts are made to ascertain the duration of the solar year, and whether there are names for the constellations, and what they are if they exist.

They weave sweet flow'rets of the desert earth,
Of finest texture and of richest worth—
Robes bright of hue as flowers which deck the mead,
Of finer texture than the spider's thread.'

Dionys. Perieg. 755.

In every nation, however barbarous, it is probable that some sort of morality exists in the sentiments of men—some notions of right and wrong—and that some practices are considered lawful and praiseworthy, while others are forbidden. The same religious impressions and the same superstitions prevail through all the branches of a widely spread race, as the superstition of the *tabú* among the Polynesians. Inquiry should be made as to all traits of this description, and all the phenomena which enter into the psychological character of a particular tribe. Among these traits are the regulations respecting marriages in different communities. In some countries a complicated, and, as it appears, a very elaborate and artificial system of rules prevails, founded on the intention of preventing intermarriages between families even remotely connected by consanguinity. The institution of the *Totem*, as it was termed among the North American nations, has its counterpart among the nations of Australia. Whether the existence of customs so similar among these widely-separated races is a result of former intercourse, or a merely accidental coincidence, it is unnecessary to inquire. In both countries it constitutes a remarkable trait in the social and moral character of the nations among whom it prevails, and it may lead us to believe that nations apparently the most savage and destitute are not always governed merely by accidental impulses and merely animal passions, but are capable of deep thoughts and reflection, and of enacting laws with a wise and well-understood import.

Inquiries should be made as to all the regulations of social life, not only among civilized nations and those who are possessed of the external appearances of civilization, wealth, and conveniences, but likewise among people less prosperous in their condition, and having the aspect of barbarism.

Where polygamy prevails, it should be ascertained, if possible, whether there is any real disproportion in the numbers of the sexes. This should, indeed, be a subject of inquiry in every tribe where statistical information can be procured, as a matter connected with the physical history of the people; but it has a particular relation to the prevalence of polygamy.

The mode of civil government should everywhere be a subject of inquiry. The more simple nations are often without any common and central government, and are in the habit of only appointing a leader in time of war, when they select for some temporary enterprise as their chieftain some individual whose fame and prowess inspire them with confidence in his guidance. Some nations have been entirely without the idea of combining for mutual aid, and the Finnish races are said to have been conquered by the Germanic nations piecemeal, one family after another falling under the yoke, till all were subdued. The Polynesian nations have princes or chieftains, according to some well-understood laws. We are not yet acquainted with the social institutions of the Papuas, if they have any such among them, which is probable. It is very desirable that the fact should be ascertained by inquiry.

The religious impressions and the superstitious practices of every tribe of men should be carefully investigated, as forming a remarkable part of the history of the particular people, and an item in the psychological history of mankind. It is probable that no human race is destitute of some belief, more or less explicit or obscure, in the existence of supernatural powers, good and evil, and likewise of a future and invisible state. But there are nations who scarcely recognise in the invisible being anything like will or power to punish the guilty or reward the good, and who do not suppose the future state to be a scene of retribution. This is the account which missionaries and other persons have given of the Polynesian superstitions.

The adoration of rude nations is generally directed towards visible objects. From this remark we must except most of the American nations, who are said to believe in the existence of a spiritual ruler of the universe. By one class of rude nations the heavenly bodies are worshipped, and the Polynesians connect this superstition with a mythology which is poetical and not devoid of ingenuity. Others, like the African nations, worship *fetiches*, or visible objects, in which they suppose some magical or supernatural power to be concealed, capable of exercising an influence on their destiny and of ensuring success in any undertaking—a superstition of which traces are to be discovered among the vulgar in many countries.

In every tribe of people among whom intelligent travellers may hereafter be thrown, it should be a subject of inquiry how far any of these observations may be confirmed and extended by the history of their superstitious belief and practices, and to what division of nations they are by such traits associated.

III. *Language, Poetry, Literature.*

As no other means have contributed so much to the increase of ethnology, and to the ascertaining of the connexions and relationship of different nations, as a comparison of languages, great care should be taken in every newly discovered country, and among tribes whose history is not perfectly known, to collect the most correct information as to the language of the people.

Among tribes of people who have any poetry or other literature, pains should be taken to obtain the best specimens of composition in their languages. Manuscripts in their languages should be procured if it can possibly be done; and it would be worth while to incur even a considerable expense rather than forego such an opportunity.

In countries where the inhabitants have no knowledge of letters, it may sometimes be found that they have preserved oral compositions, generally in some sort of verse, which they have recorded in their memory, and handed down from one generation to another. It would be very desirable in such a case to write down the most complete specimen of any such pieces, and to select any which relate to the ancient and primeval history of the people.

If no literature or compositions of any kind have been preserved, the best things that can be done will be the following :—

I. To get some intelligent person to translate into the prevailing language some continuous composition, and to copy it from his mouth with the greatest care. Get in the first place the Lord's Prayer, since this same composition has been most frequently collected already, and exists in a much greater number of languages than any other. Next to the Lord's Prayer, which does not contain a sufficient quantity of words, the Gospel of St. Luke probably exists in a greater number of languages than any other composition. The sixth, and perhaps also the seventh chapter, may be selected from this Gospel. A good translation of these two chapters will enable a person skilled in philology to furnish a tolerably complete analysis of almost any language.

II. A vocabulary should also be taken down from the mouths of intelligent natives. Care should be taken to compare the words given by one person with the testimony of others, in order to correct any defect or peculiarity of pronunciation.

It is very important to select properly the classes of words. The following should be chosen :—

1. The numerals up to a hundred or more. Ascertain how far the people of each tribe can reckon.

2. Words denoting family relations, such as father, mother, brother, sister, &c.

3. Names of the different parts of the body,—head, arm, foot, &c.

4. Names of visible natural objects, elements, &c.,—sun, moon, fire, water, &c.

5. Names of animals, especially domestic animals.

6. Verbs expressive of universal bodily acts, such as eat, drink, walk, sleep, see, hear, &c.

7. Personal pronouns,—I, thou, he, &c.

8. Prepositions,—in, from, to, &c.—if they can be obtained.

III. It would be useful, in the third place, to observe some of the grammatical rules of the language, if opportunity exists of becoming acquainted with them ; though if any composition of some length shall have been obtained, the grammatical analysis may be furnished afterwards. It will not, however, be amiss to make the following observations :—

One great feature in the grammatical structure of different languages, which distinguishes several classes of languages from each other, is the peculiar position given to auxiliary words in sentences. By auxiliary words are meant such words as have no proper meaning of their own, but tend to explain the relation of nouns and verbs. Such are prepositions in our language,—upon, in, through, &c. It should be observed what position such words hold with relation to nouns. It is a character of one great class of languages—viz., the Tartar dialect, or the languages of High Asia—to place all such particles at the end of nouns : thus prepositions become postpositions. In most African languages, as yet known, particles are placed at the beginnings of words ; and that is the case not only with prepositions, but with particles of all kinds, such as syllables which change the singular into the plural number, as Anakosah becomes the plural of Kosah. Again, in the American language, particles are as it were swallowed up by the principal words, or are inserted in the middle of them.

It may be right to observe also whether languages admit the composition of words making compound epithets by amalgamating two or more simple words.

Observe also whether the words, such as names of objects, are monosyllables, or consist of several syllables.

APPENDIX.

By the EDITOR.

In compiling vocabularies from the mouths of natives, whether of written or unwritten languages, but especially of the latter, and of languages which, though reduced to writing, are so in characters (like the Chinese, &c.) illegible to Europeans, it is of the utmost importance to secure th possibility of a reasonably faithful reproduction of the sounds from the writing when read by a third party having no personal communication with either the speaker or writer. This can only, of course, be accomplished by the adoption of a system of writing very different indeed from our ordinary English practice of spelling (which is utterly inapplicable to the purpose); fixing upon a set of letters, each of which shall express a distinct, recognised, and as nearly as possible invariable sound, and regulating their combination by simple and fixed rules.

Pending the introduction of a Phonetic character free from objection, and bearing in mind that, after all, it is only a very imperfect representation of the native pronunciation which can be so conveyed (although amply sufficient if due care be taken to render the speech of a foreigner intelligible among them), the voyager or traveller will find in the Ethnical Alphabet' of Mr. Ellis* a stock of characters prepared to his hand capable of accomplishing to a considerable extent the object proposed;† or he may adopt the following as a conventional system, in which only Roman, Greek, and Italic characters are employed, and which therefore can be at once transferred from MS. into print at any ordinary printing office. In the examples annexed the letters printed in Italic are those whose sounds are intended to be exemplified.

Vowels.

1. *u* long (*uu*) as in Engl. b*oo*t; Germ. Br*u*der; Ital. verd*u*ra; Fr. *ou*vrir:—short (*u*) as in Engl. f*oo*t; Germ. r*u*nd; Ital. br*u*tto:—very short or coalescent as in Engl. *w*ig; Germ. q*u*er; Ital. q*u*ale.

* 'The Ethnical Alphabet, or Alphabet of Nations, tabularly arranged for the use of Travellers and Missionaries, with Examples in ten Languages.'

† In thus directing attention to the 'Ethnical Alphabet' of Mr. Ellis for this special purpose, the Editor must be understood emphatically to protest against being considered an advocate of the "spelling reform" of the English language *for the use of Englishmen*, as proposed and urged by either that gentleman, Mr. Pitman, or Mr. Faulder.

2. *o* long (*oo*) as in Engl. gh*o*st; Germ. Sch*oo*s; Ital. c*o*sa; Fr. *O*r:—short as in Engl. res*o*lute; Germ. h*o*ld; Ital. d*o*lente; Fr. Nap*o*leon.
3. *v* long (*vv*) as in Engl. p*u*rse; Fr. l*eu*r? Gael. l*u*gh:—short, and very short (*v*), or in Mr. Ellis's nomenclature stopped, as in Engl. p*e*rt, c*u*t; Germ. V*e*rsuch.
4. *v̇* as in Germ. G*ü*te; Fr. Aug*u*ste.
5. *ȯ* as in Germ. L*ö*we; Fr. l*eu*r? conn*ue*.
6. *ô* long (*ôô*) as in Engl. l*aw*:—short as in Engl. h*o*t; Germ. G*o*tt? k*o*mmen.
7. a long (aa) as in Engl. h*a*rd; Germ. H*aa*r; Ital. and*a*r; Fr. ch*a*r:—short (a) as in Engl. *A*meric*a*; Germ. Burgsch*a*ft; Ital. *a*ndar; Fr. charl*a*tan.
8. *a* long (*aa*) as in Engl. w*a*ft, l*au*gh:—short (*a*) as in Engl. h*a*ve; qu*a*ff.
9. *a* as in Engl. Bank, h*a*g; Fr. Pr*i*nce; *ai*nsi V*i*n (?)
10. *ȧ* long (*ȧȧ*) as in Engl. h*ai*l; Germ. S*ee*, st*ä*dtchen; Ital. l*ie*ta; Fr. m*ê*me, f*ai*t:—short (*ȧ*) as in Engl. accur*a*te.
11. e long (ee) as in Engl. H*ei*r, H*a*re, H*ai*r, w*e*re; Germ. B*e*rg, St*ä*rke; Ital. lumi*e*ra; Fr. lumi*e*re:—short (e) as in Engl. m*e*n, l*e*mon, *e*very; Germ. b*e*sser, *e*mpor; Ital. cast*e*llo; Fr. dang*e*reux, *e*ffort, *e*loigner.
12 *i* as in Engl. h*i*ll, b*i*t; Germ. g*i*ft, g*i*tter; Ital. c*i*nque.
13. *ï* as in Engl. p*ee*l, l*ea*ve, bel*ie*ve; Germ. L*ie*be; Ital. v*i*no; Fr. qu*i*.
14. • as in Engl. peop*l*e; Germ. lieb*e*n (pïp•l, lïb•n).

Diphthongs.

15. *j* as in Eng. b*i*te; Germ. b*ei*ssen.
16. ȣ as in Engl. br*ow*n, b*ou*nd; Germ. br*au*n; Fr. s*aou*l?

Consonants, &c.

17. *s* as in Engl. *s*oft; Germ. *s*anft; Ital. *s*olo; Fr. *s*alle.
18. *z* as in Engl. *z*inc; Germ. ro*s*e; Ital. ro*s*a; Fr. a*z*ur.
19. *sh* as in Engl. *sh*arp; Germ. *sch*arf; Ital. la*sc*iare; Fr. *ch*ien.
20. *zh* as in Engl. plea*s*ure; Fr. *j*ardin.
21. *th* as in Engl. *th*ing; Span. *z*apato, na*c*ion.
22. *dh* as in Engl. *th*at.
23—31. *t, d, p, b, f, v, m, n, l*, as in the English, German, Italian, and French.
32. *rr* as in Engl. p*r*ay; Germ. *R*abe; Ital. *r*osa; Fr. e*rr*eur.
33. *r* as in Engl. smalle*r*.
34. *ρ* or *rh* as in Engl. Rhatany, Rhubarb.

35. *ν* as in Engl ha*ng*; Germ. klin*g*en; Ital. li*n*gua fra*n*ca.
36. *ν̇* as in Fr. ai*n*si, rie*n*.
37. *ν̃* The nasal sound in Æt*n*a, D*n*ieper.
38. *h* as in Engl. *h*alt; Germ. *H*exe; Fr. *H*alte!
39. *χ* as in Germ. la*ch*en; Span. *X*imenes, relo*j*; Gael. crua*ch*an.
40. *y* as in Germ. *g*elten; Gael. Lu*gh*.

Any supplemental letters may be used, if exactly exemplified and identified, for sounds peculiar to certain languages, as the Caffer and Hottentot *clicks*, &c.

Rules to be observed.

1. Do not use a running hand in writing from pronunciation, but form each letter separately; take care not to confound a and *a*.
2. For capitals use the small characters enlarged.
3. A vowel sound is understood to be prolonged by repeating its character according to the analogy of the German and Dutch. If the sound be really repeated, as in Oolite, insert a hyphen O-o thus O-o, or an apostrophe O'o. If the vowel be simply once written, it expresses the shortest sound conveying the *full vowel sound.* If intended to be very short, or to have that abruptness which has been called the stopped sound before a consonant, *double the consonant*, especially if the "stopped sound" be really perceptibly different *as a true vowel sound* from the "open," which in the English is *sometimes* the case.
4. Two different vowels coming together, when the first is intended to be shortened to the utmost possible degree consistent with the distinct audibility of its vowel character, it is to be prefixed singly to the other; as in the so-called English diphthongs *oi*, *eu* (ôi, iuu), or, as in such words as *wet*, *ye*, q*ua*ff (uett, iï, ku*a*ff). But if the vowels are intended to be separately and distinctly pronounced, as in the Italian p*au*ra, an apostrophe must be interposed, as pa'uura, or, if still more completely separated, a hyphen.
5. *h* means always a true aspiration, except in the combinations sh, zh, th, dh—for which, if any one should prefer to write *ʃ*, ʒ, *θ*, and ϑ respectively, he may do so with much advantage and with our entire approbation. The insertion of h in its true place among other consonants is a matter of much nicety, and requires an exact and discriminating ear.
6. The "obscure vowel," No. 14, represented by a large unmistakeable fullpoint, occurs only in such words as people, lieben (Germ.), &c. Its nearest representative as a prolonged sound is *υ* (in the above nomenclature); but it is a great fault to use this character, or any equivalent one, in cases where a real, distinguishable, and particu-

larly an essential etymological vowel is slurred over and obliterated by negligent and vulgar usage, as for instance, if we were to write the words *America* (Engl.), Stuf*e* (Germ.), ventur*a* (Ital.), J*e* (Fr.), or the Indian name B*e*nar*e*s, all indiscriminately with the character (*v*) appropriated to the vowel sound in the English word c*u*t. If, therefore, the necessity of imitating a *well-educated* usage require us to indicate (as no doubt it often does) a certain approach to this obscure *v*, it should be done by subscribing the point beneath the appropriate representation of the true vowel thus, Amerika, stuufe, ventuura, zhe, Benaares.

7. Compound consonants, as in *ch*ur*ch*, journal, may be resolved into their elements (tsh and dzh).
8. Particular attention should be paid to the accentuation by a single mark (′) of that syllable in each word where the prominent stress is laid in pronunciation, nor should the intonation of the voice be altogether neglected, though very difficult to reduce to any regular system of rules or signs, and rather a matter of description or musical notation than of alphabetic registry.

SECTION XIV.

MEDICINE AND MEDICAL STATISTICS.

BY ALEXANDER BRYSON, ESQ., M.D.

IT will necessarily happen, from the great proportion of the naval force employed on foreign stations, that amongst the first things which will attract the attention of a medical officer, are the effects produced on the constitution by a change of climate; and the question of the necessity or non-necessity of meeting this change by an alteration of personal habits or modes of living; whether on entering the tropics it will be prudent to continue the use of the same daily amount of food, to lessen its quantity, or to adopt a diet less stimulant as regards fluids, and more farinaceous as regards solids. Different views have been adopted on this subject, some of them erroneous, others extravagant, or only feasible were the human body a mere machine; while there is a third class founded on practical experience, and which is deserving of the most respectful consideration. Opportunities to simplify and reduce these into a more intelligible form will not be found wanting in the naval service.

In noting the meteoric changes which are likely to

affect health, there are not, it may be assumed, any great difficulties to be encountered as regards instrumental observation; in these mathematical precision, at all events, is not so essential as they would be, were the results aimed at depending on the truth of a series of arithmetical sums. There is, nevertheless, a proper degree of accuracy required in the mere registration of this kind of formulæ, as one omission may invalidate a whole set of observations,—such for example as the geographical position of the ship at the time the observations were made.

With regard to the atmosphere, the principal things to be observed are its heat, degree of humidity, and weight. That the two first greatly influence health there is not any reason to doubt; but with regard to the third, it would be hazardous to offer any decided opinion. Amongst men who have devoted much of their time and attention to the subject, there are perhaps a few who consider that it has at least some influence in disturbing occasionally the equanimity of the mental functions. Thermometrical observations with the view of noticing the influence of atmospherical heat on health, should be made several times a day, in order that the minimum, medium, and maximum in the shade may be ascertained; or even more frequently should there be a sudden rise or fall of the mercury. On board a ship under weigh, it is hardly possible, in consequence of the great variety of aspects in which she may be placed with respect to the sun, and the various currents of air rushing from her lower deck, to find a suitable place for the instrument; the black bulwarks and hammock-cloths rapidly absorb the heat of the sun's rays, and again throw it out by radiation for a considerable time even after the sun has gone down. Should the instrument therefore be placed, as has sometimes happened, contiguous to these, it will give an exaggerated view of the temperature. The under surface of the deck planks also radiates heat abundantly after the upper surface has been long exposed to the rays of the sun, consequently the temperature of the cabins and between the decks of a ship is sometimes greatly increased; this, however, if continuous in apposition with the heat in the sun's rays, and in the shade, it would be desirable to place on record, and also to state the influence it may be supposed to have on the general health of a ship's company, whether the inference drawn be of a practical or a theoretical nature. In connexion with accumulated heat from these or other causes, it would also be proper to state the space allowed to each hammock; the number of hammocks berthed on one deck, and in a general way the dimensions of the deck, together with the size and disposition of the scuttles, ports, and windsails.

Acute inflammatory diseases and fevers have most unquestionably been induced by a current of external air rushing from the lower orifice of a windsail on men sleeping close to its exit. Are we then to suppose, in the absence of all terrestrial miasmata, that these diseases are the result of the sudden abstraction of heat from the system? Simple immersion in the sea, or exposure to the external air in a state of nudity, has not, generally speaking, to the same extent, an equally deleterious effect.

These, and subjects of a like nature, are well deserving the attention of every medical inquirer; as there are still few of the doctrines respecting the origin of disease, or the manner in which the different forms of morbid action (when once established) progress, culminate, and decline, that are so clearly demonstrable as could be wished.

As a humid state of the atmosphere, particularly within the tropics, seems to exercise a considerable influence over the health of Europeans, hygrometrical observations are not less essential than thermometrical, to a full investigation into the causes and nature of any of those diseases usually denominated climatorial. Various instruments have been used for these purposes; but those which denote with ordinary accuracy the state of the atmosphere, and are the least liable to get out of repair, are the best. The appearance of surrounding objects and our ordinary sensations may be even trusted where there are no better means at hand. It will naturally occur to the observer to guard against confounding the moisture arising from any local cause, such as damp decks, or the halitus from the breath of a large body of men confined in a small space, with the natural moisture of the external air. Should the disparity, however, between the latter and the air of the deck on which the men generally congregate and sleep, be great, it will be incumbent on him when he uses an instrument to note the difference. From these data, viewed in connexion with the results of the thermometer, the necessity of a more free ventilation in all vessels of war destined to remain for years within the humid regions of the tropics may thus be made apparent.

To a dry air we are accustomed to attribute a bracing effect, to a moist air a relaxing; and there seems to be little reason to doubt the general truth of the postulate;

the first increases, and the second diminishes the amount of watery fluid in the system; the one as a general rule conduces to health, the other to disease. How far these conditions modify morbid action, it would be desirable to ascertain. That intermittent and remittent fevers are the peculiar product of moist localities, experience amply proves; and although the subject yet requires to be more fully examined, facts are not wanting to lead to the supposition that dysentery, and diarrhœa approaching to dysentery, are more frequently the result of atmospheric changes in certain dry localities within the tropics, than they are in moist localities in similar parallels of latitude.

The relative degrees of health enjoyed in vessels differing in the hygrometrical condition of the air between decks, from whatever cause (exclusive of external causes) such differences may arise, is a subject which has long engaged the attention of all classes of naval officers; and although the majority are of opinion that a dry condition is the more healthy, still there are others practically acquainted with the subject, who do not admit that the difference is appreciable, or who altogether repudiate the idea of damp decks having anything whatever to do with the health of a ship's company. As these conditions greatly depend on the modes of cleaning the lower decks, it more especially belongs to the medical officers to watch with vigilance, and report (but not without due and ample experience) the effects of dampness, whether from accident, stress of weather, or artificially produced, as well as the effects of dryness artificially maintained by swinging stoves or other contrivances.

The great difference between the appearance of men

employed in the bread-room and hold, and those who are freely exposed on deck, or in open boats, at all hours of the day, cannot escape the notice of the most superficial observer. It is therefore of importance to ascertain whether exclusion from the solar rays be not, to a greater extent than is generally believed, one reason why those men who have in consequence acquired a pale waxy look from confinement below, are more susceptible to disease, and less capable of sustaining its shocks, than are those whose blood is enriched and strengthened by the free exposure to light, heat, and air, which their different avocations ensure. The force of these remarks, however, will be best understood by those who have had opportunities of witnessing the rapid change which takes place in the human constitution by exposure for only a short time to the direct rays of a tropical sun. Why, in a state of perfect repose, the blood should acquire a brighter tinge, and an increased force of circulation, are inquiries, the value of which the observant physiologist will not fail justly to appreciate, neither will he fail, as often as opportunities occur, to follow up these phenomena, should they terminate in disease, or unhappily produce death.

Whether the stationary population at great elevations above the sea-level, differ from those habitually resident *at* the sea-level in rapidity of pulse and respiration, are questions respecting which there is still but little known. In connexion with this subject the following are the principal objects deserving the attention of medical or other travellers, particularly when opportunities occur of visiting places of great altitude:—the number of pulsations of the heart, the number of respirations per minute, and the circumference of the chest at several places. It will also be necessary to note the height, age, sex, and colour of the person examined, and whether in a standing or sitting posture at the time of examination.

The extent to which terrestrial miasmata may be conveyed by the wind has been so variously estimated, that correct information on the subject would tend not only to the benefit of the public service, but also to the credit of the medical profession. In selecting a proper position for an encampment, or for the anchorage of vessels of war, the greatest discretion and judgment are required, particularly in those countries which abound in the aërial or telluric agencies inimical to man; and although it may appear that these are matters with which the medical officer has little to do, and although necessity and the exigencies of the service may render the selection of any spot but that which is the best suited in a healthy point of view inevitable, still, dreading the suffering, loss of life, and the inefficiency that may accrue to the force from a position badly chosen, the external geological features of any coast or island off which a squadron may require to be concentrated cannot fail to attract his attention.

In connexion with terrestrial emanations, atmospherical currents, depending on local causes, together with a description of land and sea breezes, are also subjects deeply interesting to all classes of men, whether employed in Her Majesty's naval service, or otherwise engaged in maritime pursuits. It is, therefore, much to be desired that the country contiguous to any unfrequented creek or bay, or the embouchures of tidal rivers which are likely to become the resort of shipping, should be examined, and, if found to contain lagoons or marshes, mapped in such a way that those spots which are the most exposed to the malaria coming from these localities may be known, and if possible avoided as an anchorage. The nature of the soil in the immediate neighbourhood, the kind and the depth of water in lagoons, the character, depth, and consistence of swamp, bog, or marsh land, the description of plants which surround or grow from them, would greatly enhance the value of such information. These being the acknowledged sources of fever and ague, it would not escape the zeal of the inquirer to ascertain whether they were liable to irruptions from the sea, or floods from the interior; whether fogs arose from them, and if so, at what time of the day or year they were most observable; and also whether they emitted noxious effluvia. It has sometimes occurred that officers and men employed on boat service have been rendered conscious of the fact, that certain particular spots emit noxious effluvia more perceptibly than others.

The tides, by occasionally washing over or breaking down the banks of low alluvial lands, and by spreading over the adjacent country, form extensive brackish lagoons and marshes, which greatly impair the sanitary condition of a country. These circumstances, therefore, and their influence on health, should invariably be noticed under the head of topographical information.

There are few things of more importance to the naval medical officer than the origin and characters of febrile diseases, as a knowledge of the facts connected with the former may greatly bias his judgment with regard to the latter, and as the expression of his professional opinion thus influenced or formed, particularly with regard to their being of an infectious or of a non-infectious character, may involve not only the safety of the greater part of the men in his own vessel, but that of the crews belonging to other ships, or even of communities residing on shore, it will be admitted that these are not subjects, when opportunities occur, that ought to be superficially examined or inattentively reported.

Besides endemic and epidemic diseases arising from general or terrestrial sources extraneous to a ship, there are others which originate in local or personal causes existing on board. To distinguish between these is a matter of greater difficulty than seems to be generally apprehended. For instance, it has frequently occurred that fever has broken out in a single vessel of a squadron, and attacked not only the whole or the greater part of her crew, but all visitors who ventured on board, although they remained in her only a few hours. If these latter, after returning to their own ship or home, passed through the disease without communicating it to any other person, the opinion generally formed has been, that the fever was the result of exposure to some local cause unconnected with the personal emanations of the sick; but if in either case the attendants or immediate neighbours of the visitors were subsequently, that is, within two or three weeks, seized with fever similar to that of the latter and of the patients in the ship, and again other persons who had been in close communication with them were attacked, then the conclusion arrived at has been (as indeed it could not be otherwise) that the disease, if it were not in the first instance the result of personal contagion, had acquired in the course of its progress the power of propagating itself, and that in all probability it would through a series of subjects retain that power for an indefinite time. Still, notwithstanding the most careful sifting of every circumstance connected with the first cases, (it having been also ascertained that no disease of the same character had existed for several months previously in or near the locality, and that the men had not been on shore or absent in boats,) may render it necessary to conclude that it originated from some cause within the ship; it will yet remain to be determined whether that was of a local or of a personal nature, or of some peculiar combination of the two, either with or without adjunctive predisposition. In a large majority of instances it most unquestionably will be difficult, if not impossible, to decide; nevertheless, a concise narrative of the events as they occur should be committed to paper, in order that it may be made available, should it be required for any investigation in connexion with the reappearance of the fever at a future period either in the same or in a different locality.

When a fever has broken out in a vessel at sea, from a foul state of her holds, and, without her having any subsequent communication with the shore, continues to make progress, attacking man after man, how, it may be asked, is it possible to ascertain whether, as is sometimes the case, it has acquired a contagious character or not? The space is small, and the whole of the men being equally exposed to the original exciting cause, and, if such have been generated, to the personal, are there any means of distinguishing the effects of the one from those of the other, with that degree of certainty which would warrant the medical officer giving a conscientious opinion, if required by the arrival of the vessel in a port? The great similarity of all continued and remittent fevers, but more particularly of the fevers of the tropics, from whatever source they may have sprung, together with predisposition from fear, fatigue, or derangement of the digestive organs, and the utter impossibility of complete segregation, even in the most roomy vessel, will, it is apprehended, render it extremely difficult to make such a distinction; and the delivery of any opinion beyond that which may be hypothetically formed impracticable. Still experience, undoubtedly, in so critical a juncture, will greatly assist those who have had the benefit of its teaching; hence the value to be attached to all such authenticated truthful data, and the suspicion with which imperfect, speculative, or garbled accounts should be received.

If we were desired to write down "what to observe" with respect to tropical fever, the subject might be expanded into a volume, or compressed into a few words; for the occasion it may suffice to mention those things which would facilitate any inquiry for which it might be necessary to consult the medical returns of the Navy. One of the first objects of the medical officer, when an irruption of fever occurs in a ship of war, will be to ascertain, if possible, whether it arises from causes internal or external to the ship; for on this will depend the propriety of removing the cause, or removing from the cause, viz., clearing out the vessel, or quitting the locality. If it arise from causes within the vessel, these should be stated, and also the means taken to remove them; if from causes extraneous to the ship, they also, if possible, should be described, as well as the manner in which the men were exposed to their influence. The treatment of the disease will naturally rivet the attention to the symptoms; these again should lead to a more practically useful nosography than is generally adopted; the disease being placed under one of the three following distinctive heads, viz., continued, remittent, or intermittent. This, however, should it be deemed of importance, need not prevent the annexation of any other qualifying distinction, such as bilious, ardent, or yellow; but to reconcile the conflicting opinions which the writings of discrepant authors have called into existence respecting tropical fever, proof of the absence or presence of remissions, in some of its worst forms, may still be considered necessary.

As long as there is a British squadron on the sea, yellow-fever, as it is called, must claim a large share of attention; and as it is seldom brought to these shores, he who encounters it on its own domain, profiting by the occasion, will do well, while it is under his eye, to examine carefully into its origin and character. When it occurs as an epidemic its source should be looked for, its course traced, and its disappearance noted; and whether yellow suffusion be present in all the cases, or only in part of them, whether, when black vomit occurs, the disease seems to acquire generally a greater degree of virulence, and whether, in consequence of such aggravation, marked by deep yellow suffusion, dark-coloured blood, hæmorrhage, and, in the fatal cases, black vomit, it has assumed contagious properties. If the fever has commenced in a ship at sea, it will be in vain, as already noticed, to attempt coming to any decision as to the question of contagion, until it has been communicated to some person who had not been on board, from him to a second, and perhaps from the second to a third. Second attacks of this disease should invariably be noted in the returns.

In the treatment of yellow-fever there is most assuredly much to observe, and much to learn. The effects of the most vaunted remedies should be compared, without losing sight of the natural resiliency of the vital functions towards a state of health: this is a rule so essentially necessary, that not to apply it would vitiate the inferences. Blood-letting, and the nature of the blood abstracted, offer a fair field for observation, whilst the empirical modes in which we have been taught to exhibit mercury will perhaps, after some experience, induce the younger physician to reinvestigate the grounds on which his seniors recommended these questionable practices, and to compare them with the results obtained in the present day. The stage of the disease when quinine, the most valuable of all our remedies, should be commenced, and the extent to which it may be administered, are questions that have not yet been settled.

Pathological investigations have thrown but little light on the nature of simple idiopathic fever. To detect the seat of the disease, or to ascertain the cause of death, the solids have been explored in vain; the vital organs of the head, the thorax, and the abdomen, have each refused to reveal why they gave up their functions: to the fluids, therefore, armed with the additional appliances which the science of optics affords, and with a better knowledge of animal chemistry, we must resort. The blood, in connexion with respiration, will require to be examined to ascertain whether the functional derangement which exists in the organs of sanguification and respiration, in the first instance, be the effect of a chemical change in the blood; or whether it be the result of an impression made on the nervous system by some power exterior to the body; and whether such derangement of action interrupts the normal transfer of elementary principles between the external air and the blood, thereby leaving the latter so greatly altered and deteriorated as to be chemically defective in those constituents requisite for the repair of the organic structures, while it abounds in the lethal effete matters that are constantly received into it from the decay of the latter.

In the Naval service, more perhaps than in any other, there are frequent opportunities of ascertaining to a day, and even to an hour, the exact period of incubation in certain febrile and eruptive diseases: although this will also greatly depend on the disease being gradually or suddenly developed. A party of men, a boat's crew for instance, may enter a vessel, a house, or a village in which disease is raging; or they may land, expose themselves to the influence of a "homicidal marsh," and then return on board their own vessel, having inhaled a sufficiency of the poison to establish a certain specific morbid action, bearing, if of a personal nature, the exact similitude of its parent; and if of a terrestrial, that type of fever which is peculiar to the climate or locality, or to the prevailing epidemic—it will of course follow that in proportion to the length of time the patients have been exposed to the exciting miasm, so in an inverse degree will be the value of the information, as during a protracted exposure there is not any means of even approximately ascertaining when the system had acquired the requisite charge necessary to the evolution of the disease; although the latent period of endemic and epidemic diseases is a subject which is both curious and interesting; still as regards contagious diseases it is infinitely more so, as it is principally on a correct knowledge of these periods that the quarantine laws can be efficiently administered.

With a knowledge of these facts it will not be saying too much to affirm, that it greatly behoves the medical officer of the naval service to lose no opportunities of placing on record a succinct history of every case of disease, which has been contracted by exposure to any specific exciting cause for so short a period as will serve to mark the stage of incubation to a single day. There may by this means be such a mass of evidence brought to bear on the subject, as will greatly simplify our views on the doctrines of contagion, and at the same time disembarrass the laws of quarantine of many restrictive formalities, that are not only useless in a sanitary point of view, but injurious to commerce, and personally vexatious.

As the preceding observations are applicable more or less in a general way to other endemic, epidemic, or contagious diseases, it will be unnecessary to go over the same ground with respect to them. The incubative period of plague, and if it rage epidemically, proof of its having, independently of epidemic influence, acquired contagious properties which have been transmitted from one person to another, either simply through the medium of the atmosphere, or by means of fomites, are still questions of paramount interest to every nation which has communication with the shores of the Mediterranean. With respect to malignant cholera, it would be difficult to say in what particular it is most deserving attention; the causes essential to its production, the manner in which the fluids become poisoned, and the vital functions deranged, are unfortunately about as little understood as they were the first day the malady came under the observation of the European physician. This disease, therefore, together with the modes of treating it, offers a wide field for medical inquiry.

Some curious information may be occasionally obtained in distant countries relative to the modes of treating diseases amongst uncivilized tribes; not that it is likely to prove of much value, but as a matter of history it may be worth recording. It would even be interesting to know the virtues attached to charms and amulets, as well as the manner in which they are obtained, of what they consist, and how they are worn; nor would the methods of performing surgical operations be of less interest. The Albanians, it is reported, without the slightest knowledge of the anatomy of the parts, perform the operation of lithotomy with as much dexterity and success as has ever yet been reached in this country. The Marabouts of Africa, with a fallen tree for their table, may be found, with little display, performing the initiatory rites of Mahomedanism on the assembled youths of an entire village; while the Fetish man, on another part of the continent, ministering to the pride of caste, makes such fearful gashes on the faces of his patients as would astonish our boldest practitioner. How these wounds are cured might be worth knowing, as the scars sufficiently attest the excellence of the surgery.

In the central parts of Africa, and in some of the islands of the Indian Archipelago, there is reason to believe that the natives are in possession of narcotic poisons with which we are still unacquainted. An account of these, and of their modes of preparing them, would be interesting. And on all occasions the diseases most prevalent in the various foreign countries visited, and the most approved methods of treating them, together with an account of the medicinal plants, and other means in general use as remedies, should in conformity with the public instructions be invariably reported.

In preserving medical plants or seeds, or, in fact, any other object of natural history, for the purpose of bringing them to this country, it will be found no easy matter to protect them from the ravages of insects, and in damp countries from the effects of mildew. The tin cases now used for certain articles of dress are well adapted for the safe keeping of vegetable or animal substances; but when they cannot be procured, a tolerably large deal box, of a form such as will fit snugly between the beams of the small cabins allotted to gun-room officers, its seams being closed up by pasting paper inside, is the best substitute. From these all insects may be completely excluded by placing loosely in them, scattered amongst the contents, several pieces of camphor, and rags sprinkled with turpentine: the latter will require to be remoistened now and then. A few drops of the oil of petroleum may also be a very useful addition. Into a box so protected neither ants nor cockroaches will enter; and without some contrivance of the kind it will be in vain to attempt to preserve almost any object of natural history of an animal or vegetable substance; unless it be placed in spirits or dilute solution of the chloride of zinc. The latter, as it is now generally employed in all ships of war for the destruction of vermin and fetid exhalations from the holds, is not only the most available, but in other respects it is the best, the cheapest, and the most generally useful. Its preservative powers are equal to spirits, while even when the most putrid substances have been immersed in it, it remains perfectly free from the noisome odour which animal matters impart to the latter, and which renders the opening of any jar in which preparations are kept a nuisance, which few men would venture to inflict on their shipmates. The strength used by the curator of the Museum at Haslar, for fish and reptiles, or, when in good condition, specimens of morbid anatomy, is in the proportion of one part of the concentrated solution to twenty of common water; but when they are very putrid they require at first a much stronger mixture, namely, about equal parts of each. In this the preparation is allowed to remain until the smell is quite subdued, when it may be finally put up in a solution of the first-mentioned strength.

It would greatly enhance the value of the medical returns if, in addition to those now required by the printed instructions, there were an additional nosological return sent in annually, commencing on the 1st of January, or on the day the ship was commissioned, and ending invariably on the 31st of December, or, in the event of her being paid off, on the day she was put out of commission. In this the mean numerical strength of the ship's company should be noticed, together with a list of the men dead, specifying their names and the causes of death, whether from accident, disease, or suicide; whether occurring on board, on detached service, on shore, on leave, or in hospital. The name of each person dead is essentially necessary, to prevent one death being twice noticed, an error which, were it reported both from a hospital and the ship to which the man belonged, it would be difficult to avoid. By following out these plans the exact mortality of the service could be ascertained without any great difficulty or much trouble. At present this cannot be accomplished without a long and patient examination of various data, involving an unnecessary waste of time and much labour, which in the end is unsatisfactory, inasmuch as it is impossible to arrive at anything like the correctness which ought to stamp the character of all statistical details, as arithmetical facts.

Moreover, at present there is not any means of forming even the most distant conception of the relative loss of service per man from sickness in any given force, in the course of a year; nor of the number of days' sickness attributable per man to each disease separately; or, in either case, of the mean loss of service in comparison with the number of attacks. These are points of considerable importance in a statistical point of view, and unless they can be ascertained, it will be impossible to form any correct estimate of the health of the navy in comparison with other bodies of men.

The additional information required to carry out these calculations being in the hands of the medical officers of the navy, it is to be hoped the time is not far distant when it will be generally furnished. For the sake of system it will be necessary to add an additional column to the sick list, in which to state the number of days each case was under treatment. Adding these together at the close of the year, and dividing the aggregate sum by the mean numerical strength, will of course give the proportional number of days' sickness per man for the year. The relative proportion of sickness, with respect to different diseases, may also be ascertained. The additional trouble (if indeed it should be so considered) which these details would impose, divided amongst so many, would not be great, while the facilities they would afford, and the correctness they would insure in the compilation of the general details for the whole service, would be of the utmost importance.

There is still another object which would add greatly to the interest and value of the vital and medical statistics of the navy, and this cannot be effected at head-quarters unless by the employment of an extensive staff of clerks, namely, to class the whole of a ship's company by their ages into decennial periods, beginning at fifteen and terminating at fifty-five, in order to ascertain the relative degrees of sickness and mortality in each of these stages of life. By a proper arrangement of these and the preceding data, in tabular forms, the relative amount of sickness at certain ages, and from every, or, at all events, from the most important diseases, might be deduced, and the relative degrees of health enjoyed not only in different squadrons, but in different ships, ascertained by a single glance, and with a degree of accuracy which it is impossible to arrive at by the present system.

SECTION XV.

STATISTICS.

BY G. R. PORTER, ESQ.

THE population of any place or country must be considered as the groundwork of all statistical inquiry concerning it. We cannot form a correct judgment concerning any community until we shall have become acquainted with the number of human beings of which it is composed, nor until we shall have ascertained many points that indicate their condition, not only as they exist at the time of inquiry, but comparatively also with former periods.

In the section of this volume which is devoted to geographical observations, directions are given for collecting the actual numbers of the population, a branch of inquiry which properly falls within the province both of political geography and of statistics. The division of the inhabitants of any country into races, using different languages or dialects, belongs to the first named of the two sections, and need be no further noticed here.

The actual numbers of any population can never be so satisfactorily ascertained as by the interference of the government, and the first inquiries upon the subject should be for official enumerations. Where such do not exist, it may still be possible to procure data for satisfactory computations from governmental departments, and especially those connected with the taxation of the country; but it must be evident that, to render such data available, the circumstances under which it has been collected must, as far as possible, be ascertained and recorded. Where no official accounts can be made available, recourse should be had to private channels, giving the preference to such statements (if such exist) as may have been published in the country, and have thereby been subjected to criticism and correction on the part of those best qualified to form a judgment on the subject. Local registers are sometimes to be met with, where the central government has not interfered. Such were carefully kept in many parishes in England, before any government census was undertaken. From such registers, comparing births with deaths through a series of years, the population of a country may be estimated with some approach to accuracy. The rate of mortality is a fact of so much importance towards any useful knowledge of a country, that it is naturally among the subjects of inquiry that should earliest command attention. If registers of burials, which record the ages at which the deaths occur, can be obtained, they would elucidate many points of great interest as to the condition of the people and the effect of the climate, and would besides afford means, in connexion with the number of births and marriages, for more nearly approximating towards an accurate estimate of the population. Where a census has been taken, a distinction will doubtless have been made between the sexes; and if the ages also have been recorded, the tables will themselves afford means for testing their general accuracy, as it may be assumed that the proportion of adult males—twenty years of age and upwards—are *about* one-fourth of the whole population. Where no census has been taken, it may be possible to ascertain the number of *fighting men*, that is, of males between given ages. Should all other sources of information be wanting, it will then be necessary to have recourse to oral information, in estimating the correctness of which the observer must avail himself of such aids as present themselves. The question whether a community is increasing, stationary, or diminishing, may be judged from the amount of buildings in progress, or of houses untenanted or in a state of decay. If any account is taken, for purposes of taxation or otherwise, of the number of inhabited houses, and especially if these should be divided into different scales, a little personal observation as to the average number of inhabitants to be found in each will furnish valuable information concerning the population; but to do this, the inquirer must inform himself concerning the domestic habits of various classes of the people; the necessity for which caution will be made apparent by the fact, that while in all England the average number of inhabitants to each house is under $5\frac{1}{2}$, the average number in the metropolitan county exceeds $7\frac{1}{2}$; while the number to each house in Dublin is $12\frac{3}{4}$, which is double the average number in all Ireland, where the house accommodation is generally of the most wretched description.

Having ascertained, as well as circumstances allow, the numbers of the people, it becomes of importance to know how they are employed. It cannot be expected that any one, who is without the authority of the government for the purpose, can succeed in ascertaining with minuteness the numbers occupied in each of the various branches of employment, but opportunity may probably be found for ascertaining those numbers in certain great leading divisions, following in this respect the more usual course of inquiry in this country, and distinguishing individuals as employed, first, in agriculture; secondly, in trade and manufactures; and, thirdly, in all other pursuits. By knowing the proportionate number of any people who are employed in raising food for themselves and the remainder of the community, we possess a very important element towards estimating the social condition of the people. The truth of this remark is made apparent by the fact shown at the census of 1841, that while in Great Britain 251 persons raised the food consumed by themselves and 749 other persons, or while 1000 persons engaged in agricultural labour supplied the wants of 3984 persons, including their own; in Ireland, in the same year, the labour of 662 persons was required to supply food for themselves and 338 others, so that 1000 persons supplied food for only 1511 persons, themselves included. The deductions to be drawn from the like facts in other countries are liable to modification, and particularly if it shall appear that families, or any portion of them, which draw their chief support from agriculture, employ any portion of their time in domestic manufactures. Previous to the inventions of Arkwright and Hargreaves, the spinning-wheel was in general operation in cottages throughout a great part of England; and the time is yet more recent at which the shuttle might be heard in those cottages during the long evenings of winter, and at times when out-of-door labour was prevented by bad weather. Hand-loom weaving, except as the substantive occupation of the family, may now be said to have ceased in this country, and the spinning-wheel has long been wholly superseded; but this is far from being the case in many, or perhaps in most, other countries, where the females of a family are at times employed in spinning and weaving, at least for the supply of their own household, if they do not provide a further quantity of fabrics for sale to others.

Where manufactures are carried on in factories or large establishments, it will not be very difficult to obtain a tolerably accurate estimate of the number of such establishments, and of the hands employed in them. In some countries, the government requires that a patent or licence shall be taken out yearly by the proprietors of manufactories, and by this means a correct account of their number might be obtained. In the same way, the number of dealers may sometimes be ascertained, and probably classified as being wholesale or retail traders, as well as distinguished according to the branches of business pursued by them.

It is desirable to know the usual and average size of farms or holdings of land, and the system under which they are cultivated, whether by the proprietor of the soil or by tenants; and if by the latter, then upon what terms, whether by payment of an annual rent, and at what rate usually for a given measure; or by a division of the gross produce, and then in what proportion the landlord participates for the mere use of the land and farm-buildings, or whether he furnishes the stock or any proportion of it. Inquiry should be made as to the existence of what is understood by "tenant-right;" whether by law or by custom the farmer is entitled to compensation for such improvements as he may have made in the condition of the land. The number and kinds of live animals that are bred and kept upon farms should if possible be ascertained, as well as the number of labourers usually employed upon a given extent of land; the rate of wages which they receive; whether those wages are lessened by reason of their being boarded by the farmer, or whether they live and board themselves in separate cottages; and also whether there is employment on the farms for women or children, with the rates of wages paid to them; and further, if the labourers have any other advantages in aid of wages.

If it be important to know how the people of any country are employed, it cannot be less so to ascertain the result of their labour. It is especially desirable to know the proportionate quantity of each kind of food raised upon farms of a given size, or upon any known measure of land of the average degree of fertility; the quality of such of the cereal grains as may be raised will best be ascertained by learning the weight of a given quantity by measurement. While making inquiries concerning the supply of food of home growth, it must be essential to ascertain whether, in seasons of average productiveness, that supply is equal to or greater than the ordinary consumption of the country. Should it fall short of the requirements of the people, inquiries should be made concerning the quantity deficient, and the sources whence the same is ordinarily made good. On the other hand, should the home produce exceed the consumption, the amount of that excess and the usual channels employed for disposing of it should be ascertained.

Similar inquiries should likewise be made concerning the mineral productions of the country. It will not be enough to know only the number of persons employed in mining operations, since the value of such labour varies exceedingly in different countries. It was stated at the meeting of the British Association in 1844, that, according to the official reports of the French government, each workman employed in the coal-mines of France raised no more on the average than 116 tons in the year, while the average quantity raised by English miners in that time was 253 tons. Nor is this discrepancy confined to coal-mining. The quantity of iron made in Great Britain is four times that made in France, while the number of persons employed for the purpose is actually greater in France than in England: the numbers actually so employed in 1841 were, in France, 47,830, who made 377,142 tons of pig-iron, and in England 42,418, who produced 1,500,000 tons of that metal; so that the labour of each man in France produced barely 8 tons, while in Great Britain it sufficed to produce more than 35 tons.

The like inquiries should be made with reference to every branch of occupation, as occasion offers or may be found. Upon this subject it is essential to know the number of hours in the day during which, at various seasons of the year, workmen are ordinarily employed, whether the routine of their occupations is disturbed by the intervention of holidays, and to what extent such interruptions are carried in different branches of industry.

Also, whether any and what restrictions are placed by law or custom against the employment of women or children in any branch of trade or manufacture. Naturally connected with these inquiries is the share which the workpeople obtain of the value of the objects upon which their industry is employed. To ascertain this it is not only necessary to learn the usual rates of daily, or weekly, or yearly wages paid, but also the amount which a family of average industry, consisting of a man, his wife, and say four children, are ordinarily able to earn in the course of the year, including such perquisites as custom provides in aid of the ordinary wages, the nature as well as the value of which it must be interesting to know. It hardly needs to be said that a distinction must be drawn between the earnings of the skilled and those who are unskilled, those whose qualifications are the result of a previous expenditure of time and money, that is, of education, and those who bring little more than their bodily strength to the performance of their task. Neither does it need to be pointed out, that however numerically important are the classes usually understood by the term workmen, their condition does not comprise the whole of what it is desirable to know in forming an estimate of a community; the circumstances of the better educated portion of the people, including those who by their studies and acquired skill influence so greatly the general well-being, and upon whom mainly depend the progress of civilization, are to the full as necessary to be known. It will probably not be difficult to learn, as respects these, the fees paid to professional men, such as physicians and advocates, the salaries of schoolmasters and mistresses, as well as the salaries and other emoluments of men employed in the higher and in the more subordinate offices of the government. Coupled with these particulars we should endeavour to ascertain the necessary expenditure of families in the various walks of life. This is a more difficult task, and it requires much knowledge of the various conditions of the community to estimate the correctness of such statements as may be gathered, especially as regards the expenditure of the poorer classes. It is a curious fact, that in almost every case where details of this nature were offered to the Commissioners of Poor Law Inquiry in England, the expenditure as stated was found upon examination to exceed in no small degree the income of the family, although the parties affirmed that they did not run in debt. It must greatly help towards forming a correct estimate if the retail prices are ascertained of different qualities of the various articles used and consumed in families holding different ranks in the scale of society. The incomes of a very important class of public functionaries, the clergy, it may be more difficult to ascertain, especially in lands which have made a comparatively small progress in civilization, and where it is understood the priest often avails himself of the superstitious terrors of the ill-informed people to advance his own personal interest. In other countries, comparatively free from this evil, it is, however, not easy to ascertain the average rate of incomes of the clergy, which may be derived partly from one source and partly from another,—sometimes from the State, by a direct payment; sometimes from land, the profit of which will vary from year to year; sometimes from fees given for the performance of certain religious offices, such as marriage, baptism, and burial; and sometimes also from voluntary payments, or offerings, in acknowledgment for the instruction and consolation imparted. Nor will the cases be rare in which several of these sources are combined, in order to make up the income. It will be more easy to learn the number of the clergy, and to ascertain the manner of their appointment, whether by election on the part of the people, or by nomination on the part of the government or of individuals; and an estimate may be made of their general incomes by observing the class of the community among whom they usually live upon a footing of equality.

There is no subject which will so well enable us to judge concerning the progress and probable future condition of any people, as the state and degree of instruction which is provided for the youthful among them. The inquirer will therefore endeavour to learn, not only the number of educational establishments and of students attending them, but also the nature and quality of the instruction imparted; the proportion of schools connected in any way with the State, and of those established and supported by private means. It will not be difficult to judge from observation, and also through conversation with the inhabitants, how far the means provided have been effectual in former times in rendering the people intelligent and in forming their characters. The cost of instruction should also be learned, and whether in any and what degree that cost is borne by the government; also whether any and what degree of proficiency in the usual branches of knowledge is requisite to enable any person to take upon himself any official duties, or to authorise him to assume certain responsibilities in society, where the fortunes, the happiness, and it may be the lives of others, will be intrusted to his charge.

Closely connected with this subject is the state of crime in every country. The number of prisons, the amount of accommodation which they afford, and the number of inmates usually to be found within them, should, if possible, be obtained, as well as some acquaintance with the system of punishments pursued and the treatment of prisoners. The number of executions that have taken place within a given number of years, and the nature of the crimes for which that extreme punishment has been inflicted, should be ascertained. If any more general record of offenders can be had, it would be well to inquire the prices of food during the particular years to which those records relate, in order to judge correctly concerning the moral character of the people under one of its most important aspects,—its tendency towards criminal courses. To know the nature, generally, of the offences committed will give us an insight into many subjects of interest, provided the people have made any considerable advances in civilization; but if the country should be very backward in this respect, many crimes will go "unwhipt of justice." If the criminal records of England existed for any period further back than half a century, we should probably search them in vain in order to learn the number of pickpockets; not that the offence of picking pockets was unknown, but that when the offender was detected the mob took his punishment into their own hands, and by pumping upon him, or dragging him through a horsepond, or by some other more convenient summary proceeding, satisfied their views of justice, and let the culprit go. It is very desirable to know among what classes of people offenders are chiefly found; whether among labourers in agriculture, or handicraftsmen, or others; and also whether educated persons add in any, and in what degree, to the list of culprits. It is highly important to draw a distinction between male and female offenders, since their proportionate numbers will throw light upon the general character of the community in some of its features. In the early part of the present century there were 40 females to each 100 males committed for trial in England and Wales; but during the 10 years from 1838 to 1847 inclusive, the average proportion has not been quite equal to 24 in each 100, indicating a change in condition, manners, and morals, favourable to the present day. It is equally desirable to know the proportionate numbers of juvenile offenders, classing under that head all under 15 years of age, or such other period of life as, under the influence of climate or any other cause, may determine the date at which the youths of the country generally assume an independent position and provide for their own support. It will be well to distinguish the sexes of these young offenders.

By making inquiries of intelligent residents it may be learned whether, with the progress of time, criminality has increased or diminished in the country. The criminal records, if such exist, will by no means furnish data upon which reliance can be placed for judging upon this point, since it often, or it might be said, most frequently happens, that with advancing civilization the police regulations of a country are more strict; besides which, increasing population, and increasing wealth, may lead to a greater number of offenders, without really adding to the criminality of the community, since the nature and quality of the crimes committed may have become less serious. The number registered in the calendars will be increased if two cases of petty larceny shall have taken the place of one murder, and yet no one would thence affirm that crime has increased in the country.

The provision made for the indigent generally, and especially for the sick and the aged among them, will naturally call for inquiry. The number and extent of establishments answering to our union-houses, alms-houses, hospitals, dispensaries, and lunatic asylums, should be sought for, with every particular that can be gathered concerning the manner in which they have been established and are supported, and the number succoured. It would be a service rendered to an important branch of science if the numbers, in proportion to population, are ascertained of lunatics, of blind persons, and of the deaf and dumb.

The length and condition of the public roads should be inquired into, as well as the system under which their repair is provided for, whether by the State, or by tolls collected from passengers, or by the money or labour contributed by residents in the districts through which the roads are carried. The modes of travelling, as well as the nature and number of public carriages; and whether, as in some countries, they are the property of the government, or, as in England, the result of private enterprise, hould also be ascertained, as they easily may be. The means for internal navigation, whether by rivers or by artificial canals, it may not be difficult to learn; recording the direction and the length of each, and the size of vessels in which the traffic can be conveyed. In the case of canals, it will be interesting to know the date of their construction, and, if possible, their cost, as well as the nature and amount of goods conveyed upon them (and upon the rivers also), the rate of toll, and the degree in which their construction has answered, both for the advantage of the community and the profit of the owners. The interest which attaches to railroads, in most places where they have been introduced, has been such as to cause every publicity to be given to their statistical conditions, and printed accounts, in which every question that it may be necessary to ask concerning them may be easily procurable, and should be secured.

The manufacturing industry of a country will naturally claim attention from every inquirer, who, in the probable absence of all precise information concerning its extent, will endeavour to supply its place by means of such circumstantial information as he can bring to bear upon the subject. With respect to such branches of manufacture as depend for their raw material upon foreign supply, it will not be very difficult to arrive at a tolerably close approximation to the truth, in regard to the quantity of such material used. Such cases are comparatively few, however, and with regard to those branches of industry which derive their material partly or wholly from the native soil, the person who visits any country must usually content himself with such statements as he can draw from trustworthy persons, preferring those accounts, if any such there be, which, having been published to the world, have stood the test of local criticism. The cotton and silk manufactures of England are examples of the first-mentioned of these conditions, while our linen and woollen manufactures sufficiently explain the other class. To ascertain merely the quantity of raw material used would go but a little way towards determining the value of any manufacture to a country. This will be plainly seen if we call to mind the familiar instance of the chain-cable and the watch-spring, both of which are products of the same material; while one, by reason of the amount of labour bestowed upon it, is many thousand fold more valuable, weight for weight, than the other. The cotton manufacture is open to the like difficulty, although in a minor degree; even the yarn, which is the result of a preliminary process, sells according to its degree of fineness, from a few pence to as many shillings per pound. The inquirer will, therefore, feel it necessary to ascertain the increased value that is ordinarily imparted by processes of manufacture to the materials used, and whether any and what changes are going forward in this respect. The information here suggested may partly be gathered by comparing the prices of given weights and measures of the materials with those of the average qualities of finished goods; but it must be evident that little more can be done in this branch than to apply to men of intelligence and respectability for such information and opinions as they may be able to impart. It need hardly be pointed out as desirable to know in what degree the general population shares in the use or consumption ot home-manufactured articles—whether any part of them falls to the lot of the working classes, or whether they are wholly engrossed by the high-born and wealthy. It is desirable to know whether any, and what, branches of manufacture are carried on by foreigners to the exclusion of native workmen: also, if women and children find employment in manufactures; and further, if the degree of comfort in which the various classes of the manufacturing population live, is greater or otherwise than the comfort enjoyed by those who follow other occupations. It is of importance to learn whether any manufactories are maintained or assisted by the government, and in what form that assistance is given; whether by direct money-payment or by the grant of privileges or monopolies; and in case any such system is followed, then whether in the branches thus favoured there is found a greater amount of success than ordinarily attends the employment of capital and skill in the country. The seats of the several manufactures should be indicated; and where any mechanical power is employed, the nature of the same should be explained; and also the degree of proficiency attained in the production of machinery, when it is made in the country, and if it is brought from abroad, then the places whence such machinery is derived.

The foreign commerce of a country is matter of especial interest to every other country, and more particularly to England, so much of whose prosperity depends upon its commercial relations. Among the earliest inquiries to be made on this head will be the amount and description of the shipping under the national flag, and whether the same be increasing or otherwise; whether any, and what, privileges are accorded to the native marine. Then, what other flags frequent the ports, distinguishing those which participate most largely in the trade, and whether they mostly or entirely trade with their own ports, or engage in the carrying trade from foreign countries.

The description and quantities of goods imported and exported may usually be learned without much difficulty from intelligent merchants, or, what is better, from the accounts of custom-houses. Distinction should, as far as possible, be drawn between goods imported for use and those brought in transit, dividing them, in both cases, into raw materials and finished articles, and classifying them according to their nature, distinguishing food, clothing, metals, &c. The like statements should be obtained, and distinctions made, in regard to exports. It is desirable to know whether goods are imported directly from the various countries of their production, or indirectly from third markets, and in this latter case the reasons should be sought why the apparently less desirable course is followed. The rate of customs-duties can always be procured, and in most cases in a printed form. This will serve to show whether any differential or preferential duties are levied, to the hurt or advantage of particular countries. In regard to duties upon exports, their amount and nature should be sought equally with those charged upon imports and consumption.

The home trade of countries, unless they be of such extent as to include different climates, and consequently to yield different products, is usually comprised in transmitting imported articles from the ports to towns in the interior and to country districts, or in transferring articles of home growth from the country districts to the different towns and ports. Besides this there will be, in manufacturing countries, the transmission of goods from the seats of manufacture to the towns and villages, for the supply of their inhabitants and of the neighbouring districts. A traffic of this kind it must be at all times difficult to register, and the most that can be done by a stranger or visitor is to learn the general nature and course of the trade, and to collect opinions as to its amount and condition at various periods of time. I. any internal duties, answering to our excise-duties, are charged upon home productions brought into consumption, their nature and amount should be ascertained.

The subject of currency and banking is of very high importance, and every information concerning it that can possibly be had should be carefully obtained. The nature and value of coins in use, their weight and denominations, should be noted, and whether means are used to prevent their exportation by laws passed for that purpose, or by the coins being made to contain any considerable portion of baser metal. If any auxiliaries to the use of coin should be established, such as bank or government notes, or transfers in books of public account, as practised in some trading cities in Europe (Hamburg, for instance), the nature of such should be described. Until a recent period, the chief, if not the only method used for making payments in France was by the transmission of silver coin; and it often happened that public carriages, passing between two places in opposite directions, conveyed at the same time tons weight of five-franc pieces. This inconvenient and expensive practice has of late years been in some degree remedied by the more general establishing of banks of issue, whose notes are transmitted by post, as well as by the extended use of bills of exchange.

The weights and measures in use should be stated, with the proportions which they bear to those in this country, or to other well known standards: and, connected with this subject, it is well to know what articles of general use are sold by weight and what by measure, and whether different weights or measures, or different usages in regard to them, are adopted in different parts of the same country, as was at one time the case in different parts of England.

If any joint-stock associations are in operation for trading purposes, their nature and the extent of their capital; the peculiar privileges, if any, that they enjoy, and the effect they are judged to have upon the general interests of the community, should be carefully gathered. There may be other associations not strictly trading and yet closely allied to trading interests, which should equally be the objects of inquiry; such as docks, insurance offices, and the like.

The public revenue and expenditure of countries, when published at all, are put forth by the government; and all statements of this kind should be made objects of enquiry, with a view to obtaining the same. If the government should not think fit to publish information of this kind, it will seldom be of any use to seek for it in any other quarter.

In every country, having any claim to civilization, it will be possible to procure maps, and by conversation with men of intelligence the visitor may get to know the degree of reliance that is to be placed upon their accuracy.

The limited space that can be given in this volume to the subject of statistical enquiries has necessarily confined the recommendations which are offered to the more leading or important objects, which are also noticed with the utmost brevity. To persons of intelligence who visit other lands, many peculiarities will present themselves which they will think worth recording, although nothing should have been said concerning them in these pages. One caution it appears desirable to offer; it is, that no fact shall be disregarded as without value by reason of the incompleteness of the information it yields, since it may well be that this very fact may supply a link in the chain that will give value and completeness to former or to future observations.

APPENDIX.

Plan and Regulations of the Establishment and Adjudication of Two Prize Medals for the Encouragement of the Medical Officers of the Royal Navy, and the Improvement of Physic and Surgery in that Department of the Public Service. Founded by Sir Gilbert Blane, Baronet, First Physician to the King, F.R.SS., Lond., Edin., Gött., Member of the Imperial Academy of Sciences of Russia, of the Institute of France, &c.

1. The Founder—considering how much it will conduce to the advancement of the Public Service that emulation should be excited among the Medical Officers of the Royal Navy by honorary distinctions for professional merit—has vested the sum of three hundred pounds in the three per cent. consolidated Bank Annuities, in the corporation of the Royal College of Surgeons of London, in trust with the dividends which shall be from time to time receivable, for the purpose of conferring, once in two years, Two Gold Medals of equal value on two medical officers, surgeons of ships of war in commission, or assistant-surgeons of King's ships in commission not bearing surgeons, who, in the time required, shall have delivered into the proper office, Journals, evincing the most distinguished proofs of skill, diligence, humanity, and learning in the exercise of their professional duties: these journals to be delivered in the form in which they have been kept from day to day, stating the symptoms as they shall have occurred at the time; but without prejudice or hindrance to their making such observations, practical or theoretical, as they may judge proper to annex to them.

2. The first selection to be made by the Medical Commissioners on the 12th of August, 1831, from the Journals delivered between the 12th of July, 1827, and the 12th of July, 1831. All future selections to be made on the 12th of August, at the interval of two years from each other, from the Journals delivered in the two pre-

ceding years, up to the 12th of July immediately preceding such selection.

3. In the selection of these Journals the Founder proposes that the Medical Commissioners of the Navy shall, out of the whole Journals delivered to them in the course of the intervals above specified, make choice of such as in their judgment possess the highest degree of merit, in number not more than ten nor less than five, which shall be transmitted to the Founder during his lifetime, for his selection out of the number so sent, of two, or one in case there should not be another of sufficient merit, the authors or author of which, in his judgment, may be most deserving of the prizes. And after his decease the said Journals to be conveyed to the President of the College of Physicians, who, after due examination, is to communicate them to the President of the College of Surgeons, and after proper deliberation the said Presidents are to call to their assistance the Senior Medical Commissioner of the Royal Navy, and jointly with him select from the said Journals one or two. the author or authors of which, in the opinion of the majority, possess the highest merit, and become thereby entitled to the Medal or Medals. The Medal or Medals when adjudged are to be put into the hands of the attending Medical Commissioner, to be by him presented to the successful candidate or candidates. All the Journals of the first selection to be returned into the custody of the Medical Commissioners.

4. In case of the impossibility of performing the before-mentioned duties, through the illness or unavoidable absence of the parties described, the duty is to devolve on the next in rank; that is, on the Senior Censor of the College of Physicians, the Vice-President of the College of Surgeons, or the Junior Medical Commissioner.

5. In case it should happen at any of the periods of adjudication that in the opinion of the Founder, or of the two Presidents after his decease, there shall not be found a Journal or Journals of adequate merit to entitle any candidate to the prize, the Medal or Medals shall be withheld until the next period of adjudication, and the unadjudicated Medals are to be conferred on such as may possess sufficient merit over and above those subject to adjudication at that period. But this regulation is to be so construed and limited that no more than four prizes shall be adjudicated at any one period; and if the unadjudged

Medals should exceed this number, their value in money is to be given to the Supplemental Fund for the Children of Medical Officers.

6. In case at any time the Founder or the two Presidents shall omit to make the adjudication for a longer period than three months, they shall be considered as having forfeited their right, and the ultimate selection shall devolve on the Medical Commissioners, who, in case of difference of opinion, may call in such a referee as they may judge necessary or advisable.

7. The Founder shall provide and deposit with the Royal College of Surgeons the Die engraved for the Medal, from which they will cause the Medals to be struck at the prescribed periods, and to be delivered to the Medical Commissioners to be presented by them to the successful candidates.

8. No successful candidate to be admitted as a competitor a second time.

9. The Presidents of the Royal College of Physicians and Surgeons and the Senior Medical Commissioner to be considered as guardians of the fund and its equitable administration.

10. In case any of those surgeons whose Journals have been given in, should have been paid off previous to adjudication, or should they have been appointed to an hospital, or any other situation on shore except that of Medical Commissioner, such surgeons shall still be deemed eligible candidates for the Medals in case of adequate merit.

11. After a lapse of not less than ten years from the decease of the Founder it shall be competent for the Presidents of the two Royal Colleges and the Medical Commissioners of the Navy to hold an interview for the purpose of consulting whether any and what additions or alterations would be advisable in the preceding plan and regulations, and to adopt them in case of their being unanimous for the adoption: subject, nevertheless, to the approbation of the Lord High Admiral, or the Commissioners for executing the office of the Lord High Admiral.

RECOMMENDATIONS AND SUGGESTIONS.

THE Founder, with all deference to the high professional authorities who are to adjudge the Medals, begs to suggest and recommend as follows:—

1. That a book be kept in the custody of the Medical Commissioners of the Royal Navy, wherein is to be transcribed the plan and regulations, and to serve also as a record of the periodical adjudications, and wherein not only the names of the successful candidates may be inscribed, but also of all those of the first selection; among whom it cannot be doubted that there will be found tokens of merit which may go without their due reward, from the limited number of Medals, and all of whom will, of course, possess a considerable share of merit above the unselected, and be deserving of consideration.

2. That there be transcribed into this book of record such remarks as may have arisen out of the examinations, deliberations, and discussions of those appointed to adjudge the Medals, and which may prove a source of much valuable information not only for the interests of the Navy, but of the community at large, while it will open a source of liberal and useful intercourse between the members of the different public professional institutions of the empire, provided some degree of publicity should be given to them.

THE END.

London: Printed by WILLIAM CLOWES and SONS, Stamford Street.

Printed in the United States
By Bookmasters